The Prussian Army

1640-1871

Jonathan R. White

University Press of America, Inc.
Lanham • New York • London

Copyright © 1996 by
University Press of America,® Inc.
4720 Boston Way
Lanham, Maryland 20706

3 Henrietta Street
London, WC2E 8LU England

Library of Congress Cataloging-in-Publication Data

White, Jonathan Randall.
The Prussian Army, 1640-1871 / Jonathan R. White.
p.   cm.
Includes bibliographical references and index.
1. Prussia (Germany). Armee--History. 2. Prussia (Germany)--
History, Military. 3. Military art and science. I. Title.
UA718.P9W5   1996   355'.00943'0903--dc20   95-39549 CIP

ISBN 0-7618-0205-3 (cloth: alk: ppr.)
ISBN 0-7618-0206-1 (paper: alk: ppr.)

℗™The paper used in this publication meets the minimum
requirements of American National Standard for information
Sciences—Permanence of Paper for Printed Library Materials,
ANSI Z39.48—1984

Dedicated to:

Dencil and the memory of Crystle White;

and

Ken and Helen Medendorp

Blessed are the peacemakers,
for they will be called
Children of God.
Matthew 5:9

Table of Contents

TABLE OF CONTENTS

Table of Maps

Maps

Prussia 1640

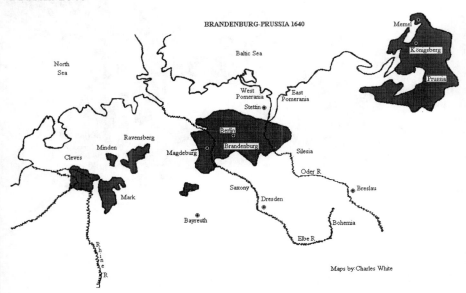

BRANDENBURG-PRUSSIA 1640

Memel

Königsberg

Prussia

Baltic Sea

North
Sea

West
Pomerania

East
Pomerania

Stettin

Berlin

Brandenburg

Ravensberg

Minden

Magdeburg

Silesia

Cleves

Oder R.

Mark

Saxony

Breslau

Dresden

Bayreuth

Bohemia

Elbe R.

Rhine R.

Maps by Charles White

Prussia 1740–1806

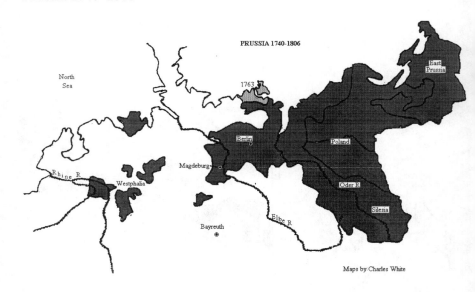

Maps by Charles White

Prussia 1815–1871

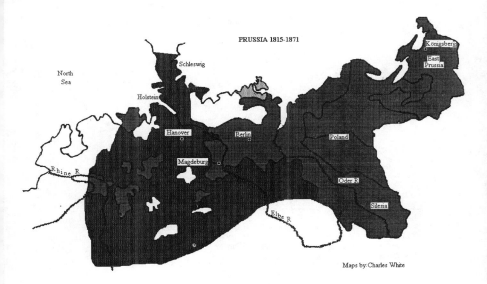

PRUSSIA 1815-1871

Maps by:Charles White

Maps of Battles Described in Text

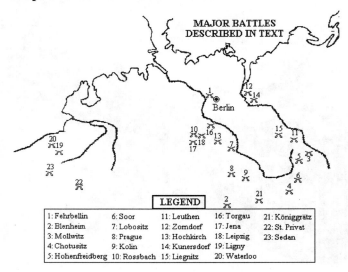

MAJOR BATTLES
DESCRIBED IN TEXT

Berlin

LEGEND				
1: Fehrbellin	6: Soor	11: Leuthen	16: Torgau	21: Königgrätz
2: Blenheim	7: Lobositz	12: Zorndorf	17: Jena	22: St. Privat
3: Mollwitz	8: Prague	13: Hochkirch	18: Leipzig	23: Sedan
4: Chotusitz	9: Kolin	14: Kunersdorf	19: Ligny	
5: Hohenfreidberg	10: Rossbach	15: Liegnitz	20: Waterloo	

The Hohenzollerns
1640–1871

Frederick William, the Great Elector
1640–1688

Frederick III (Later Fredrick I, King in Prussia)
1688–1713

Frederick William I
1713–1740

Frederick II (Frederick the Great)
1740–1786

Fredrick William II
1786–1797

Frederick William III
1797–1840

Frederick William IV
1840–1858

William I
Prince Regent 1858–1861
King of Prussia 1861–1871
Kaiser of Germany 1871–1888

Introduction

In the summer of 1987 I attended a seminar sponsored by the Carnegie Endowment for International Peace at the University of Wisconsin, Madison. It was a multi-disciplinary workshop designed for professors who teach classes in peace-making and problems with militarism. During one of the breaks, I spied an abridged version of Clausewitz's *On War*, and I absent-mindedly began leafing through it. Aware that someone was looking over my shoulder, I looked up to see one of the conference participants beaming at me. "That guy was a German general," he said. I started to nod, but my friend interrupted. "I discovered him last summer. Imagine, that fellow wrote an entire book on the philosophy of war. I think he's dead now, but it's still a good book. You should get a copy."

I smiled and watched as the professor returned to our seminar room. At first, I wondered if he could ever teach his students *anything* about military affairs, but then I remembered my own naïve introductions to abstract concepts. For example, I "discovered" Norbert Wiener's work on cybernetics in the late 1970s and bored everyone in my family for weeks with my excitement. It has been over 150 years since Marie von Clausewitz published her late husband's work, but my companion was consumed by the passion of his discovery. When I saw him later in the day, I thought it best not to tell him that a few other people had also "discovered" Clausewitz. I am sure he knows by now.

This book is written for those who want to make a discovery. It is a text designed to introduce American college students to military affairs through the spectacles of Prussian military history. It is specifically written for undergraduates who have no background in the development of Europe. Gordon Craig, Christopher Duffy, Peter Paret, John Kegan, Robert Asprey, Michael Howard, and others have told this story before, but students who have not studied history are not familiar with their work. This book is designed to give students the background to read other works. The purpose is to give non-historians the ability to understand Craig and the other masters.

In technical terms, *The Prussian Army* is narrative history based on secondary sources. Less formally, it is a set of selective introductory stories about some of the world's most professional soldiers. The work focuses on military leadership and the progress of Prussia's battles. References are primarily in the English language, and I have tried to cite works that are available in typical American liberal arts college libraries. The major purpose of the footnotes is to guide students to further reading. References include standard European historical surveys, classic English-language works on the Prussian army, and citations of popular literature.

Although much of my formal training is in the discipline of history, I have purposely violated almost every principle I had to learn. In a world gone astray with militarism and violence, peace is the most desirable outcome of international relations. Yet, if we want to teach peace—and I pray that is the purpose for which this book is used—we need to conceptualize both the attraction and the process of war. The story of Prussia serves as a great beginning, and I have tried to make this book an entertaining introduction to the methods of war by telling the story of an army. Making no attempt to be professionally objective, I have tried to spin a whopping good yarn with a reasonable respect for factual material.

Many people assisted with the project, and I would like to give them full credit. My immense gratitude is expressed to my editor, Dr. Benjamin Lockerd, and my copy editor, Craig Stapert. Additional critical review came from a dear friend and true renaissance man, John Schneider. Mr. Schneider provided valuable historical and structural comments as well as long-standing friendship. I also appreciate the editorial review from my colleague, Dr. Agnes Baro, and the moral support from the faculty and staff of the School of Criminal Justice, Grand Valley State University. Special thanks goes to Cindy Driesenga for providing top-quality secretarial support. I also appreciate Ms. Helen Hudson and the entire UPA staff for their assistance with the project. The computer-generated sketch-maps were produced by Charlie White. (Thanks, Charlie, for many hours devoted to dad's new book.) Finally, I would like to express my appreciation to my wife Marcia and my daughter Katie for their support and love throughout the months it took to write the book. (I *really* appreciate you putting up with me on the long days!)

Jonathan White
East Grand Rapids, Michigan

Chapter 1

The Elector Builds an Army

In the beginning, there was no Prussian army. Although being "Prussian" would come to epitomize martial professionalism, the army germinated inauspiciously during the last phase of the Thirty Years' War in the small north German state of Brandenburg. In 1640 the embryonic formations that would grow into the army were composed of a ragamuffin group of miscreants who were no match for the major powers of the day. The army's combat strength was maintained by purchasing the services of mercenaries, its leadership was questionable, and its finances were in disarray. To make matters worse, the new leader of Brandenburg, the twenty-year-old Elector Frederick William, found himself at war with Sweden, a nation that commanded one of the most powerful land forces in Europe. To fight Sweden, Frederick William needed money and an army, yet creating an army was far beyond the young elector's means. With every card in the deck stacked against him, the new elector decided to do the impossible. Since Brandenburg's only salvation could come through military force, Frederick William decided to build an army.

The Electorate of Brandenburg

The story of the Prussian army begins in the north German state of Brandenburg, a territory in the northern European plain with almost no natural defensive barriers. In the early 1600s Brandenburg was only one state among many small German governments, and the Germany of the seventeenth century was not the Germany of the modern nationalistic era. One commentator has gone so far as to describe the divided region as a concept, not a nation. It was a geographical conglomeration composed of German-speaking people. There were a handful of states the size of Brandenburg and nearly 300 smaller independent political entities. Seven principalities were called electorates because the royal leaders of these states could cast a vote for the emperor of the Holy Roman Empire. The house of Hohenzollern held the throne in Brandenburg and claimed ownership of lands that stretched from Prussia in Poland to provinces along the Rhine. The leader of Brandenburg also held an electoral vote in the empire.[1]

Austria was the most powerful state among the German-speaking people, and the electors usually united in support of the Hapsburg throne. The emperor, therefore, was Austrian and the most powerful figure among the German princes. The emperor's power, however, was primarily titular. Although he exerted influence on the few hundred provincial rulers, he did not exercise direct control. The seven electors granted the emperor power, but they gave him only nominal allegiance. They were quick to ally with him when they believed it to be in their best interests, yet they held no long-lasting loyalty. The four most important electorates were Austria, Brandenburg, Saxony, and Bavaria. These regions had been ideologically divided since the Protestant Reformation of 1517. The electors of Saxony and Brandenburg were Protestant, while the emperor and the Bavarian elector were Catholic. Actually, the empire was a loosely bound political confederation of Germans who were divided by religion, politics, and geography. The emperor ruled a divided house.[2]

Austria still managed to maintain some control of the empire, and the electors were subject to the emperor's call for provincial troops from time to time. Yet, religion served as the most divisive political issue in the Holy Roman Empire, and it eventually drew the Germans into a devastating war. At the same time, Austria faced problems outside German borders. The Ottoman Turks were forever threatening Austria's southern flank. During most of the sixteenth and seventeenth centuries, Austria's political attentions were constantly torn between the Protestants of the north and the Ottomans of the south. When France threatened expansion into Germany in the 1600s, it added to Austria's woes. It was a difficult balancing act for the Hapsburgs.

The house of Hohenzollern had been in control of Brandenburg since the Middle Ages. Wresting control of the sandy and often swamp-infested

---

[1]H. W. Koch, "Brandenburg-Prussia," in John Miller (ed.), *Abosultism in Seventeenth Century Europe* (New York: St. Martin's Press, 1990), 127–28.

[2]Geoffrey Parker, *The Thirty Years' War* (London: Routledge and Kegan Paul, 1984), 13–24.

ground from the Slavs, the Hohenzollerns had slowly moved eastward since 1400. Several generations of family members gradually added territory to the royal domains. They gained Prussia in Poland while purchasing lands along the Rhine. Many of the inhabitants of Brandenburg were Lutheran, but members of the royal house and many other subjects were Reformed. While religion was a major issue for the Hohenzollerns, the more important question of government focused on geography. The lands of the Hohenzollerns spread over a 400-mile region, and the borders did not interconnect. The Hohenzollerns ruled from Berlin, but each of their holdings maintained its own capital and governing body. In addition, Prussia was officially under control of the king of Poland even though its inhabitants were German. It was a strange geographical configuration.

Brandenburg was not an extremely important political entity in seventeenth-century Europe. Austria was not only the dominant political force in the Holy Roman Empire, it was one of the more important military powers in Europe. To the east, the medieval kingdom of Poland was breaking up into political chaos, causing Russia to cast a longing eye toward the Polish steppes. Poland did not present a political threat to Brandenburg, but foreign intervention in Polish affairs was dangerous. Sweden was the dominant military power of northern Europe, and it held Western Pomerania adjacent to Brandenburg's northern boundary. To the west, the English and the Dutch competed with one another to control the high seas. They presented no threat to the Hohenzollerns, and they offered the possibilities of alliance. Spain had been in strategic retreat since the defeat of its fleet—the Armada—in 1588, but it remained a far greater military power than Brandenburg. The most powerful country of the age was France. Indeed, the last part of the seventeenth century would become known as the age of Louis XIV. The scattered, poverty-stricken lands of the Hohenzollerns were no match for such major European nations.[3]

---

[3]See John B. Wolf, *Toward a European Balance of Power: 1620–1715* (Chicago: Rand McNally and Co., 1970), 1–28, and John A. Lynn, "The Pattern of Army Growth, 1445–1994," in John A. Lynn (ed.), *Tools of War: Instruments, Ideas and Institutions, 1445–1871* (Urbana, Ill.: University of Illinois Press, 1990).

The Thirty Years' War

Central Europe was engulfed in a tragic struggle known as the Thirty Years' War from 1618 to 1648. Just as Germany was not a nation in the modern sense, the Thirty Years' War was not a conflict in twentieth-century terms. It is best to see the war as a series of conflicts involving a variety of nations for a myriad of different purposes. The war began over a religious dispute in the Holy Roman Empire, but it quickly turned into a political struggle. During the final phase of the war, all pretense of religion was forgotten, and the Roman Catholic powers of France and Austria fought for control of the continent. While Brandenburg was not a major character in the war, like many north German states it played a critical role. It sat on the crossroads of the major powers.[4]

The war started with an incident in the Bohemian city of Prague. After the Archbishop of Prague ordered the destruction of a Protestant church, the Protestant city leaders of Prague decided to close the Catholic cathedrals. The Catholics of Prague appealed to the emperor in Vienna. He sent a delegation to order the Protestant city leaders of Prague to reopen the cathedrals. The Protestants responded by throwing the imperial delegation from the second story window of city hall. Angered, the emperor responded with troops. After two years of fighting, the Austrians gained control of the entire state of Bohemia, and the Protestants were ordered to become Catholic.

Appealing to their Reformed and Lutheran cousins in northern Europe, the Bohemian Protestants found sympathy from Denmark. When the Danes entered the war, the Austrians saw an opportunity to expand their influence to the north. Other countries joined the fray, and for the next thirty years a conglomeration of strange armies fought across Germany. The war was dominated by the Danes from 1625 to 1629, the Swedes from 1630 to 1635, and the Swedes and the French from 1635 to 1648. The Thirty Years' War ended with the Peace of Westphalia in 1648.

Military historian Michael Howard has placed the Thirty Years' War in the proper perspective for the Germans. Professional armies, Howard demonstrated, would come into being after the war. From 1618 to 1648, however, Howard said the armies roaming across Germany were akin to swarms

---

[4]Parker, *Thirty Years' War*, 48-61.

of locusts.[5] They fed on the German people. Ragtag groups of adventurers and mercenaries terrorized the German population at will. No little German state could stop these armies. By the end of the war, they had wreaked havoc upon the cities and countryside. The German states lay in ruins. Outside of Austria, half of the German-speaking people were dead, most of them killed by famine.[6] Some German leaders concluded that their only hope of protection would come when they had their own military forces. Among those leaders was the new elector of Brandenburg, Frederick William von Hohenzollern.

Frederick William in Königsberg

As a young boy, Frederick William had watched his father, George William, attempt to protect the lands of Brandenburg from a position of weakness. The policy had not worked. Young Frederick William noted that foreign armies roamed at will through all the Hohenzollern provinces. At age twelve he left Berlin to study in Holland, only to find a completely different way of life. In Amsterdam there were no foreign armies marching forth and back across Dutch territory. The future elector concluded that Dutch power and wealth were responsible for this state of affairs. At eighteen he returned to Berlin, hopeful that Brandenburg might imitate the Dutch and drive the foreign intruders away. He hoped in vain.

George William could not place his faith in power and wealth; he could only wish for them. Brandenburg was bankrupt. Massive foreign armies moved wherever they wanted, and George William could only pray that they would leave his subjects alone. By 1638 prayer was not enough. Swedish troops violated the very heart of Brandenburg, leaving George William in a lopsided war. Fearing for his family, George William moved the court to the relative safety of Prussia, the Hohenzollern province on the northern coast of Poland. Königsberg, its capital city, was far from harm's way. George William appointed Johan von Schwarzenberg as the provincial governor in Berlin, and Schwarzenberg established himself as a virtual military dictator

---

[5]Micheal Howard, *War in European History* (Oxford: Oxford University Press, 1976), 29. See also J. R. Hale, *War and Society in Renaissance Europe: 1450–1620* (New York: St. Martin's Press, 1985), 179.

[6]Gustav Freytag, "The German Catastrophe," (orig. 1882) in Theodore K. Rabb, (ed.) *The Thirty Years' War* (Lexington, Mass.: D. C. Heath and Co.: 1972), 3–7.

in the elector's absence. George William never returned to Berlin. He died in 1640, a leader in exile. Twenty-year-old Frederick William was proclaimed "The New Lord" far away from home.

Frederick William believed that diplomacy and promises were little more than a façade. Brandenburg needed power to protect itself, and power could only come with military force. Military force, in turn, could only come with money. The new elector needed money. Optimistically ignoring Brandenburg's bankruptcy, Frederick William told his advisors that he intended to raise money and an army. In 1640 this seemed a most impossible task. The new elector was impoverished, exiled, and had almost no military power at his beckoning. He was, as his American biographer Ferdinand Schevill so aptly stated, a beggar on horseback. Yet, Frederick William devoted himself to the task of creating an army, and he never veered from the course for the next forty-eighty years.[7]

The new elector seemed keenly aware of both problems and opportunities. He was hardly in a position to march into Berlin, but he could back diplomatic negotiations with the few troops he had. The key, Frederick William believed, was to end the war with Sweden and entice the Swedes to withdraw from his domain. This was difficult because his minister in Berlin had charted a different course, tying Brandenburg to Austria, endorsing the war with Sweden, and exhibiting open Catholic sympathies. The Austrians liked this, but they were of little help. They could not dislodge the Swedes, who still controlled Brandenburg. Frederick William wanted to remove both Sweden and Austria. Before he could put his plans into operation, he had to find a method for dealing with Schwarzenberg's love affair with Vienna.

Frederick William had one trump card. While Brandenburg did not have substantial military power, Schwarzenberg had managed to raise a meager army of local peasants and mercenaries. When inside a fortress, the ragtag army was strong enough to hold it. In fact, this army could hold several fortresses at one time. Despite all its weaknesses, Brandenburg had good fortresses, especially along the southern approaches from Austria. Schwarzen-

---

[7]Ferdinand Schevill, *The Great Elector* (Chicago: University of Chicago Press, 1947), 52.

berg had taken full advantage of the situation.[8] Without bothering to consult the Privy Council, the governing body under the elector, he raised local taxes to support the small army and garrisoned as many fortresses as possible. This policy brought a tremendous financial burden to Brandenburg, and it increased the military importance of Brandenburg in the eyes of the surrounding armies. Schwarzenberg used the fortresses to keep Brandenburg in the war. Frederick William had other objectives in mind.

According to Schevill, Frederick William wanted immediately to negotiate a peace with the Swedes.[9] At first glance, this would seem to imply a contradiction to Frederick William's desire to build an army. The elector's primary negotiating pawn was the ability to withdraw from the Brandenburg fortresses. In other words, he wanted to reduce the army, thus curtailing the immediate military threat of Brandenburg. Frederick William knew that this was a short-term step, because no army would remain outside the realm on good faith alone. Yet, he hoped to accomplish two difficult objectives. Since the loyalty of Schwarzenberg's mercenaries was suspect, he wanted to dissolve the existing force. In addition, he wanted to secure enough finances to build a regular army. Offering Sweden a temporary reduction of his power did not contradict his stated purpose of military aggrandizement; it was the a foundation for future growth of his army. Therefore, he began negotiations with the Swedes and discussed the possibility of disbanding the Brandenburg army. All of this remained a secret to Schwarzenberg.

Reduction of his meager forces did not mean that Frederick William would completely abandon the fortresses. Protection would come from garrisons assigned to critical crossroads. Frederick William thought that this option would provide a strategic defense, and he knew it was something the tight-fisted merchants of the province could afford. He also knew that Schwarzenberg would never agree to such a plan. In response, the new elector proved to be somewhat devious, if not outright ruthless. The elector sent to Schwarzenberg a message that detailed nothing of his true intentions.

---

[8]See Christopher Duffy, *Siege Warfare: The Fortress in the Early Modern World, 1494–1660* (London: Routledge and Kegan Paul, 1979), 174–82. (Duffy has an excellent discussion on the role of fortresses in the Thirty Years' War.)

[9]Schevill, 105–8.

Frederick William sent word to Berlin, telling Schwarzenberg to continue the pro-Austrian policy. In the meantime, Frederick William called upon a crafty colonel named Konrad von Burgsdorf. Burgsdorf was a crude man given to hard drinking and gambling, but he was a professional soldier. He resented Schwarzenberg's authority and mistrusted his Catholic leanings. Of all the colonels in the makeshift army, only Burgsdorf had refused to obey Schwarzenberg. Each time the provincial governor issued an order, Burgsdorf cleared it through Frederick William before taking action. He was just the person the elector needed. Frederick William summoned his colonel to Königsberg and indicated his actual intentions.

In February 1641 Burgsdorf commanded the fortress of Küstrin on the Oder River. It sat along the right flank of the Austrian army, protecting it from Swedish advances. From Königsberg, Frederick William ordered Schwarzenberg to inform the Austrian court of Frederick William's Viennese sympathies and to offer Küstrin to the Austrian army as a gesture of good will. At the same time, the elector ordered Burgsdorf to hold the fortress and not to admit Austrian troops. Sweden was the actual target of the elector's diplomacy. By denying the fortress to the Austrians, Frederick William destroyed Schwarzenberg's credibility and exposed the Austrian right flank to the Swedes. He hoped the Swedes would accept the gracious offer. For his part, Burgsdorf was delighted. He had always believed that war with Sweden had been a mistake. Now he had a chance to rectify the situation. As Schwarzenberg approached the Austrians with news of the elector's supposed Hapsburg leanings, Burgsdorf actually exposed the Austrian field forces. The plan worked. Schwarzenberg and the Austrians were caught completely by surprise.[10]

Frederick William's faith in the Swedes proved to be naïve. Schwarzenberg contracted a fever and died, while the elector made a conciliatory gesture to Sweden from Königsberg. Despite Burgsdorf's horrified protests, the elector virtually surrendered to Sweden in an effort to obtain peace, implementing the plan he had in mind all along. Frederick William agreed to disband the major units of the Brandenburg army in exchange for peace with Sweden. Burgsdorf's concerned reaction came because the elector agreed to disband the forces *before* the Swedes left the elector's territory, not *after*

---

[10]Herbert Tuttle, *History of Prussia, Volume I: 1134–1740* (New York: AMS Press, 1971; orig. 1884), 145. See also discussion by Schevill, 107–8.

they had departed. Burgsdorf begged the elector to keep the army intact until the Swedes were gone, but the elector placed his trust in a Swedish promise to leave. The 9,000-man Brandenburg army was reduced to a standing force of 2,000 stationed at Küstrin, Spandau, and a few other fortresses. In desperation, Burgsdorf pointed out that the Swedes might be more inclined to abandon Brandenburg if the elector maintained a small field army until all the conditions of the treaty were fulfilled. Frederick William brushed his colonel's opinion aside. It proved to be a mistake.

Burgsdorf became a prophet without honor. No sooner had Frederick William dismissed his field force than the Swedes began to hedge on their promise to leave Brandenburg. Claiming that they would leave as soon as the Austrian threat disappeared, the Swedes maintained garrison forces in southern Brandenburg as well as a sizable field army. By the standards of the Thirty Years' War, the local country folk suffered the most as the Swedish army lived off the land. Frederick William remained exiled from his capital, surprised and disappointed. Burgsdorf, now elevated to a position on the Privy Council, was not so surprised. As a career soldier he believed that diplomacy worked only when it was backed by force, and with only a handful of troops locked up in garrisons, the elector had no muscle behind his diplomacy. With his less than stately manner, Burgsdorf explained this lesson to Frederick William. Although the elector would develop a reputation for cashiering advisors who had served their usefulness, he listened to Burgsdorf. The Swedes took their time in leaving, and the elector never forgot the lesson.

Denmark unintentionally saved Frederick William. In the last phase of the Thirty Years' War, alliances changed quickly, and they focused on power politics, not religion. In 1643, Protestant Denmark reentered the war, primarily because it was concerned about expanding Protestant Swedish power. The Swedes, quickly retreating through Brandenburg to meet the threat, now grew anxious to seek Frederick William's favor. Swedish generals sent word to Königsberg, indicating that Austria was no longer a threat and that they would be quite happy to remember their promise to withdraw. They agreed to pull their troops out of Brandenburg and even gave Frederick William some of their fortresses. It was not a statement of brotherly love. The Swedes wanted their southern flank protected. Frederick William took advantage of the opportunity. He began to solicit funding for a field army, and with Burgsdorf's support, he began reconstituting the Brandenburg military force. Although it would not be proper to label this new force as the Prussian army, the nucleus of that force was spawned by the actions of Frederick William and Burgsdorf. On March 4, 1643, three years after he had been proclaimed the Elector of Brandenburg, Frederick William finally entered Berlin under the protection of Burgsdorf's army. If he wished to remain, he needed a stronger army.

Building an Army

Three administrative tasks faced Frederick William in the construction of the army. First, he had to find means to finance it. Second, there was a need to establish an organizational structure. Third, Frederick William had to create a force capable of completing its mission; that is, he needed a group of fighting soldiers who could protect Brandenburg. He threw himself into all three tasks with fierce energy.

The means of financing the army was of utmost importance.[11] The scattered Hohenzollern domains had semi-autonomous governing units that had the exclusive authority to levy taxes. These local assemblies, or diets, were very concerned about their own defenses, but they indicated almost no concern for collective defense. Most of all, they saw no reason to protect Brandenburg. Diets outside of Frederick William's capital were quite happy to let the Brandenburgers fend for themselves. To understand this attitude, it is necessary to look briefly at the cultural and geographical composition of the Hohenzollern states. Germans in Brandenburg, Prussia, Cleve, Mark, and the other scattered domains were perfectly willing to acknowledge Frederick William as their sovereign. What they refused to acknowledge, however, was their German identity. For example, most people in Cleve believed people from Brandenburg to be foreigners. By the same token, Brandenburgers thought that the Germans of the Hohenzollern town of Königsberg were aliens. The people of the Hohenzollern holdings spoke German, they were primarily Reformed or Lutheran, and they were German. Yet, they were geographically and ideologically separated, and they behaved accordingly. They were reluctant to commit tax revenue for the protection of a

---

[11]E. J. Feuchtwagner, *Prussia: Myth and Reality: The Role of Prussia in German History* (Chicago: Henry Regnery Co., 1970), 26–27.

foreign government. In the Hohenzollern lands outside Brandenburg, Berlin was a foreign capital.

Frederick William approached the problem of financing in a number of ways, adapting each method to the situation in each local province. He rented some of the family estates and sold others. While the diets controlled most tax revenues, the elector began charging tolls to use rivers and roads. He also imposed a tax on commerce. He revamped the tax system and created a new excise tax. He even went to the diets and asked for permission to raise taxes. Sometimes he simply tired of dealing with the diets and gathered funds illegally. No matter which method he used, Frederick increased and centralized his power, and he placed his new-found money in the royal war chest.

Frederick William's efforts met with success. By 1644 troops were increasing at a slow pace. He maintained a personal guard of musketeers and increased their ranks to 500 men. In addition to the 2,000 garrison troops in Brandenburg, he forced each province to raise a small standing army. Once each army was established, he moved it among the provinces at his discretion. When compared to his potential enemies, this was not much of a force. It was, however, a beginning.[12]

The problem of organization was a different issue. Frederick William wanted to create an army that would be loyal to the Hohenzollern family. It would serve no purpose to have an officer corps that held more loyalty to its local province than to its sovereign, so the selection of each regiment's officers was crucial to the elector's goal. As Frederick William formed local regiments, he paid particular attention to the selection of officers. At the outset of his reign, most of Frederick William's military efforts were aimed at simply gathering enough troops to protect his scattered borders. As he selected leaders for these troops, he began to lay the foundation for the most professional officer corps in history. Frederick William began to solicit experienced officers who pledged absolute loyalty to the Hohenzollerns.

The first person to come into his orbit was Colonel Burgsdorf. The role of a colonel in the seventeenth century was important and merits explanation. Most military units in the Thirty Years' War fought as autonomous commands. The base unit was the regiment and its leader, a colonel. Often little more than a commander of mercenaries, the colonel was responsible

---

[12]Schevill, 114.

for commanding, supplying, and committing the troops. He also served as paymaster. There were any number of methods to become a colonel and an equal number of creative ways to raise a regiment. While such informal practices may seem strange to a culture acclimated to the bureaucratic regulation of everything from families to private industry, the power of the colonel was the norm in the seventeenth century. The colonel was the primary authority figure in the basic military unit. He selected his own officers and employed his men when and where he liked. Generals, dukes, and kings depended on the loyalty of their colonels. This made Burgsdorf an ideal candidate for Frederick William.

Konrad von Burgsdorf believed in the abilities of Frederick William and was quick to pledge his loyalty to the elector. After helping the elector depose Schwarzenberg, Burgsdorf accepted his next task with enthusiasm. When the elector realized that the Swedes were not going to live up to their promise of leaving Brandenburg, Frederick William charged Burgsdorf with the task of gathering troops for an army. Burgsdorf assembled the garrison troops in Brandenburg and raised another 1,000 in Prussia. By 1644 the army was growing. Frederick William made one amendment to the common practice of officer selection. As Burgsdorf developed candidates for regimental command, Frederick William took an active role in selecting not only the colonels but every other officer in the regiment. As a result, every officer was selected and commissions by the elector.

The elector found another commander cast in the mode of Burgsdorf from the western provinces. General Johann Norprath was charged with assembling an army in Cleve. While Norprath faced the same fiscal constraints as Burgsdorf, he began scouring the countryside. Officers were relatively easy to recruit from the nobility or, on rare occasions, from talented peasant soldiers. The non-commissioned ranks were gathered from the least fortunate of society and citizens who volunteered for local militias. Once again, Frederick William selected all the officers. Norprath was able to raise nearly 2,000 troops virtually from thin air. By the end of the Thirty Years' War the elector commanded roughly 5,000 troops. It was a small army, but a beginning, and the officers owed their positions solely to the elector.[13]

---

[13]See comments by Gordon A. Craig, *The Politics of the Prussian Army: 1640–1945* (Oxford: Oxford University Press, 1968), 6–7.

The elector's final task was to create an effective fighting force. After all, even if his army was well financed and organized with efficient, loyal officers, it would be virtually worthless unless it could fight. Frederick William modeled his army after the Swedish army, decreasing the ratio of pike men to musketeers due to lack of personnel. Cavalry and artillery units were necessarily small due to their tremendous costs. In order to understand how this structure affected the army's ability to wage war, it is necessary to discuss the nature of seventeenth-century warfare.

Neoclassical Warfare

After the Renaissance, European armies fought in lines like the ancient Romans and Greeks, with one major difference—European armies fought with gun powder. Essentially, armies engaged in two types of fighting, sieges and field battles. Fortifications dominated the central European countryside through much of the seventeenth and eighteenth centuries. Any army moving through an area faced the necessity of reducing or neutralizing fortifications. Therefore, one of the most common forms of military conflict involved laying siege to an enemy fortress. Conversely, the construction and defense of fortresses literally became a military science. This style of fighting was known as siege craft and fortification.

Most major towns in Europe were constructed to double as fortresses. Control of a city was immensely important because it covered supply and communication lines.[14] When armies foraged, such as the armies of the Thirty Years' War had done, cities could also serve as supply bases. In defensive operations, fortresses were of immense importance. A small group of defenders could often hold out against an army many times its size. Most cities were held by local garrison troops, even in peace time. Defenders held most of the advantages. It took many more soldiers to attack a fortress than it did to defend one. Defending forces had three major tasks. First, they maintained the fortress walls, ensuring that an attacking enemy would be subjected to cannon and musket fire. Second, they raided enemy trenches. Third, they dug counter trenches and counter mines when enemy trenches, or works, came close to the walls. Trench digging was boring, dangerous work, usually reserved for inferior troops.

---

[14]Duffy, *Siege Warfare,* 247–64.

The job of the attacking commander was to breach the city walls and force its capture. The methods for doing so were prescribed by tactical traditions and the physical range of fortress artillery, and the attacking commander usually needed a group of talented engineers. Some of the best masters of siege craft, such as the French Marquis Sebastien Vauban, were outstanding engineers. Storming a fortress wall could be done, but it was costly. Most attackers preferred to surround a city and cut it off from any supporting troops. Attacks were usually limited to artillery bombardment and digging. Attacking forces dug trenches to bring their artillery closer to the fortress walls. Counter-mining involved attempts to disrupt the attackers' digging.

Using the three basic tactics, the defending commander simply had to wait out the siege. If the defenders had an adequate supply of water and food, and if the sanitary system inside the besieged city remained intact so that disease was contained, a defending commander could usually count on success. If the attacking commander was able to dig a series of trenches that allowed him to reach the city wall with artillery, the story might be different. In siege warfare the goal of the attacking commander was to compel a fortress to surrender. If the city was poorly supplied, the attacker merely had to wait until starvation and disease caused the defenders to raise the white flag. If the fortress was well-stocked, the attacker was usually forced to dig toward the wall. Digging fell to the sappers, the ancestors of modern military engineers.

The sappers or miners employed two basic tactics. The most common was to approach the fortified wall in a series of trenches. The purpose was to bring the trenches progressively closer to the enemy's wall so that cannon shot could eventually breach the wall. By the rules of seventeenth-century warfare, once a wall was breached, the defenders were expected to surrender with full military honors. That usually stopped the wholesale slaughter of defending military forces. The second tactic was a variation on the first. Rather than relying on cannons to breach the walls, miners would dig underground toward the enemy wall. Once they were under the wall, a number of techniques could be employed to break it. Sometimes a large cavern was constructed and supported by wooden beams. The miners then placed air holes along the top of a cavern and set the beams on fire. When the beams burned away, the ground beneath the wall collapsed, usually taking the wall with it. A variation on this theme was to fill the cavern with gunpowder and then blow a hole in the wall. Regardless of the technique, the goal of the attacking force was to break the wall while the defenders tried to keep the wall intact.

Although the terrain around fortifications varied, the primary method of attacking a fortress was fairly standardized. The attacker dug the first trench, called the first parallel because it ran parallel to the fortress wall, about 600–700 yards away from the fortress. This matched the limit of fortress artillery. If the fortress was elevated, it was necessary to dig the first trench further away from the wall to account for the increased range of heavy artillery pieces. Digging commenced by slowly carving out areas where heavy siege artillery could be used to fire toward the city. These siege batteries became small fortresses of their own.

As digging continued, sappers moved to a second trench called the second parallel. This trench was important. It allowed the heavier guns of the besieging army to strike the walls of the fortress directly. In addition, it provided an area for the sappers to begin a direct approach to the city walls. Using a series of smaller oblique trenches, sappers gradually moved forward to the final trench, the third parallel. All arms, including muskets, could be brought to bear from the third parallel. Sappers also dug mines from this area to destroy a defending wall. The infantry played an important role in the last two parallels, defending the sappers and the artillery from raids out of a defending camp. In 1648 it would have taken nearly all of Frederick William's military resources to conduct a siege. On the other hand, he could have defended several fortresses with his modest army.[15]

The other style of fighting was known as the field battle, and this suited the structure and numerical strength of Frederick William's army. The Thirty Years' War reintroduced Europe to Greek and Roman styles of battle. European armies fought in a series of linear formations. In standard formations the infantry gathered at the center, wings of cavalry units were assigned to each flank, and artillery was positioned directly in front of the in-

---

[15]Christopher Duffy, *The Fortress in the Age of Vauban and Frederick the Great: 1660–1789* (London: Routledge and Kegan Paul, 1985), 78–81. For an excellent case study see Simon Pepper and Nicholas Adams, *Military Architecture and Siege Warfare in Sixteenth-Century Siena* (Chicago: University of Chicago Press, 1986). For an English summary of Vauban's comments see Geoffrey Symcox (ed.), *War, Diplomacy, and Imperialism: 1618–1763* (New York: Walker and Company, 1974), 166–76.

fantry. The primary tactic involved frontal assaults, and the opposing forces generally spent their efforts in attempts to punch a hole in their enemy's line. If this was achieved, the attacker attempted to flood the gap with cavalry or infantry.[16]

The infantry was composed of two types of soldiers. The most common soldier in Europe was the pike man. Pike men carried long wooded spears from twelve to eighteen feet long, and they fought from three ranks. When threatened by an enemy advance, especially from the cavalry, the pike men rushed forward and buried the blunt ends of their spears. The front row knelt and buried the butt of the pikes in the earth. A second row stood and provided a complementing row of sharpened points. A third rank stood behind, ready to take the place of any fallen comrade. In reality, the job of the pike man differed little from that of the spear man in a Greek phalanx 2,000 years earlier.[17]

King Gustavus Adolphus of Sweden brought a new role for the pike men during the Thirty Years' War. Instead of relying solely on the three ranks of pikes to defend, Gustavus massed his Swedish pike men in a single formation. When an enemy's line appeared to be in danger of collapsing, the Swedish pike men would rush forward in a massive charge. Gustavus also covered the ends of the wooden pikes with metal spear tips. This protected them from damage by enemy cavalry who frequently rode parallel to pike formations to cut off the points of pikes with their swords. The Swedes were known for their ferocity and their ability to break enemy formations with the pike.

The second type of foot soldier was the forerunner of the modern infantry, the musketeers. At the beginning of the Thirty Years' War, they carried large, thirty-pound weapons. The cumbersome guns were fired by igniting a charge of powder with a smoldering wick attached to the trigger mechanism. The weapons were called "matchlocks" after their firing mechanisms, and they often failed to operate in poor or rainy weather. The heavy weapons required a stand when pointed at the enemy. A good musketeer could

---

[16]Geoffrey Parker, *The Military Revolution: Military Innovation and the Rise of the West, 1500–1800* (Cambridge: Cambridge University Press, 1988), 16. (Parker has an outstanding discussion on the organization and purpose of field battles.)

[17]For a full discussion see Martin van Creveld, *Technology and War: From 2000 B.C. to the Present* (New York: The Free Press, 1989), 81–97.

fire about one shot every sixty to ninety seconds. The rounded bullets were so inaccurate that infantrymen fired in volleys—that is, several weapons in a line firing at once—to ensure that at least some of their shots would strike the target. There was no standard caliber or cartridge. After a musket was fired, the musketeer relied on his comrades with pikes for protection.

Gustavus Adolphus changed the role of the musketeer. Always thinking of the offensive, Gustavus streamlined the muskets of the Swedish army until they weighed slightly more than eleven pounds. He also introduced paper cartridges. Each musketeer received a number of pre-measured rounds easing the complicated process of reloading. The Swedes experimented with different types of firing mechanisms. While the matchlock was the most common type of musket, other locks improved the rate of fire. The Swedes became so proficient that they could fire one volley per minute. Other European armies followed suit.

The cavalry had three basic functions: scouting, protecting the flanks of the infantry, and exploiting weaknesses in the enemy line. Occasionally cavalry units were used for direct assaults against infantry, but this was a costly and risky venture. Horses were expensive and difficult to train. While hard-charging cavalry commanders usually threw caution to the wind, army commanders tended to employ cavalry units with caution. In the initial stages of a battle, the primary purposes of the cavalry were usually limited. Cavalry units harassed any enemy infantry in the area and defended their own battle line from enemy cavalry. If the enemy's line weakened, the cavalry's role increased dramatically. Massed cavalry units would be ordered to attack the weak point. As a thunderous herd of horses closed toward the enemy, the front line of the attacking cavalry would be compressed. It was not unusual for many of the horses and riders in the midst of such a charge to be lifted from the ground and carried along in the air. The enemy line would hopefully be broken by "shock action" as hundreds of cavaliers smashed into a line of infantry at twenty to forty-five miles per hour.

Linear infantry formations not only allowed the musketeers to fire in volleys, they afforded some measure of protection from the cavalry. As long as the infantry line remained intact, even the shock action of a cavalry charge could be stopped. Volleys of musket fire and cannon shot could also stop cavalry formations in their tracks. Infantrymen utilized two other forms of protection, spiked wooden barriers and caltrops—metal triangle spikes thrown on the ground to penetrate horses' hoofs—to protect themselves from cavalry assaults. In the early 1600s, the role of the cavalry was crucial, but by the end of the century its importance was beginning to wane. The infantry's muskets were gradually asserting their superiority over the cavalry.

Three types of cavalry were used in the seventeenth century, and they remained on the battle field until well after the Napoleonic period. Heavy cavalry, or cuirassiers, formed the mainstay of cavalry troops. These heavy cavalrymen were used both to shock enemy troops and to exploit any breach in an enemy's line. When the cuirassiers charged, most of the other cavalry units followed suit. Cuirassiers usually rode large horses, and they wore armored breastplates and helmets. Their uniforms were designed for protection against pikes. They were directly descended from the medieval knights, and they wore their title proudly.

The second type of cavalier was a mounted infantryman known as a dragoon. Dragoons had a twofold purpose. On a moment's notice, they could perform the role of any type of cavalry unit. By the same token, the dragoons could dismount and fight as infantry. Carrying short-barreled carbines, they could speedily ride to an area along the front, dismount, and fire a rapid volley. In theory, this made them more mobile than the infantry. In practice, cavalry fire was notoriously inaccurate. The dragoons were no excetion. They usually fought as heavy cavalry.

The final type of horse soldiers was the light cavalry. True to their name, these cavaliers were generally short in stature and they carried a lean frame. They had to be small because their purpose was speed. The light cavalry was used to scout, harass, and forage. Heavy cavalry was designed for frontal assault. The light cavalry was designed to cover ground. As light cavalry evolved, everything from horses to uniforms was geared toward speed. They became masters of feinting and reconnoitering. The Hungarian hussars became the most notable light cavalry units in Europe. Other armies imitated their tactics and even copied their name. Eventually, the term "light cavalry" became synonymous with "hussar."

The role of the artillery increased during the seventeenth century. In the early years of the Thirty Years' War, most commanders felt that artillery was indispensable. At the same time they recognized that using the artillery was a cumbersome and burdening process. Creating an artillery train, the logistical supply force needed to bring the artillery into the field, was one of the nightmarish administrative tasks for seventeenth-century commanders. The job of moving and supplying six cannon, for example, involved a tremendous amount of organization. It would entail moving hundreds of men, scores of horses, and dozens of supply wagons. The role of the artillery was limited because it was cumbersome.[18]

Artillery was usually deployed to the front of a battle line, and three types of pieces were available to the armed forces.[19] The most common weapon used in the field battle was the cannon or gun. The cannon was a direct-fire weapon, and it was designed to shoot straight at the enemy. There were no standard calibers until Gustavus intervened, but the projectiles were grouped into variations of a single theme. For long range, cannons used solid ball shot. For shorter ranges, they would used shot with chains attached or two cannon balls held together with a single bar. For close encounters, artillerists used canister or grapeshot. This method of firing involved placing a canister of musket balls in the barrel of a cannon and firing it directly into an advancing enemy. The shot spread in a pattern similar to a shotgun blast. If the enemy advance was threatening to overrun a position, artillery men would often simply improvise grapeshot by placing musket balls and other suitable objects in the barrel. Grapeshot was seldom effective beyond 100 yards. Within its effective range, however, it was deadly.

Mortars complemented cannon by employing indirect fire. Firing at high angles, the mortar could be used to loft heavy shells high into the air. They were excellent for firing over mountains or fortress walls, but they were almost impossible to utilize in a field battle because of their tremendous weight and complete lack of mobility. Howitzers provided an acceptable alternative. Unlike cannons, they fired at an angle, producing a trajectory much like that of a football pass. Yet, they were mounted on wheels and were much more mobile than mortars. The howitzer also frequently fired another type of shot, a fused shell full of gunpowder. Upon being fired, the fuse would ignite and the cannoneer hoped it would explode somewhere in the vicinity of the enemy. One of the main duties of artillery men was to trim fuses to time an explosion properly. This was hardly an exact science, and rainy weather or damp ground would extinguish the fuse. Indirect fire

---

[18]David Chandler, *The Art of War in the Age of Marlborough* (New York: Hippocrene, 1976), 141–57.

[19]O. F. G. Hogg, *Artillery: Its Origin, Heydey, and Decline* (Hamden, Conn.: Archon Books, 1970), 25–27.

weapons were available to field commanders, but the direct fire of the cannon dominated the tactical employment of artillery on the battlefield.

Frederick William's New Army

As Frederick William tried to create an effective fighting force, his main challenge was to build an army that could fight in the field. Small units could be used to garrison his fortresses, and garrison duty required little training or skill. Second-rate troops could generally do an adequate job when asked to defend a fortress. Field battles were another issue. Frederick William needed mobility and commanders who could seize the initiative. As he increased the size of his army, he needed to have the mobility to get to any potential battle site in his scattered domains and to bring the full weight of combined infantry, cavalry, and artillery to bear on his enemies. In other words, his army required speed and coordination if the Hohenzollern lands were to be defended. Frederick William began building his army to meet these requirements.[20]

The core of the elector's army became his field force. By 1645 he had 5,000 men under arms, and he concentrated his efforts on the infantry. The elector slowly built artillery units, but cannons were expensive. Cavalry units were also costly, and the elector's subjects had few horses to spare. Therefore, the Brandenburg cavaliers were mounted on horses that were assigned to double duty. Most of the cavalry sat astride farm animals. When not employed in a military assignment, the horses pulled milk wagons and plows. Faced with shortages of cavalry and artillery units, the Brandenburg infantrymen also compensated by developing tactics to cope with situations without support. Frederick William's infantry began to grow very proficient. Although his soldiers could not know it, they were beginning a tradition that would become legendary.

---

[20]Schevill, 115. See comments by Hans Delbrück, *History of the Art of War within the Framework of Political History, Volume IV, The Modern Era* (trans. by Walter J. Renfree, Jr.) (Westport, Conn.: Greenwood Press, 1985; orig. 1920), 245.

Chapter 2

The Elector Employs His Army

Frederick William fancied himself an army commander, and he conceived his role as head of state in military terms. It is not surprising that the elector was not content simply to build an army. While it would be unfair to portray Frederick William as a steed champing at the bit, he did not hesitate to employ military force when he thought it was advisable. He did so even before the end of the Thirty Years' War in a religious dispute over the city of Jülich. He went to war again in 1655, albeit rather reluctantly, when Sweden invaded Poland. After two frustrating encounters with France in the 1670s, Frederick William refocused his military energies on Sweden and the conquest of western Pomerania. Although he assumed the throne in a position of weakness, he left the throne with power. In the end, he lived up to his self-image. He was an excellent general.

The Jülich Interludes

When Denmark rejoined the Thirty Years' War in 1643, it was obvious that the major powers were no longer concerned with religion. The two Protestant powers of the north, Sweden and Denmark, fought with each other, while Catholic France waged war against Catholic Austria. The Swedes were unapologetic when they handed Brandenburg's fortresses back to Frederick William and hastily retreated from his land. The Swedish commander, Lennart Tortensson, voiced nominal concern for the Lutheran church. His intended to eliminate the Danes.[21] Frederick William was caught in the midst of these diplomatic struggles. Loyal to the Reformed faith, he followed a dual standard in the practice of statesmanship. Personally, he followed the moral code of his faith. Diplomacy was another matter.

---

[21]For an analysis see J. V. Polišenský, *The Thirty Years' War* (trans. by Robert Evans) (Berkeley, Calif.: University of California Press, 1971), 257–63.

He would lie, cheat, and steal to protect Brandenburg.[22] Yet, there was something deeper. Frederick William was one of the few minor German princes would still fight for his faith, even when the encounter threatened Brandenburg.

With Tortensson gone, the elector took full advantage of the Swedish retreat. Not only had he learned a valuable lesson about backing diplomatic promises with military muscle, but he vowed never again to be caught in a position of such utter weakness. After forming the 5,000-man army, the elector sought to strengthen it. He knew Tortensson's word was only good while the Danes threatened the Swedish rear. If that threat were eliminated, the Swedes might well return to face the Austrians. The only solution the elector saw to the security problem involved the reinforcement and expansion of the Brandenburg military. The elector cared neither for Austrian Catholicism nor Swedish Protestantism in the realm of diplomacy. Going to war for the sake of religion would seem to have been the furthest thing from his mind.

For the next two years, the elector focused on taxation, not religion. He was in a familiar vicious circle of revenue gathering, and his first military policies were more suited to economics and political philosophy than the drill field. He needed to revamp the tax system and convince the scattered Hohenzollern states that they had an obligation to provide for a common defense.[23] Rhenish merchants were no more receptive to Frederick William's ideas than the Junker farmers of Prussia. They controlled the purse strings and, therefore, much of the elector's military power. Under these circumstances, it is difficult to think that any cause could motivate the elector to military action. To everyone's surprise, the elector mobilized in 1646 under the banner of Protestantism.

---

[22]Ferdinand Schevill, *The Great Elector* (Chicago: University of Chicago Press, 1947), 52. See also E. J. Feuchtwanger, *Prussia: Myth and Reality: The Role of Prussia in German History* (Chicago: Henry Regnery Co., 1970), 22, and discussion by Herbert Tuttle, *History of Prussia, volume I: 1134–1740* (New York: AMS Press, 1971; orig. 1884), 52.

[23]Schevill, 187–95. See also F. L. Carsten, *The Origins of Prussia* (Oxford: Oxford University Press, 1954), 129 ff. (Carsten has a detailed summary of the elector's taxation policy.)

Wolfgang Wilhelm was the Catholic Duke of Neuburg and a claimant to the territory of Jülich, a minor city state along the Rhine River. The Hohenzollerns had a legitimate claim to Jülich based on a previous treaty. Under the agreement, Jülich was to be a free city where the rights of Protestants and Catholics were to be protected. The Hohenzollerns were to enforce the edict. Frederick William had not paid much attention to the small burg until the summer of 1646 when the Protestant residents claimed that they were being harassed by the Catholic duke. The elector sent word to Wolfgang Wilhelm, ordering him to let the Protestants live in peace. The duke completely ignored both Frederick William and the Hohenzollern claim to Jülich. Frederick William decided to act. If the duke would not accept the elector's claim to Jülich, perhaps he might accept a show of force. Frederick William mobilized a good portion of his army.[24]

Selecting General Norprath to command the force, Frederick William dispatched troops to the Rhenish province of Cleve and strengthened them with the local garrison. Comprising 3,000 men, the army was not strong enough to shake the confidence of Sweden and France, but it certainly frightened the 500 men in Wolfgang Wilhelm's army. The duke took a different approach. Believing that Frederick William would not use military force without the permission of a major power, he ordered his small army to stand defiantly in Jülich. Wolfgang Wilhelm believed that he would find an ally to thwart the elector's threats. To his consternation, no one responded. In November Frederick William ordered the army to cross the duke's borders. With no ally to support his cause, Wolfgang Wilhelm took the most logical action: he ran.

The Duke of Neuberg's small army escaped to a fortress before Frederick William's men could strike it in the field. Frederick William had proven his point, but the powers in Europe were no longer anxious to see the Thirty Years' War continue. France spoke of peace, and Austria, Sweden, and Denmark hastened to engage in deliberations. As peace negotiations opened, the major powers had no desire to divert their attention to Jülich or to see the local dispute flare to a wider conflict. Frederick William was forced to withdraw under strong diplomatic pressure. Under his protest, the territory of Jülich was divided, and the elector was given the county of Ra-

---

[24]Schevill, 122–23. See also John Stoye, *Europe Unfolding: 1648–1688* (London: William Collins Sons and Co. Ltd., 1969), 37.

vensberg as compensation. Wolfgang Wilhelm agreed to stop persecuting the Protestants in Jülich, and the two German royals signed a peace treaty on April 8, 1647. All was not settled, however. The elector had quite effectively demonstrated that he had an army and that he was not afraid to use it.

The entire affair displeased Frederick William. His army had failed to achieve its objective, and the elector was dissatisfied with the manner of mobilization. Blame fell on Konrad von Burgsdorf. While Burgsdorf's crude, soldierly manners made him a popular field commander, they were not suited to organization, supply, and movement. His talents fit the saddle, not the desk. Norprath had performed well once the invasion started, but Burgsdorf had not provided the necessary troops and supplies. Frederick William believed that his army should have been able to catch Wolfgang Wilhelm before he ran away, and he exhibited little sympathy toward the loyal Burgsdorf. Frederick William believed that the mobilization had been lethargic and disorganized. In the elector's eyes, Norprath had accomplished his mission; Burgsdorf had not. Frederick William would not tolerate failure, even from his most trusted soldier.

Burgsdorf incurred Frederick William's wrath once again in 1651 when Wolfgang Wilhelm began harassing the Protestants of Jülich for a second time.[25] After sending notes of protest, the elector mobilized the army, giving Burgsdorf instructions not to fail. The elector's actions were not popular among his advisors. The relative peace following the Treaty of Westphalia was good for the Hohenzollern lands. The Privy Council was appalled by Frederick William's order for mobilization and strongly recommended against military action. Burgsdorf joined in the crescendo of protests, stating that there too many risks associated with another Jülich invasion. Frederick William dismissed such worries, and he noted Burgsdorf's reluctance to support his policy. On June 13, 1651, Frederick William ordered 4,000 men to invade the duke of Neuberg's territory. Mobilization and supplies were problems.

The Brandenburgers captured border towns without a great deal of effort and within days surrounded Wolfgang Wilhelm in Düsseldorf. Their numbers were far too small to lay siege to the city, so the Brandenburgers were forced to camp outside the range of Düsseldorf's fortress cannon. The duke immediately appealed for help from the neighboring German states,

---

[25]Ibid., 139–44.

and his Catholic colleagues responded with 7,000 troops. Realizing that he was on the verge of a major war, Frederick William responded by mobilizing his entire army and sending it forward. As the army marched away, he then came to see the logic of the Privy Council. With the bulk of his army far away at Düsseldorf, his lands lay unprotected. War over Jülich was unwise. Unfortunately for Burgsdorf, the elector would not accept blame for the failed policy.

Two reluctant armies stood near Düsseldorf in the summer of 1651, neither wishing openly to challenge the other nor feeling that retreat was possible. It was a logistical debacle. Relief surged through the men-at-arms when the Holy Roman Emperor intervened from Vienna. Assuring Frederick William that he would force Wolfgang Wilhelm to honor the conditions of religious tolerance, the emperor brought the duke and the elector to the bargaining table. The crisis fizzled out. On October 11, 1651, all the combatants signed a peace treaty. Frederick William was not completely mollified. Once again he held Burgsdorf responsible for the failure. Frederick William saw the weakness of his trusted colonel. Better organization could have prevented the debacle at Düsseldorf, but he perceived that Burgsdorf was incapable of providing such organization. Burgsdorf became the sacrificial lamb. Frederick William dismissed the colonel and ordered him never to return to his beloved Brandenburg. Broken in spirit, Burgsdorf left for his family estate. Frederick William tolerated no failure in service to Brandenburg, and he never blamed himself.

The elector learned another lesson from his adventures over Jülich. Realizing that organization was the key to employing the army over scattered provinces, Frederick William created a military office, the quartermaster general, to supply and support all military actions. While the title of the office shifted over the years, a general officer held the position. In the 1600s it was common to refer to all the generals as the "general staff," but the function of the quartermaster general was geared toward staff operations. Frederick William selected General Otto von Sparr to supply, support, and organize all the logistics of the Brandenburg army. For all intents and purposes, Sparr was a general staff officer. When Frederick William next went to war, he wanted to make sure that the army was organized. Sparr was responsible for doing so.[26]

The Polish War 1655–1660

Frederick William's system of military organization was tested again in 1655 when the elector was reluctantly drawn into a conflict on his eastern border.[27] Seventeenth-century Poland was a country still languishing in the Middle Ages. While western Europe was caught in the grips of the Scientific Revolution, Poland disavowed the principles of Copernicus. Its agrarian mode of production blended with the feudal distribution of land. Serfs were tied to the farms and subject to the jurisdiction of their nobles. The once-strong Polish army had dominated eastern Europe. By the 1600s its methods of fighting were outdated and ill suited to deal with the power of armies using musketeers armed with gunpowder. The feudal style of warfare made Poland a tempting plum for a modern army. Many of Poland's neighbors looked at the county's military backwardness as an opportunity for political expansion—expansion to be won at the expense of the Polish people.

Poland's most aggressive and opportunistic neighbor was Sweden. Charles X, King of Sweden, envisioned himself as the new Gustavus Adolphus. Perhaps he overestimated himself. Far from the mold of the politically astute Gustavus, Charles was a short-sighted militarist who excelled in battle and failed at the affairs of diplomacy. In 1655 he saw in Poland an unbridled opportunity. Knowing that Poland was threatened by Russia and the Ottoman Turks, Charles believed the time was ripe to strike at its undefended front—the Baltic coast. Armed with a desire to turn the Baltic Sea into a Swedish lake, Charles devised a plan to destroy Poland.

Charles developed a two-front strategy for confronting Poland. One wing of his invasion would move by land along the Baltic coast. This would allow Sweden to seize the Baltic states—Latvia, Estonia, and Lithuania—and attack Poland from the east. The other wing would comprise two groups united under Charles's personal command. The first group, under

---

[26]Walter Goelitz, *The German General Staff: Its History and Structure: 1657–1945* (trans. by Brian Battershan) (London: Hollis and Carter, 1953), 3.

[27]Schevill, 162–75.

Count Arfwid von Wittenberg, was to originate in Swedish Pomerania. Wittenberg was to cross East Pomerania and move into Poland. Charles was to cross the Baltic Sea with another force and link with Wittenberg in northern Poland. Together they would move toward Warsaw. This posed a political problem for Frederick William. East Pomerania was a Hohenzollern territory.

Oblivious to the niceties of diplomatic negotiation, Charles began the invasion in the summer of 1655. He asked Frederick William for permission to cross East Pomerania, and he also asked for seaports in Prussia to land the troops from Sweden and to serve as their base of supply. The elector was not eager to cooperate. For one thing, he was a vassal of John Casimir, the Polish king. Prussia was Polish territory. Frederick William saw no political advantage in confronting his suzerain, and he feared several potential disadvantages should Charles fail. In addition, Charles had eyes on two seaports, Memel and Pillau, for supply bases. Frederick William realized that he would virtually have to surrender these ports if he acquiesced to Charles's demands. Frederick William quietly approached the Dutch to ask for protection from the sea. At the same time he began to assemble an army in Königsberg.

Frederick William was not without friends. His wife was from the house of Orange, and the royal family of the Netherlands was openly sympathetic to Brandenburg. If it had been up to the Dutch king, Holland and Brandenburg would have been allies. Frederick William had hoped for this alliance since 1640, but the tight-fisted merchants of Holland had other ideas. Alliances meant armies, and armies meant expenses. The Dutch traders had no intention of paying for a military alliance with Brandenburg prior to 1655. Their attitude suddenly changed when Charles's expansionist dreams threatened Dutch profits in the Baltic. If Sweden controlled all the Baltic ports, it would decrease Dutch revenues. The potential loss of income worried the Dutch estates. They agreed to protect Frederick William by sea, if he would agree not to surrender Memel and Pillau. It seemed to be a good idea, but it came too late. Brandenburg signed an alliance with the Netherlands in July, a few weeks after Charles and Wittenberg joined forces in northern Poland.

Charles did what the Dutch could not do. He defeated himself. Enraged with Frederick William's failure to cooperate, Charles simply plunged into

Poland in late July. The Poles fought hard, but they were overmatched by a force of 32,000 Swedes. Charles moved south, seizing several Polish cities along the way. By September 8 he had taken Warsaw, and by September 25 Charles was in Cracow. John Casimir was forced to flee to Silesia while Charles relished his victory. The celebration was to be short-lived.

While Charles was busy in Poland, Frederick William managed to gather a number of troops in the north. Moving through Polish Prussia, a province to the west of Hohenzollern-owned East Prussia, the elector signed defensive alliances with the Polish nobles. He gathered a number of troops at Königsberg and prepared for war against Sweden. The policy was not completely successful. Charles returned from Cracow and attacked East Prussia. He also drove deep into Brandenburg and prepared to lay siege to Berlin. Frederick William consulted with his generals, who urged him not to fight. For his part, Charles wanted an alliance with the elector, not a war. The elector reluctantly saw the futility of the situation. In January 1656 he formally allied with Sweden. Charles did not have the insight to know that he had enlisted an unwilling ally.

Disaster was not far away for Charles. The Polish army might have been outmoded, but a sense of nationalism burned in the hearts of the average Pole. Polish peasants were infuriated as they saw Swedish troops— Protestant troops—occupy their most sacred cathedrals. Stories of Swedish atrocities circulated from village to village. Altars were desecrated and the land laid waste. Men and women were slaughtered under the swords of the merciless Swedes—so the stories said. It was more than the peasantry could bear. Armed with popular anger, the Polish peasants began a spontaneous winter uprising. They recalled John Casimir from Silesia and placed him at the head of a guerrilla army. Even the militarist Charles had to take note. His long supply lines were in jeopardy. King Charles rejoined the main body of his army in Cracow only to find that he needed to retreat. His once-strong army began to trudge north through the snow.

Frederick William was aware of Charles's difficulties. While Swedish strength was being frittered away in a long winter retreat, the elector kept building his force in Königsberg. When spring came he had several thousand troops in Prussia. This was much better than the Swedish position. Charles had retreated to Warsaw but found little relief. The Polish peasant army surrounded him, and the fierce winter had destroyed all but a handful of his original army. Faced with impending disaster, Charles left Wittenberg in Warsaw with 4,000 troops and escaped to the north. This time he approached Frederick William with a little less bravado than he had exhibited while dictating his earlier terms. One of the elector's advisors, George Frederick Waldek, suggested that the elector take advantage of Charles's situation. Frederick William told Charles that Brandenburg would assist Sweden, but the elector had some new conditions.

The elector was wise to assist the Swedes for two reasons. First, Charles was wounded, but the wounds were not permanent. To be sure, he lost an army in Poland, but the king could produce a new army. The second Swedish army, for example, was well on its way through the Baltic states. Second, Frederick William had been a vassal to John Casimir, and, albeit under duress, he had broken his feudal oath to him. Prussia was now under the control of Sweden and Charles, for better or for worse, was the elector's new suzerain. It would not have been wise to abandon Charles at this juncture. If all the factors were held constant, the Swedes seemed to be the favorite candidate for victory. On Waldek's advice, the elector agreed to help Charles with more than 8,000 troops, provided that Frederick William retained autonomous control over the Brandenburgers. Charles agreed to the condition; he had little choice.

The elector and his new king marched south from Prussia in the early summer of 1656. Wittenberg's garrison surrendered Warsaw in June before Charles and Frederick William could arrive. As the Poles turned to meet the new army, Frederick William prepared to fight his first major battle. On July 28, 1656, 8,500 Brandenburgers and 12,000 Swedes attacked 50,000 Poles just outside Warsaw. The battle lasted for three days, resulting in a victory for Charles and the elector. At a council of war, Charles demanded that the combined armies push south. The elector was not so sure. He was reluctant to follow the same policy that had resulted in the destruction of the first Swedish army, and he was alarmed by reports of Lithuanian cavalry raids into Prussia. After a heated exchange, Frederick William retreated north to Prussia. Helpless to stop him, Charles moved south to Cracow.

If Charles was unable to read the nuances of diplomatic language, Frederick William was a master at it. His movement back to Prussia proved fruitful. The Swedes in northern Poland were in siege around Danzig while the elector concentrated his forces in East Prussia. At long last, the Dutch navy came on the scene. They sailed into Danzig and lifted the siege. Frederick William, who had never wanted to fight the Poles, saw a diplomatic opportunity. He courted the Dutch and the northern Polish nobles in hopes of withdrawing from the war. For Charles, things only became worse. The nationalistic Poles would not quit. He found himself in another winter retreat. To make matters worse, the Austrians had become concerned with the Swedish presence along their border. The emperor sent an army under General Raimondo Montecuccoli to assist King John Casimir. After losing a second army, Charles found himself back in East Prussia in November 1656.[28]

The once-proud Charles was now forced to beg Frederick William for help. The elector listened to the Swedish king's plea while searching for his own opportunities. With Waldek and 10,000 troops at his command, the elector agreed to help Charles on one condition: Charles must release Frederick William from his vassalage and recognize him as the autonomous Duke of Prussia. Charles was angered, but since he needed the elector's help, he agreed to Frederick William's terms. The elector sent a small army under Waldek to assist Charles, ordering Waldek not to support the Swedish cause enthusiastically. The Brandenburgers might shift sides again in the future. The sly advisor understood the implications perfectly. Both men knew that diplomatic shifts had been kind to Frederick William.

Charles' last effort in Poland faired no better than his first. Then, in May 1657, the Austrians entered the war in full force. Even though the Poles were virtually destroyed by subsequent fighting against Russia and Transylvania, Charles was on the run. By summer he was forced into Pomerania, where he found an even greater problem. The Danes, knowing Sweden to be weakened by the Polish disaster, brought a new war against Charles. Rather than seeking a negotiated settlement, Charles threw himself into an all out attack on Denmark. Frederick William approached the Polish court and asked John Casimir to accept an olive branch. The Austrians lobbied hard for the elector because they wanted to use Brandenburg forces against the Swedes. John Casimir was in no position to remind Frederick William that East Prussia rightfully belonged to Poland. Instead, he recognized Frederick William's independence. As a result, the Hohenzollerns in-

---

[28]See Tuttle, vol I, 177–80.

creasingly began to identify with their autonomous state of Prussia and less with the electorate of Brandenburg. For the army, the remaining portion of the Polish-Swedish war, the War of the North, accomplished the elector's original goals.[29] Frederick William drove the Swedes from the Polish Prussian and Pomeranian coasts. If the elector had gained control of the Swedish fortress of Stettin, he could have claimed uncontested control of Swedish Pomerania but he was unable to do so. When Charles died in February 1660, all sides sought peace. Louis XIV, the rising Sun King of France, offered to negotiate a settlement. Since Frederick William had failed to take Stettin, Louis told the elector to withdraw from Swedish Pomerania.[30] In compensation, Louis told the powers of Europe that Frederick William was the new Duke of Prussia, beholden to no one. France was the most powerful country in the world, and Louis's authority was unquestioned. The Duke of Prussia returned to Berlin, casting a wary eye toward the Swedish army encamped in West Pomerania.

Curt Jany cited this period as the formal birth of the Prussian army.[31] The Brandenburgers of the Thirty Years' War were the nucleus, and the staff and command system after Jülich brought the necessary administrative controls. The Polish war brought all of the factors together. Far from being a beggar on horseback, Schevill's appropriate term for Frederick William in 1640, the elector had assisted with the defeat of Sweden. He had also raised and commanded over 20,000 troops. When the war ended, the elector reduced his army to a skeletal force but kept the structure intact so that soldiers could be recalled in an emergency. The war gave the elector an autonomous state, Prussia, and the elector had given all the Hohenzollern domains an army. It was, to say the least, an a miraculous turn of events.[32]

---

[29]F. L. Carsten, "The Rise of Brandenburg," in G. N. Clark et al. (eds.), *The New Cambridge Modern History, Volume IV, The Ascendancy of France: 1648–88*), 546.

[30]Duffy, *Siege Warfare*, 188–89.

[31]Curt Jany, *Geschichte der königlich Preussischen Armee, Volume I* (Berlin: 1928–29), 116–27.

[32]For a summary of European military changes during this period see Richard Bunney, *The European Dynastic States: 1494–1660* (Oxford: Oxford University Press, 1991), 345–51.

The Dutch Wars

The Treaty of Oliva brought an end to the elector's involvement in Poland. Frederick William still desired to remove the Swedes from Pomerania, but he focused his attention on domestic affairs. Little did the elector know that he would soon be drawn to the Rhine River to confront an ambitious French policy of expansion. Louis XIV wanted to move French boundaries toward their "natural frontiers." In the south this meant the Pyrenees. In the west and north this meant the Rhine River. Any move toward the Rhine would force Louis to deal with the Spanish Netherlands and the United Provinces. Louis's best military option would be to cross the Rhine and strike the low countries from the east.[33] This would necessitate an advance through Germany and, much to Frederick William's consternation, movement through Hohenzollern lands. By 1665 the elector began casting cautious glances toward Paris.

Frederick William's sympathies were clearly with the House of Orange, the ruling family of the Netherlands. Louise Henrietta, his beloved Dutch wife, provided the familial relationship with the Dutch king, while the elector's Reformed faith served as an ideological attachment to the United Provinces. These sympathies, however, were not enough to commit an army. The Hohenzollern provinces, however, were another matter. The Rhenish provinces, Mark, Cleve, and the protected city of Jülich lay along the Rhine River. These small states and other properties formed the western section of the Hohenzollern lands. In effect, they made the elector a western prince. When the Rhenish German leaders appealed to the emperor for help, Frederick William was firmly in their camp.

Louis XIV made his opening move in 1667 by invading the low countries during the so-called War of Devolution. Claiming the Spanish crown by marriage, Louis sent troops across his northern border, only to encounter strong allied opposition. The Dutch attempted to secure their border by joining forces with the English and the Swedes in a Triple Alliance. Frederick William, who was greatly alarmed by Louis's attack, joined the Triple Alliance with Austrian support. It was too much for the Sun King. Realizing

---

[33]For an excellent summary see Paul Sonnino, *Louis XIV and the Origins of the Dutch War* (Cambridge: Cambridge University Press, 1988).

that he was isolated, Louis beat a hasty retreat and turned to diplomacy.[34] Louis understood that any move against the United Provinces could only come when the Dutch were isolated. The prospect of alliance seemed to be a more potent French weapon than gunpowder, given the situation.

The soldiers of the United Provinces were brave, but their army was small. Dutch military strength was vested in two areas outside the army. The wealth of the merchants and the strength of the navy were the highest ranking cards in the hand of the House of Orange. Louis hoped to trump these cards by building new alliances. Stated simply, an isolated Dutch army could not stand indefinitely against superior French numbers. The key, Louis believed, was to force the Netherlands to stand alone. Accordingly, the French king cautiously approached the Swedes and advised them that if they would abandon their alliance with the United Provinces, he would turn a blind eye to any moves they made in Poland. The Swedes quietly accepted Louis's offer. Louis then focused on England. Knowing that Charles II, King of England, desired to bring England back into the Roman Catholic fold, Louis signed the secret Treaty of Dover with Charles. The treaty promised French support for re-Catholicizing England. In return, the English were to allow Louis to have a free hand with Holland. With the Triple Alliance falling into disarray, Louis thought the time was right for a second move against the United Provinces. In 1670 he again declared war against Holland.[35]

If Louis had counted on German neutrality, his hopes were in vain. By treaty, the Dutch stood alone, but when Louis's troops crossed the Rhine River, the German princes rallied to the Netherlands. Frederick William was first among them, becoming their official leader in the absence of the emperor. The elector also led the local German princes in an appeal to Vienna. The emperor, Leopold I, responded with a polite but cold shoulder. The court at Vienna had more concerns with expanding Turkish power than with French violations of the Rhine. The Turks were in a position to threaten Vienna, and the Hapsburgs were trying to maintain a tenuous peace on their

---

[34]Ibid., 35–37.

[35]For summaries of the Dutch position see E. H. Kossman, "The Dutch Republic," 287–96 and G. Zeller, "French Diplomacy and Foreign Policy in the European Setting," 210–20 in G. N. Clark et al. (eds.), *The Cambridge Modern History, Volume IV, The Ascendancy of France: 1648–88* (Cambridge: Cambridge University Press, 1961).

Balkan flank. The emperor had little desire to send his army off to another front while the Turks were at his back door.

Yet, Leopold was not completely unsympathetic to the plight of his Rhenish subjects. Despite his concern with the Turks, he wanted to listen to Frederick William with an empathetic ear. After all, he was the emperor and the protector of the Germans. Sentimentally, this meant that he must resist the French. In addition, Louis's claim to the Spanish throne sent shudders through the Hapsburg family. Leopold was not eager to see France expand to the south. On the other hand, the Turkish threat was real. As much as Leopold wanted to help his German cousins, he had to defend against the threat to Vienna. He worked for a face-saving compromise. In an effort to appease the Germans, Leopold called on his best general, Raimondo Montecuccoli, ordering him to take an Austrian army to the Rhine.[36] In private, however, the emperor was of another mind. He also issued confidential instructions to Montecuccoli, instructions purposely kept secret from the elector.

Had Frederick William understood the emperor's true intentions, he might have approached the war a bit more hesitantly. The emperor, anxious to be able to recall Montecuccoli quickly, hobbled the Austrian army from the beginning. He secretly ordered Montecuccoli to avoid contact with the French unless he was assured of a decisive victory at no risk. Montecuccoli's real purpose was to show the flag. In addition, by keeping his general independent of the elector, the emperor effectively placed a single army under two commanders. Frederick William, as the ranking nobleman, voiced his objection, but the emperor would agree to no other terms. Montecuccoli was to retain an autonomous command. This effectively eliminated any plans Frederick William could develop for an offensive thrust. Decisions for joint actions had to be negotiated with a less-than-enthusiastic partner. Montecuccoli played the part well. He marched laboriously to the Rhine and moved only grudgingly when Frederick William summoned his assistance. The Hapsburg general suspected that his real battle would soon come against the Turks.

---

[36]See Gunther E. Rothenburg, "Maurice of Nassau, Gustavus Adolphus, Raimondo Montecuccoli, and the 'Military Revolution' of the Seventeenth Century," in Peter Paret et al. (eds.), *Makers of Modern Strategy: From Machiavelli to the Nuclear Age*), 58.

The campaign began in the summer of 1672, and it was an absolute disaster. The combined armies did not arrive until the autumn, and Montecuccoli's hidden agenda frustrated the elector. Frederick William spent much of his time watching the Austrian army march back and forth along the Rhine, avoiding all contact with the French. When the elector would march to an advantageous position, he would ask Montecuccoli to join in an attack. Inevitably, the Austrian general would respond with an excuse about the terrain, the weather, or supplies while marching in the opposite direction. Since the elector was insufficiently strong to fight the French alone, he would find himself retracing his footsteps in an effort to join Montecuccoli. The process so infuriated Frederick William that he took matters into his own hands.

Without the emperor's authority, Frederick William took direct command of the army, bluffing the Austrian colonels into submission through the use of his electoral rank. No longer able to dupe the elector, Montecuccoli agreed to Frederick William's ploy to avoid embarrassing the emperor. It all worked to Montecuccoli's advantage, however. When the combined armies moved toward the French in December, the French retreated in haste, and as the cold weather ended the campaign season, the French army slipped away entirely. Montecuccoli declared that the French were defeated. He abandoned the elector, establishing winter quarters for the imperial troops. With winter snows covering the German roads and the French nowhere to be found, the frustrated Frederick William took up winter billets for his own men. He was impotent without Montecuccoli.

The spring of 1673 brought Frederick William scant promise of victory. Montecuccoli left the theater entirely for a return trip to Vienna. After all, the imperial general maintained, the French were on the run and the combined German forces had been victorious. The elector, outnumbered by the French, was exposed with all his forces at risk. His position was fruitless. Realizing the danger, the elector bowed to Leopold and signed a treaty with the French at Vossem in June 1673. The elector agreed to pull his army back to Brandenburg. At least Louis was generous with the terms. Frederick William was in a position to be humiliated, but the French wanted the House of Orange, not the holdings of the Hohenzollerns. Louis promised to leave the German states and abandon the Hohenzollern provinces that he had invaded. Leopold agreed to recognize Louis's claim to the Spanish Netherlands.

Louis had little intention of honoring the terms of the Treaty of Vossem, and both the elector and the emperor probably realized that. The problem for Frederick William was that the emperor was not concerned about it.

Louis XIV began a new offensive against Holland in the summer of 1674 with three of his best field generals: Condé, Turenne, and Vauban.[37] The main assault was to come through the Low Countries under Louis de Bourbon, the Prince of Condé. Condé, victor over the Spanish at Rocroi in 1645, was a vain man with a brilliant reputation and an equally brilliant mind. He was Louis's favorite general. Supporting Condé's flank on the German side of the Rhine was one of the best commanders in military history, Marshal Henri Turenne. Turenne was equal to Condé, if not better, but he did not enjoy the same favor in the court.[38] Turenne was the master of maneuver, and his presence gave Louis a tremendous one-two punch. This deadly combination was complemented by Louis's engineer *extraordinaire*, Marshal Sebastien de Vauban. Not only was Vauban the greatest siege-master and engineer of the age, the entire method of siege craft would become known as the "Vauban system." He prepared thirty-three new fortresses to supply the invasion and laid siege to the Dutch fortresses along the Rhine. Louis seemed to have an unbeatable combination.

Henri Turenne moved along the Rhine with fierce determination. Much to his dismay, he was under special orders from the French king. Not only was he to protect Condé's flank along the river and move toward Holland, but Turenne was given the objectionable task of reducing the German ability to wage war. He was to scorch the earth, destroying people, villages, and farmlands. He protested mightily, but Louis wanted no German threat to his territorial ambitions. The destruction of the Rhineland began in the summer of 1674. French troops plowed under German farmlands, burned villages, and destroyed cities. Turenne found the duty distasteful, but he was efficient. When his soldiers crossed into Hohenzollern lands, the policy of destruction continued.

The German princes flooded Vienna with protests and begged for relief from the emperor. Frederick William also sent envoys, and this time he meant business. He wanted sole command of an allied army that would

---

[37]Theodore Ropp, *War in the Modern World* (New York: Collier, 1962), 40–42. (Note Ropp's comments on the abilities of these leaders.)

[38]See comments by Sonnino, 140.

move toward the Rhine. Bending under pressure, the emperor again called on Montecuccoli and placed him at the elector's disposal. At last Frederick William had a united command. He crossed Germany with his Austro-German force in the fall of 1674 ready to do battle with the forces of the Sun King. Unfortunately, his opponent was Henri Turenne.

Frederick William did not intend to go into the annals of military history with his second Rhine campaign, at least, not as history recorded it. Outmanned and outgunned, Turenne retreated in the face of Frederick William's superior army. The elector was completely frustrated. Marching and counter-marching, Turenne managed to avoid battle by retreating into France. The elector was able to stay on the offensive, but he was not able to bring Turenne to grips. Crossing the Rhine into French territory, the elector decided to stop the sly French general by cutting his supply lines. In a brilliant series of maneuvers, Turenne turned the table threatening Frederick William's rear and forcing the elector to move back across the Rhine. By autumn Montecuccoli, ever worried about the threats to Imperial troops, sought to abandon the elector. Frustrated again, the elector sought winter quarters in January 1675. The Austrians seemed to have lost what little enthusiasm they had, and he could not capture the French. Therefore, he decided to settle for the safety of winter quarters. Safety was the furthest thing from Turenne's mind.

On January 5, 1675, the elector awoke to a great surprise. Frederick William looked outside his camp to find a French army—an army armed and ready for battle—in front of his winter quarters at Turkheim. Turenne was supposed to be several miles away in his own winter quarters, but here he was, on the outskirts of Frederick William's camp. Alarmed, the elector realized speed was the essential. He hastily assembled a makeshift line, hoping for enough time to rally all his forces. Turenne had no intention of giving Frederick William that time. As the elector's army assembled, the French struck. They broke the elector's position, and Frederick William spent the remainder of the day attempting to avoid a complete rout. As evening fell, the stunned elector gathered his army and retreated to avoid annihilation. Turenne, whose campaign is still studied by military com-

manders, returned to his winter quarters with the Rhineland firmly in
French hands.[39]

Frederick William's problems were just beginning. Although he still
had a small command, Montecuccoli decided to return to Vienna. The west
German princes were completely loyal to Frederick, but they lacked the
military strength to defeat the French. As the January snows deepened, the
elector's mood grew darker. He had been outmaneuvered, outfought, and
outwitted. When it seemed that the elector's situation could not possibly
grow worse, a messenger arrived from Berlin. A Swedish army had crossed
the Brandenburg border. Berlin was in jeopardy.[40]

The Battle of Fehrbellin

Frederick William decided to remain in winter quarters. Dejected, he
realized that if he moved he would lose of troops in a winter march, troops
he would need to meet the Swedes. In the twentieth century, winter campaigns are normative. In the seventeenth century, they were seldom used
and possibly disastrous. Troops had no tents, so soldiers had to find whatever cover was available on the roadways. For the horses, winter movement
was a death sentence. Roads were not designed for heavy winter traffic, and
supplies were limited. The movement of armies in the winter was slow. A
commander could figure on losing twenty to forty percent of his troops in a
winter march without even fighting a battle. Given these facts, the elector
made a courageous decision. Frederick William saw no sense in leaving the
Rhine for a hasty winter march against Swedish troops, even though his
capital was in danger.[41]

The circumstances were different for the Swedes. Since they did not
face substantial opposition in Brandenburg, they could afford a leisurely
winter offensive with minimal risk to their forces. They assembled a large
army in Pomerania and moved slowly into Brandenburg to minimize the
hardships of a winter's march. The Swedish cavalry fanned out to forage

---

[39]Nineteenth-century German historians often glossed over Turenne's victory. See
Leopold von Ranke, *Memoirs of the House of Brandenburg during the Seventeenth and
Eighteenth Centuries, Volume I* (New York: Greenwood Press, 1968; orig. 1849), 76–78.

[40]Schevill, 311.

[41]Ibid., 314–20. (The description of the battle has been taken from Schevill.)

and supply the army on the backs of German farmers and peasants. Un-threatened by any Brandenburg army, the Swedes could work with rela-tively small units and disperse their army over dozens of miles. So, the winter campaign was not costly to them. The only action the Brandenburg-ers could take was to lock themselves in their fortresses and wait for the return of their army. The Swedes continued a rather lethargic campaign of rape, murder, and pillage.

Karl Wrangel, a famous general from the Thirty Years' War, was the Swedish commander. He was old, not too energetic, and plagued by gout. He spent much of the campaign riding in an oxcart, being knocked about on the rutted roads. No longer the man who could command a lightening cam-paign, Wrangel was well-suited for the task of leading a lethargic invasion. Wrangel's deputy commander was his younger brother, Waldemar von Wrangel. Karl took the lead of the invasion column, while Waldemar com-manded the rear. With their forces dominated by cavalry units, the Wrangels scoured the countryside and raided to the very gates of Berlin.

The Swedes were little concerned with Brandenburg resistance. Resting for the winter in an extended line, the Swedes slowly gathered their forces when spring came. Because Frederick William had not yet appeared, Karl determined that the Swedes could move at their pleasure. In fact, there were no Brandenburg troops to be seen, save the garrisons in the fortresses. This reinforced Swedish complacency. Continuing with his slow pace, Karl moved east. Waldemar moved south of Berlin as his brother continued the eastward trek. By June the Swedish army was spread over a ten-mile front. Karl moved the vanguard of the army to Havelberg at the junction of the Havel and Rhin Rivers. He hoped to watch for Frederick William from this position. He kept a small force of dragoons in Rathenow, a small village about five miles south of Havelberg, to guard his flank and keep contact with Waldemar. Karl sent Waldemar to Brandenburg with a sizable contin-gent of troops. The Wrangels probably planned to assemble the army in ei-ther the area of Havelberg or Brandenburg when the elector finally ap-peared.

Far away in a camp near the Rhineland, Frederick William had patiently waited for winter's end. When the mud on the spring roads slowly turned to the hard dirt of summer, the elector was at long last ready to move. His re-luctance to make a winter march did not reflect his overall intentions. Using

the time profitably by persuading the Dutch to send a battle fleet into the
Baltic, he effectively isolated Swedish Pomerania. On June 6, when the
roads were good, Frederick moved toward Brandenburg with 15,000 troops.
The troops were well-rested and well-fed. They moved quickly, and by the
night of June 23 they arrived in Magdeburg. Frederick William's caution
reaped a reward. Karl Wrangel had no idea that the elector had moved so
close to the Swedish army.

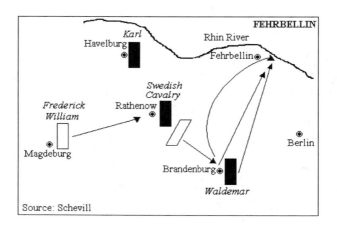

Source: Schevill

Frederick William grasped the situation by mid-afternoon on the next
day. The Brandenburg scouts reported the dispersal of the Swedish forces
and that the Swedish army was not moving. This could mean only one
thing. The Swedes were unaware of Frederick William's proximity. The
elector realized that the time was right for attack, but Magdeburg was too
far away from the Swedish positions. If the elector tried to move his entire
army, Swedish scouts would undoubtedly locate him and all surprise would
be lost. Yet, a successful assault on Rathenow would split the Swedish
army. There was an opportunity with a risk. The question the elector faced
was how to get to Rathenow before the Swedes could assemble their forces
and present a united front. There was only one answer: the elector had to
strike quickly—more quickly than his force could move by foot.

Assembling a council of war, Frederick William decided to gather all of
the horses and all the supply wagons in the center of Magdeburg. Loading

the supply wagons with 1,200 infantry, he decided to leave 7,800 men behind. With the infantry in wagons and 6,000 men on horseback, Frederick William moved toward Rathenow with all possible speed in order to surprise the Swedes.

In addition to Frederick William, two subordinate commanders played critical roles. The general in charge of the operation was an old cavalry general, Georg von Derflinger. Derflinger, a Protestant, had been lured from Swedish service at the time that the elector was building his army. A commoner, he had risen through the ranks of the elector's army on sheer ability. Ferdinand Schevill described the old war-horse well: Derflinger was sixty-nine years old, but the smell of gunpowder reduced his age by thirty or forty years. Colonel Joachim Henning commanded the vanguard unit. Henning's job was to find the enemy, and he was suited for the task. A peasant native of the area, who had risen on merit, he was familiar with the back roads and farms of the swampy Brandenburg area. Frederick William had personally selected Henning to lead the advance guard. Tactics in war generally rely on combined arms. Frederick William had sacrificed the wisdom of the age for speed and mobility. It was a risk. If it worked, the elector and his commanders would surprise the Wrangels. If it failed, he had condemned his small force to defeat. As the sun set on Magdeburg, a makeshift group of cavalry and cart-borne infantry galloped into the night.

The Brandenburgers arrived on the outskirts of Rathenow around midnight. Seeing no signs of Swedish urgency, Derflinger sent a squadron of dragoons to deal with the guards. The Swedes were eliminated in minutes and the attack began. It appeared that the Swedes were taken completely by surprise. A group of Brandenburg pioneers moved to the south end of the city and crossed a stream. They forced the city's south gate and opened the way for the cavalry to ride through. A second gate was opened by storm. As the Swedes tumbled out of bed, they were slaughtered on the streets. A small group of Swedish soldiers managed to form in the town square, but the Brandenburgers proved to be too strong. At dawn the survivors surrendered and Rathenow was in Frederick William's hands. The citizens of the village cheered the elector's victory.

Realizing that he had broken the Swedish center, Frederick William did not want to lose momentum. If the Swedes concentrated their forces, Karl Wrangel could turn the course of events. The best course of action, Frederick William reasoned, would be to eliminate the Swedish flanks. As the elector searched for a plan, he received word of Swedish movements. The group in Havelberg was retreating north while the group in Brandenburg was moving toward Potsdam. Apparently, both Karl and Waldemar believed that Frederick William had a sizable force in Rathenow. They wanted to move to the north of the Rhin River and assemble there. The plan for eliminating the southern Swedish flank became clear as the elector assessed the situation. Frederick William decided to strike the Brandenburg group before Waldemar could escape.

The exhausted Brandenburgers galloped to Brandenburg only to find that Waldemar was gone. He was moving to the village of Fehrbellin and a bridge that crossed the Rhin. The elector called his officers together. He argued that it was necessary to leave the infantry in Brandenburg and move speedily to catch the Swedes. Derflinger disagreed. Without infantry, the old general said, the Swedes would have numerical superiority. In addition, the horses and men were spent. They had been riding and fighting for an entire night. Frederick William would not be dissuaded. Leaving his infantry, the elector decided to pursue the Swedes with only his cavalry units and light cannons. Having lost the point, Derflinger threw himself into his pursuit task with enthusiasm. He took command of the vanguard and rode forward.

Frederick William knew that Waldemar's only point of retreat across the Rhin River was a bridge near the village of Fehrbellin. There the river is shallow, though extremely swampy. It would be impossible for a large force to cross without some type of bridge. Frederick William believed that if the bridge were destroyed, the Swedes could be stopped and he could engage them. The Swedes, however, had a head start on the road to Fehrbellin. The question was how to get to the bridge before them.

Frederick William called on Colonel Henning. Henning, the Brandenburg native, assured the elector that he could cut through the swamps and beat the Swedes to the bridge. The swamps, full of quicksand, mud, and bogs, were a death sentence to those unfamiliar with the territory. Henning, however, knew the area like the back of his hand. He dashed through the marshy ground with 300 troopers and made it to the bridge by early afternoon. Realizing that he had beaten the retreating Swedes, Henning set fire to the planks of the bridge and destroyed the gateway.

Waldemar's troops arrived at the bridge to find smoke billowing from the southern section. Without hesitation Waldemar ordered his cavalry forward to chase Henning's troops away. Outnumbered by several thousand, Henning's small detail retreated into the village of Fehrbellin. Assured that the problem with Henning was ended, Waldemar turned his attention to the bridge. He needed to repair it to continue his retreat. When Swedish engineers doused the fire and began gathering wood for repairs, Henning decided to harass Waldemar. Riding out of the village, he tried to divert the Swedes and slow their progress. Waldemar kept the engineers at their places by dispatching Swedish cavalry units to chase the Brandenburgers away. For the next few minutes, Henning played a dangerous game of cat and mouse with the Swedish cavalry. His tenacity was rewarded. By early afternoon the cavalry of Frederick William's army came riding up the road. The aged Derflinger, riding at the head of fifty advanced guards, attacked the entire Swedish army.

Derflinger's troops were easily thrown back, but they were soon joined by the remaining cavaliers, including the elector himself. This confused the Swedes, causing them to assume they were under attack by the entire Brandenburg army. Deploying for battle, Waldemar drew his infantry into formation in front of the bridge. He set out the artillery and recalled what cavalry he could. As a line of battle formed in front, he ordered the engineers to make haste. Frederick William regrouped and assaulted. The Brandenburg army smashed into the Swedish line and fought an engagement with both the elector and Derflinger in the thick of the fighting.[42] Despite the element of surprise, the Swedish line was too strong. The elector could not break it without the use of infantry, and the Brandenburg infantry was stretched between Magdeburg and the village of Brandenburg. Reluctantly, the elector moved his cavalry to a small hill overlooking the bridge.

Derflinger did not wish to lose the initiative. Moving to the highest point above the battle, the general brought up the cannon the cavalry had dragged along on its ride to Fehrbellin. As the Swedes sat in their battle line, Derflinger fired a salvo of terribly inaccurate cannon shot at them. It was ineffective, but it drew attention. Waldemar decided to respond by at-

---

[42]See C. Edmund Maurice, *Life of Frederick William: The Great Elector of Brandenburg* (London: George Allen and Unwin, Ltd., 1926), 131. (Both Schevill and Maurice attest to Frederick William's presence in the fighting.)

tempting to silence the guns. He launched both wings of his cavalry toward Derflinger's make-shift position. After a brief fight, the Swedes took the Brandenburg guns and forced Derflinger back. The elector and all his troops were caught in the midst of a fierce hand-to-hand battle. Eventually, the Brandenburgers drove the Swedes back and recaptured the guns. To their amazement, the Swedes had done nothing to disable the guns; Derflinger and Frederick William began another volley of cannon fire.

The Swedes repeatedly charged the hill, only to be repelled in hand-to-hand fighting. After the third charge, the elector struck back, catching his adversaries by surprise. Although they outnumbered the Brandenburgers, the Swedes lost the initiative. They scrambled down the slope, exposing their infantry to Frederick William's pursuing horses. To an untrained eye, it looked like a breakthrough. Waldemar remained calm. While his infantry kept the Brandenburg cavalry from penetrating the formation, he continued to repair the bridge. Frederick William returned to the high ground and began bombarding the Swedish infantry again. Whether Waldemar toyed with the notion of an infantry assault is not known. What historians do know is that his engineers made the necessary repairs to the bridge. With his line of retreat open, Waldemar moved across the bridge.

The Brandenburgers sat on the hill too frightened and exhausted to recall the days of the Thirty Years' War when the Swedes or other foreign troops moved through Germany at will. As time wore on, however, their recollection of the symbolic nature of the battle grew. On this day in June, the fate of Brandenburg and northern Germany seemed to have changed. The Brandenburgers looked toward the bridge and saw one of the finest armies in the world retreating in front of them. No doubt they were relieved that the fighting had ended, but while caught in the euphoria of the moment, a wave of adrenaline-induced excitement washed through the troops. It began to dawn on them that they had beaten the mighty Swedish army. They were the victors of Fehrbellin. That evening, as a tired Frederick William rode into the gates of Berlin, his subjects cheered and spontaneously gave him a new title. They called him the Great Elector, a title associated with his name for more than 300 years.

Over the next four years, Frederick William drove the Swedes from northern Germany and Poland. After a long siege, he even captured the fortress of Stettin. When Louis XIV ordered the elector to return all lost territory to Sweden at the end of the war, Frederick William threatened to go to war with France. He sent an army to the Rhine but sued for peace as soon as a large French army crossed the river. France waited another decade before moving against Germany, leaving the elector in peace during his final years. Frederick William died in 1688. He had come to the throne a "beggar on horse back" but departed from it as the Great Elector. He also left a legacy to the scattered domains of Brandenburg-Prussia, an army of 30,000 men.

Chapter 3

Prussia's Rent To Own

In 1688 Frederick III—later King Frederick I—became the Elector of Brandenburg and the Duke of Prussia. It was soon obvious that he was not cut from the same cloth as his father. Frederick the Great would say of the new elector that he was large in small things and small when dealing with large issues. Frederick neglected political and military affairs, favoring the aggrandizement of the court. As with many other German princes in the late 1600s, Frederick behaved as if he were Louis XIV. Despite all of this, he made a promise to his father and he kept it. The new elector vowed to maintain and expand Brandenburg's military forces. The army developed and expanded under Frederick's rule, and he used its power to become a king. This chapter will discuss three aspects of the army under Frederick: (1) the changing structure of the officer corps, (2) the quasi-mercenary activities of the army, and (3) the "Prussian Corps" in Marlborough's service.

The Army and the Junkers

Frederick III came to the throne in 1688 carrying a grudge against his father. After he married a woman against the Great Elector's wishes, Frederick turned his back on the Hohenzollerns and fled Berlin. He claimed that the Great Elector was seeking his life. This was absurd, but Frederick remained absent from Berlin for several months after his wedding. As the Great Elector lay on his death bed, Frederick made up with his father. Both men knew that the time for family squabbles had passed as the territorial ambitions of Louis XIV became clear once more. Although fatally ill, the Great Elector correctly predicted that Europe would soon be going to war over Louis's desire to move French borders toward the Rhine. The only method to stop this, the Great Elector believed, was to link Brandenburg-Prussia to Vienna. As father and son reconciled in the time of death, Frederick saw the merit of his father's point. Frederick promised to approach the emperor and to keep the army ready for use.

Brandenburg-Prussia was hardly a major power after the Great Elector's death, but it had become an important factor in German affairs. Aside from territory and autonomy in Prussia, the Great Elector's most notable legacy was a sizable army. With Europe on the brink of war, the new Elector of Brandenburg-Prussia realized that the Hapsburgs needed his army, and it presented him with an opportunity for economic growth. While the Great Elector used military force for the aggrandizement of Prussia, Frederick III would utilize it to enlarge his own prestige. Unskilled at statecraft, he submerged his court in opulent ceremonies and sophisticated rituals. He imported scholars and artists to the Berlin Academy to enhance his reputation among intellectuals, even though Frederick's mental prowess was limited. His goal was to make Potsdam another Versailles, and it would not be unfair to cast him as "Frederick the Ostentatious." Yet, Frederick had no economic base to finance his extravagant tastes, save one natural resource. Shortly after assuming power, he realized that the army could provide an excellent source of income.[43]

Military organizations usually reflect the societies that spawn them, and Brandenburg-Prussian was no exception. Brandenburg-Prussia was poor, but it was highly productive and known for its efficiency; the character of the army mirrored these values. The army gained a reputation that matched the country whose farmers scratched out a living from sandy soil. It was small, but efficient. In addition, while the new elector did not take the field, his officer corps was inundated with competent commanders. They were quite capable of promoting discipline, efficiency, and the new social concept of military professionalism. To his credit, Frederick recognized the importance of having such talented commanders, and when Vienna needed these troops for the war with France, he steered clear of direct military command. He left military leadership to the emerging officer corps.[44]

Frederick's decision to leave most of the army's internal affairs in the hands of his generals resulted in several important developments. Among

---

[43]David Kaiser, *Politics and War: European Conflict from Philip II to Hitler* (Cambridge, Mass.: Harvard University Press, 1990), 180–81.

[44]Russell F. Weigley, *The Age of Battles: The Quest for Decisive Warfare from Breitenfeld to Waterloo* (Blommington, Ind.: Indiana University Press, 1991), 126.

the most significant was the change in the structure of the officer corps. The Great Elector selected officers primarily on their merit and personal loyalty. He had played a role in the selection of junior officers, but he allowed his senior officers freedom to select junior officers from any social class. Under no circumstances could the Great Elector's army be characterized as a "noble army," and the Junkers had no monopoly on the command positions.[45] Frederick III began to change this. Although the "Junkerization" of the Prussian officer corps would be completed after his son and grandson came to power, Frederick began to recruit more and more nobles into the officer ranks.

Poor economic conditions provided the major incentive for Junkers to seek a career in the profession of arms. The noble estates of Brandenburg-Prussia were characterized by their relative poverty, and most nobles could only claim the sandy soil of the north German plain, their serfs, and their titles, but little else. Most families had limited incomes, and very few were wealthy. Poor agricultural conditions posed a special problem inasmuch as parents could rarely provide for all their children. Daughters could be married to men of other houses, but the sons had to survive from the estates. Frederick III offered a solution. He encouraged young Brandenburg-Prussian noblemen to seek their fortune in the army. Over the next three decades more and more Junkers accepted this call, receiving commissions in the army. The officer corps gradually became dominated by so few noble families that Walter Goerlitz could say the history of the general staff could be linked to the history of a handful of Prussian clans.[46]

Technology also played a role in the transformation of the army.[47] As musketry increased, the need for pikemen decreased. Musketeers found that they could slip a solid sharpened piece of steel down the barrel of their weapons, changing their muskets into makeshift pikes. This metal shaft, called a plug bayonet, was long, cumbersome, and clumsy, but it stopped cavalry. When Frederick came to the throne, the plug bayonet had virtually

---

[45] Martin Kitchen, *Military History of Germany from the Eighteenth Century to the Present* (Bloomington, Ind.: Indiana University Press, 1975), 4.

[46] Walter Goerlitz, *History of the German General Staff: 1657–1945* (tran. by Brian Battershaw) (New York: Frederick A. Praeger, Publishers, 1962), 1–7.

[47] For an outstanding summary see David Chandler, *The Art of War in the Age of Marlborough* (New York: Hippocrene Books, Inc., 1976), 65–92.

replaced the pikeman. Its only detriment was that it was once inserted, the musketeer could no longer fire. By plugging the barrel of his musket, the infantryman had essentially transformed his weapon into a spear. French general Sabastien Vauban developed a solution.

Vauban wrestled with the problem of the plug bayonet. The striking power of the musket was its ability to fire. The bayonet was important, but it played a secondary role, and with the barrel plugged, the firing mechanism was useless. The infantryman could act either as a musketeer or a pikeman, but he could not play both roles simultaneously. Vauban created a bayonet that would fit over the end of the barrel, leaving the weapon free to discharge. Rather than plugging the barrel with a solid core, Vauban decided to place a locking mechanism on the end of the barrel and a hollow core to fit over it. The result was the socket bayonet. It fit over the end of the musket, allowing the infantryman to load and fire while the bayonet was fixed on the end of the barrel. European armies scrambled to introduce the new weapon to their arsenals. As a result, armies drastically increased their fire power while simultaneously protecting themselves from cavalry.

A change in the musket's firing mechanism also increased the use and value of muskets. Wheel locks and matchlocks had been the hallmark of seventeenth-century fire power. The eighteenth century saw a new development, the flintlock musket. Using a piece of flint locked in place on a hammer and a steel striking plate, the flintlock reduced the cumbersome process of firing the musket. Much more reliable than its predecessors, the flintlock allowed weapon manufacturers to lighten the overall weight of a musket while shortening the time necessary to fire and reload. Disciplined armies could fire much more rapidly with flintlock muskets, and the Brandenburg-Prussians began to excel in their ability to reload and fire. They increased their rate of fire to approximately one volley every twenty to thirty seconds under combat conditions.

With improved fire power and the ability to maintain fire even while bayonets were fixed, the infantry began to utilize new formations. In the mid-seventeenth century, infantry units fought from blocks of musketeers flanked by pikemen. By the late 1600s these unwieldy formations gave way to the infantry line. The line, usually three men deep, became the standard formation for infantry units. While stretched in a line, infantry formations could attack other infantry formations or fire at advancing cavalry. If the opposing cavalry proved to be too close, the infantry could simply set up a bayonet wall from a linear formation. The key to success in a linear formation was precision in movement. If soldiers remained in position—that is, if they stayed in a solid line while moving—they were relatively safe and extremely effective. Marching, drill, and discipline became the three key factors in this process. The Brandenburg-Prussian officers were among the first to realize the importance of formations. Under Frederick III, they became increasingly proficient on the drill field.

Frederick's soldiers also developed another tactic based on the volleys from rapid-fire muskets and the "steel wall" of socket bayonets. The Brandenburg-Prussian army was always short of cavalry for one simple reason: horses cost quite a bit of money. Accordingly, the Brandenburg-Prussian cavalry was notoriously weak, and it provided little protection from the enemy horse. To deal with this deficiency, the officers began to develop tactics that would allow the infantry to confront the cavalry directly. Officers theorized that disciplined infantry could accept a cavalry charge. If the fire were controlled, the cavalry could be stopped before it reached the infantry line. If musketry failed, tight bayonet formations served as a second defense. Many of the officers concentrated their efforts on perfecting these anti-cavalry tactics.

Military preparation was not an abstract exercise in Brandenburg-Prussia. While Frederick was rather hesitant to assume direct command of his army, he was quite willing to allow others to use his forces. Units from Brandenburg-Prussia were engaged in combat throughout his reign. They distinguished themselves in Italy under Hapsburg command and in western Germany under both Austrian and British leadership. They gained a reputation as first-class fighting men, drawing respect from the best captains of the age.

King in Prussia

To understand the developments of the Prussian army during Frederick III's reign, it is necessary to reflect briefly on the elector's goals.[48] Frederick III desperately wanted to be like Louis XIV. In this sense, he was no different from many other German princes who moved through the countryside

---

[48]Max Beloff, *The Age of Absolutism: 1660–1815* (London: Hutchinson University Library, 1964), 105.

collecting taxes to build palaces, collect art, and import philosophers. But the major problem facing Frederick was one of ennoblement. As much as Frederick wanted to be like Louis, Louis was a king and Frederick was merely an elector in Brandenburg and only a duke in far away Prussia. Such status galled a man like Frederick. Although he wanted art, wisdom, and prestige, his most important goal was power. Frederick wanted to be a king, and the only person who could bestow the title was the emperor in Vienna.

Frederick III cannot be vilified for seeking elevation to king. During his reign, the Duke of Saxony was made a king, and George, the Elector of Hanover, would soon become King of England. Brandenburg-Prussia was higher in the pecking order than either of these states, and Frederick was already an autonomous duke. Since his ruling cohorts were being elevated, Frederick's ambition is understandable. In addition, most small German rulers were attempting to emulate the French court. Many lesser princes built miniature palaces and placed their states in debt to finance their tastes. Brandenburg-Prussia was more important than most of the other German states, and Frederick was quite within his rights to demand such recognition for the Hohenzollerns.

Despite its sandy soil and cold north-German winters, Brandenburg-Prussia had one natural resource that could help Frederick gain the crown: the army.[49] An opportunity presented itself that allowed him to use the army to his best advantage. War broke out in 1688, and Emperor Leopold I wanted Brandenburg-Prussian troops. The Turks, still present on the southern flank of the empire, drew the emperor's wary eye. While the French moved through the Low Countries and into the area of the Rhine, the emperor faced the age-old dilemma. He needed to protect the empire on two flanks. As the German princes begged for protection against the French, the emperor was quite willing to court Frederick's favor. The Hapsburgs needed the Hohenzollerns.

In 1688, Leopold appealed to Frederick for troops, and Frederick seized the opportunity. He agreed to heed the emperor's call but pleaded poverty. Funding for the army, the elector said, would need to come from Vienna. Leopold was sympathetic. After negotiating with Frederick, the emperor agreed to pay for approximately 50 percent of the cost of the Brandenburg-Prussian army. Frederick was elated. The negotiated price with the emperor

---

[49]See Tuttle, vol. I, 262–77.

was one thing; the actual income was another. Frederick realized that the fiscal support provided by Austria far exceeded his direct costs. He dispatched about 8,000 troops to the Rhine River while gathering Austrian money to enrich his court. The army was a natural resource that could generate income.

Frederick also acted on his other goal. Since Leopold needed troops, Frederick cautiously approached Vienna with a proposal. If Leopold would agree to elevate Frederick to the position of king, the monetary support for the Brandenburg-Prussian army could possibly be reduced. Frederick's vain suggestion sent Leopold into a rage that was calmed only by his advisors. After an outburst of anger, the emperor sent a response—hiding his rage in the language of gentle diplomacy—agreeing to take the matter under consideration. Leopold had no intention of following through, but Frederick was so elated that he failed to grasp the emperor's true intentions. "Considering" an action, Leopold understood, was considerably different from "taking" action. Unfortunately for Frederick, when peace came in 1697, the emperor was still considering the elector's request to become a king.

The Nine Years' War, or the War of the League of Augsburg, lasted from 1688 until 1697. Aside from the fiscal benefits for Frederick, the war had two notable effects on the Brandenburg-Prussian army. First, Frederick's forces had served with distinction. They drew positive attention from Dutch, English, and imperial commanders. The second effect had as much to do with the Hapsburgs as it did the Hohenzollerns. During the war, Eugene of Savoy rose through the ranks of the imperial forces while gaining direct contact with the Prussian infantry. He had cut his teeth fighting the Turks, emerging as the best commander in the emperor's service. Frederick had relegated his 8,000-man corps to a supportive role, implying that his troops would fight under a foreign commander. Eugene, the imperial Prince of Savoy, was destined to be that commander. In future years the combination of Eugene's leadership and Brandenburg-Prussian military abilities would prove a formidable combination.[50]

---

[50]For modern biographies see Nicholas Henderson, *Prince Eugene of Savoy* (New York: Frederick A. Praeger, 1964) and Derek McKay, *Prince Eugene of Savoy* (London: Thames and Hudson Ltd., 1977). See also Paul Frischauer, *Prince Eugene, 1663–1736: A Man and a Hundred Years of History* (New York: William Morrow and Company, 1934).

This combination of leadership and application was tested almost immediately. No sooner had the War of the League of Augsburg ended than a new threat appeared on the horizon. The Hapsburgs had no heir for the vacant Spanish throne, and the aging Louis felt that the time was right to place a Bourbon in Spain. As a result, England, Holland, and Austria stirred to action. By 1700 it appeared that armies would once again be marching around the Rhine. Leopold again appealed to Frederick, but this time Frederick dallied. Although his soldiers served as the major source of revenue for the court, he decided he would not send troops until the emperor agreed to confer the title of king on him. Frederick reminded Leopold of his promise to consider the elevation while Eugene reminded his emperor of the need for the Brandenburg-Prussian troops. Irritated, Leopold felt that he had little choice. In the fall of 1700, a reluctant Leopold agreed to elevate Frederick—with one little insult. He dubbed Frederick "King in Prussia" instead of "King of Prussia." The man who was great in little things paid no attention. He was, at long last, a king.[51]

The results of this elevation have been the bane of first-term European history students for many years. As with other titles in European nobility, Frederick's elevation causes much confusion to American students. Since Frederick was now the first King in Prussia, his numerical designator changed. He assumed the throne in 1688 as Frederick III because he was the third Frederick von Hohenzollern to become the Elector of Brandenburg. In 1700, however, he became the first Frederick von Hohenzollern to become a king. Therefore, Frederick III, the elector, became Frederick I, the king. Since he was King in Prussia, Frederick immediately redubbed all Hohenzollern holdings as "Prussia" and the use of the title "Brandenburg" faded into the past. Frederick III was Frederick I and the Brandenburg-Prussians as well as the inhabitants of all the other Hohenzollern domains became Prussians. And, despite all the confusion associated with these changes, King Frederick I gained both a title and a suitable income. The army had given him the means to do so.

---

[51]See Maurice Ashley, *A History of Europe: 1648–1815* (Englewood Cliff, N.J.: Prentice-Hall, Inc., 1973), 90.

The Army in Marlborough's Service

France and Austria were drawn into war in 1702 over control of the Spanish Netherlands and the Spanish throne. Lured by the promise of subsidies, King Frederick was now anxious to commit troops against Louis. As an imperial army took the field, Frederick supplied it with a well-prepared contingent of Prussian troops. The 8,000-man force was under the control of Frederick's best field general, Prince Leopold zu Anhalt-Dessau. Leopold, in turn, reported directly to Eugene. The two soldiers developed mutual respect for one another, an admiration that would last for the rest of their lives.

The Prussian troops became part of a great allied, multinational army. They could not properly be considered mercenaries because they were in the service of the imperial cause and Frederick was a vassal of the Holy Roman Emperor. The Prussians were directly attached to Eugene's imperial army, and, in fact, they became the mainstay of Eugene's infantry. The imperial army was attached to the main allied army, comprising Dutch, English, and German-speaking soldiers. Eugene's title gave him the right to command the allied army, but Prince Eugene had a character trait that has rarely been displayed in military history: he had his ego under complete control. Recognizing the abilities of another allied commander, the Prince of Savoy voluntarily settled for the role of deputy. He handed over all command of the allied army to John Churchill, the Duke of Marlborough. The infant Prussian army was destined to be weaned under Marlborough's tutelage.

The War of the Spanish Succession lasted from 1701 to 1714. It was the first global war in modern history, with battles in Europe, India, the Americas, and on the high seas.[52] It pitted a group of European nations in the Grand Alliance against the forces of France and its allies. Prussian participation was limited to operations in western and central Europe. In March 1701, the French moved to occupy fortresses in the Spanish Netherlands. As a result, Prince Eugene assembled the imperial army in Tyrol. He began operations against the French in Italy during the summer of 1701. On Septem-

---

[52]John B. Wolf, *Toward a European Balance of Power: 1620–1715* (New York: Rand McNally and Co.), 97 and 156–78. For a general political summary see Jeremy Black, *The Rise of the European Powers: 1679–1793* (London: Edward Arnold, 1990), 51–59.

ber 7, 1701, England, Holland, Austria, Prussia, and most of the western German states (with the exception of Bavaria) officially formed the Grand Alliance against Louis. On May 15, 1702, John Churchill became the military commander of the allied forces.

Marlborough spent two years operating along the Rhine and the Meuse Rivers.[53] Much of his attention focused on the fortresses Vauban had constructed a quarter of a century earlier. When not engaged with the French, Marlborough spent his remaining time trying to hold the allied coalition together. Each allied contingent had a different political mission, and it was no small task to keep attention focused against France. Marlborough succeeded until September 1702 when the rules changed completely. The Bavarian Elector Maximillian had long sought the imperial crown. As the Emperor Leopold's age increased, Maximillian made his bid for power. He joined the French, declaring war on Marlborough's Grand Alliance. The French sent to Bavaria an army that both outflanked Marlborough and posed an ominous threat to Vienna. Marlborough was politically paralyzed. He needed to face the new threat, but the Dutch and the Rhenish Germans would not follow him if he left the lower Rhine. He watched the growing Bavarian threat all through the next summer, politically helpless to move against it.

By April 1704, the French and Bavarians were presenting a sizable threat to Austria. The Elector Maximillian and his French ally Marshal Ferdinand Marsin concentrated a force of 55,000 Franco-Bavarian soldiers around Ulm in southern Germany. The French Marshal Camille Tallard was nearby with 30,000 more troops, and he was moving to assume command of a united army. The Austrians had approximately 30,000 troops around Vienna, while Eugene held an additional 10,000 troops south of Ulm. Eugene desperately needed Marlborough, but the Dutch would not consent to let Marlborough's force move south toward the Danube. The situation called for careful political innovation. Informing the Dutch officials that he planned to attack French fortresses from the south, Marlborough secretly left the lower Rhine and moved toward his real objective, the Danube. In a series of forced marches he transferred an entire army from one theater of

[53]For a solid introductory summary to Marlborough's actions see Russell F. Weigley, *The Age of Battles: The Quest for Decisive Warfare from Breitenfeld to Waterloo* (Bloomington, Ind.: Indiana University Press, 1991), 73–104.

war to another. By the time the Dutch forces in Marlborough's army determined the actual purpose of the march, it was too late to leave. Their political leaders remained in the Netherlands, blissfully ignorant of Marlborough's location. It was a superb piece of political insubordination. By August Marlborough had joined Eugene on the Danube.

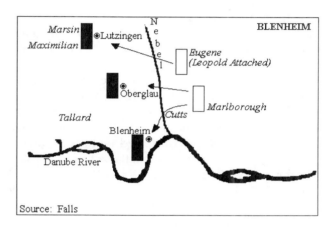

It is beyond the scope of this book to discuss Marlborough's campaign in detail, but Marlborough's actions at Blenheim merit a brief review.[54] The battle at Blenheim should not be construed as a Prussian victory, but the Prussians played an important role, similar to their role more than a century later while assisting another Anglo-Dutch force at Waterloo. The Prussian forces under the command of Eugene of Savoy performed in an outstanding manner at the Battle of Blenheim, and Leopold of Anhalt-Dessau's infantry literally saved the northern flank of the Grand Alliance. Although they were relegated to imperial service and their function was defined by foreign commanders, the Battle of Blenheim demonstrated the abilities of the Prussian infantry.

---

[54]For excellent summaries see Henderson, 91–114, and Winston S. Churchill, *Marlborough: His Life and Times* (New York: Charles Scribner's Sons, 1968), 403–17. For a complete summary see Peter Verney, *The Battle of Blenheim* (London: B. T. Batsford Ltd., 1976).

The army of the Grand Alliance that gathered on the banks of the Danube River in August 1704 was a hodgepodge of European military forces. Eugene was responsible for two major units in this menagerie, the imperial cavalry and the Prussian infantry. The Prussians were under the direct command of Prince Leopold of Anhalt-Dessau and attached to Eugene. Marlborough held the main body of the allied force. His contingent included large elements of German-speaking soldiers, most of the Dutch army, and red-coated English soldiers carrying—for the first time—the flintlock "Brown Bess" musket. The total strength of the allied army was about 56,000.

The main French positions were just to the west of Marlborough's forces. The French Marshal Camille Tallard was in command with approximately 60,000 men under arms, including ninety guns. Some historians have questioned the military prowess of Tallard, but he was a good commander. Marshal Ferdinand Marsin served as Tallard's assistant, while cooperation from Maximillian further strengthened Tallard's position. Although Tallard's force comprised both French and Bavarian units, it was not nearly as diverse as Marlborough's army, and the Bavarian elector had placed himself at the disposal of French leadership. Tallard's force was politically cohesive, and its leadership was solid.

The topographical position of the Franco-Bavarian force was also excellent. The left was anchored by a virtually impenetrable set of woods and hills just to the north of the village of Lutzingen. The right flank was guarded by the high banks of the Danube River. Three villages served to orient the four-mile front, Lutzingen to the north, Oberglau in the center, and Blenheim along the banks of the Danube. The front of the position was covered by a natural fortification, the Nebel River, which ran in swampy grounds just below the Franco-Bavarian position and emptied into the Danube just below Blenheim. Even though Marlborough had just conducted one of the most magnificent marches in military history—300 miles in thirty-five days—Tallard seemed to have few worries. He was on the high ground and he expected Marlborough to retreat. Marlborough did the opposite.

By all military logic, Marlborough and Eugene probably should have retreated or, at least, held their positions. Instead they devised a daring plan of attack. Although they were outnumbered—the Grand Alliance forces numbered approximately 56,000 men and sixty guns—Marlborough and

Eugene believed they could achieve surprise. Their plan involved splitting the army and conducting three simultaneous assaults against Tallard's strong points, Lutzingen, Oberglau, and Blenheim. Marlborough would take the Anglo-Dutch forces down through the swampy ground of the Nebel and move up the hill toward the French. A British unit would be dispatched to assault Blenheim, and the remaining forces would concentrate on the French center at Oberglau. Eugene was to move north through a heavily wooded area with the imperial cavalry and the Prussian infantry. His objective was the village of Lutzingen. If the allies could break any of the three Franco-Bavarian strong points, they could turn Tallard's line and force him to retreat. The plan had its risks. Marlborough would have his back to the Nebel and Eugene might be caught in the woods. If all three attacks failed, there would be no place to retreat.

Early on the morning of August 13, 1704, the army of the Grand Alliance arose to the sound of drums and began marching. French scouts reported the movement to Tallard, and he assumed that Marlborough must be retreating. He wrote King Louis that it appeared the allied forces were moving toward Vienna. By 7:00 A.M., Tallard wished he could retrieve the message. His scouts reported that allied forces were marching toward the Nebel. Still, Tallard saw no reason to panic. He moved a substantial portion of infantry to Blenheim and placed another group with artillery support at Oberglau. His main striking weapon, 20,000 French cuirassiers, took a position behind Oberglau out of Marlborough's line of sight. Although he felt the woods protected his northern flank, Tallard sent Marsin and Maximillian to Lutzingen with most of the Bavarian army and some French reinforcements. Even though Marlborough was attacking, Tallard still held the high ground.

Skirmishing began around the Nebel River with long-range cannon shot, as Marlborough's engineers began the tedious process of building bridges to cross the swampy ground. Eugene of Savoy, who broke away from Marlborough at mid-morning, marched into the wooded area and into a tremendous surprise. The woods included virtually impassable undergrowth and rocky ground. Eugene had only a few miles to go, but the ground almost prevented movement. In the meantime, Marlborough sat in the midst of a long-range artillery duel, believing all to be well with his imperial ally. Nothing could have been further from the truth. Eugene's imperial cavalry and Prussian infantry began hacking at the underbrush in an attempt to move. The August temperature rose, creating a human mire around the sweating troops. The expected quick two-mile march bogged down, and Eugene was not able to reach his position until shortly after noon, hours after the French cannon had started bombarding Marlborough's force on the other side of the Nebel.

Around midday, a courier arrived from the imperial forces to inform Marlborough that Eugene was finally in position. Without further delay, he ordered his commanders to their posts and began crossing the Nebel. If all went well, Marlborough could hope to break the French strength along one of the villages in the Franco-Bavarian line. All did not go well, however. The initial assault on the village of Blenheim failed, and as Marlborough's infantry and cavalry moved toward Oberglau, nearly 20,000 armored French cuirassiers gathered in front of the village to attack his bridgehead. Marlborough could only wonder about the fate of Eugene as he fought for survival.

Despite the unexpected turn of events, Marlborough's forces faired well. While his initial assault on the village of Blenheim failed, it achieved another strategic result. The French troops inside the village were trapped within a British defensive parameter. The British could not break in, but the French could not break out. This gave Marlborough a decisive advantage. His southern flank was secured. Things also went well in the center. Using an innovative combination of infantry and cavalry in a single line, Marlborough was able to break the charge of the French cuirassiers. As he advanced on Oberglau after a four-hour fight, a second cuirassier charge met with complete frustration. The French troopers stopped their mounts in midcharge and fled from the field. Many of them drowned in the Danube River while trying to escape. By late afternoon, Marlborough could claim success on the center and in the right.

All of this was unknown to Eugene. Believing that Marlborough depended on his assault at Lutzingen for success, Eugene sought to engage the Bavarian and French forces outside the town. After taking stock of the Franco-Bavarian positions, Eugene summoned the imperial cavalry. The Bavarian cavalry was prepared to assault. Although the imperial horses and soldiers were exhausted after hacking their way through the thickets, Eugene explained to his men that they must be the first to attack. The imperial cavalry galloped forward to meet the Bavarians. Their bramble wood march, however, had been too exhausting. After an exchange of volleys and a clash of swords, the Austrians retreated to their original position.

Eugene realized the gravity of the situation. If his cavalry could not stop the Bavarian threat, he had no place to hide. With thickets and foothills to his rear, Eugene would be at the mercy of the Bavarian cavalry and the French infantry that supported it. In a personal, emotional appeal, Eugene anxiously explained the situation to the exhausted troopers, and, again, ordered an Austrian cavalry charge. The Bavarians held, leaving Eugene to retreat again. A third attempt to charge only met with half-hearted results as most of the Austrian soldiers simply sat and watched the first rank ride away. For all practical purposes, the imperial cavalry had been neutralized. Refusing to accept this situation, Eugene moved from trooper to trooper in an attempt to rally his cavaliers. Then, to his horror, he looked up the hill to Lutzingen to see the Bavarian cavalry mounting for an assault. The Prince of Savoy thought his entire command was doomed.

Dramatically swearing at the imperial troopers, Eugene grabbed an imperial standard and dismounted. Screaming that he was not willing to die with horsemen who would not fight, Eugene moved in front of the Prussians vowing to die like an infantryman. He defiantly stood in front of the ranks of the gathering Prussian soldiers and looked at the oncoming Bavarians. Had Eugene looked the other way, he might have formed another opinion. The Prussians, perpetually short on cavalry and always ready to develop infantry tactics to meet cavalry assaults, had no intention of dying that day. Leopold of Anhalt-Dessau methodically lined his forces up in a slanted line with the forward elements gradually fading back to the rear elements. When the Bavarian cavalry charged forward, Prussian musketeers began blazing away. No cavalry on the field at Blenheim could have reached the Prussian infantry that day, and the Bavarians could not withstand such a rain of fire. After several minutes, the Bavarian cavalry stopped; then it turned to retreat.

If Eugene was in shock, he recovered quickly. With the Bavarian cavalry retreating, the prince ordered Leopold to move on the supporting infantry. The Prussian infantry moved forward with splendid discipline and began exchanging volleys with the French infantry. The exchange of musket fire lasted for nearly an hour when the French forces started to give way. By late afternoon, Leopold's Prussians had penetrated the village and were forcing the French to reestablish their lines. The French north completely collapsed when word came to Marsin and Maximillian that the French center at Oberglau was in trouble. As they tried to reorganize their forces, Eugene sent Leopold forward again. The French broke, and the Prussians turned south. Eugene hurriedly marched to assist Marlborough at Oberglau. His arrival in the early evening ensured the complete collapse of the French army.

The Battle of Blenheim was a victory for Marlborough, and the Prussians had played a critical role while demonstrating their mettle. Leopold's Prussian corps was a far cry from the ragamuffin group of miscreants assembled by the Great Elector sixty-four years earlier. The Prussian forces continued to play an important role in Marlborough's army throughout the War of the Spanish Succession. They accompanied Eugene on a campaign in Italy in 1706, and in 1708 they marched from Italy to unite again with Marlborough for the last phases of his Oudennarde campaign. Prussian troops also participated in Marlborough's bloody victory at Malplaquet in 1709. The soldiers won Marlborough's respect to the extent that he twice traveled to Berlin to obtain Frederick's assurance that Prussian troops would stay in the army of the Grand Alliance.

The War of the Spanish Succession ended in 1714. During the war the Prussians won the admiration of Europe's finest generals. The participation of the Prussian corps in Marlborough's army also achieved Frederick's life-long dream: he became King Frederick. He did not live to see the end of the war, but he left his son a crown. The war also had another effect on the Prussian army. Frederick's son, the Crown Prince Frederick William, did not share his father's tastes for the arts and sciences. As soon as he was of age, he joined Leopold and spent his time with the Prussian corps. While Frederick built the Academy of Arts and Sciences, the crown prince defied death at the Battle of Malplaquet. The young prince even won Eugene's admiration and became fast friends with Leopold. It was rumored that the crown prince often remarked that he loved war.

Chapter 4

The Soldier King

The Prussian army became an internationally respected organization under Frederick I despite his failure to use the army for national policy. His son, Frederick William I, wanted to take that reputation further. Assuming the crown in 1713, Frederick William was the antithesis of his father. The new king had one objective: he wanted to provide Prussia with the best army in the world. To accomplish this objective, Frederick William would transform everything in the kingdom to support military growth. Prussia became, as one British diplomat observed, an army with a state. No observation could have pleased Frederick William I more. He was a soldier king.

The Road to Militarism

Martin Kitchen described militarism as the subjugation of all social attitudes to military values.[55] This definition epitomizes Frederick William's approach to life. The new king ruled from 1713 until 1740, and his rule was dominated by a fanatical desire to expand the army. Under his authoritarian hand, the army grew from 38,000 men to 82,000 men. Frederick William completely organized the Prussian state to support the military, and he was as hard on civilians as on the army. If the Great Elector had moved toward the goal of centralized authority, Frederick William made the centralized state the source of all power. The Great Elector's grandson believed that the entire essence of the state was conveyed in its military power. Determined to bring this belief to fruition in Prussia, his army drilled, trained, and ex-

---

[55]Martin Kitchen, *Military History of Germany from the Eighteenth Century to the Present* (Bloomington, Ind.: Indiana University Press, 1975), 4.

panded. In the process Frederick William created the fourth largest army in the world.[56]

Frederick William had the personality of a military king. During his childhood he exhibited a strong temper. Tutors had no success in controlling the tempestuous child, who struck them when he became bored with lessons. On one occasion he even attempted to strangle one of his teachers. As a child he exhibited little appreciation for the finer things in life. He had a keen mind, but little interest in learning. Robert Ergang said that the young crown prince's intelligence was "strictly bounded by the utilitarian."[57] Frederick William not only failed to foster cultural affairs, he actively attacked them. His interest in the state went only as far as it served his army, and he had no interest except efficiency.[58]

Frederick William confirmed his childhood tendencies as soon as he assumed the throne. His first priority was to cut what he considered the frivolous waste of money on art and sciences. He sold paintings, discharged musicians, reduced court salaries, and began renovations to transform the royal palace in Potsdam into a hunting lodge. He intended to close his father's Royal Academy of Science, but left it open on the advice of his chief surgeon. The doctor convinced him that the Academy of Science would be useful for the medical training of military physicians. Not completely convinced of the utility of all the sciences, he did dismissed the master of the Royal Academy and replaced him with his favorite court jester.

A meticulous Christian who tolerated Catholicism, Lutheranism, and Reformed doctrines throughout his reign, the king ordered Lutheran and Reformed theologians to work out their differences in the name of efficiency. The army needed a simple religion devoid of theological complexities. Frederick William's fervent Christianity was probably the result of his childhood where he received rigorous indoctrination in Reformed Calvinism. He dismissed theologians with terse notes, telling them that their doc-

---

[56]For an excellent synopsis on the impact of Prussia's military growth on Europe see R. J. White, *Europe in the Eighteenth Century* (New York: St. Martin's Press, 1965), 81–83.

[57]Robert Ergang, *The Potsdam Führer: Frederick William I, Father of Prussian Militarism* (New York: Octagon Books, 1972; orig. 1941), 28.

[58]See R. W. Harris, *Absolutism and Enlightenment: 1660–1789* (New York: Harper and Row, 1966), 167–74.

trines were the same, and despite his harshness and orientation towards militarism, Frederick William believed with all his heart that he was a practicing Christian. The king also fostered several educational reforms, but his purposes were entirely utilitarian. He thought that his subjects should receive enough education to practice Christianity and to serve the state. He established a cadet school for officers and promoted military learning. Beyond the military and the Bible, Frederick had little use for scholarship or education. He openly dismissed most learned men as fools.

Frederick William's Christianity had an interesting impact on the military. He believed that each unit should be staffed with chaplains and quite happily provided chaplains of Reformed, Lutheran, and Roman faiths for his soldiers. The Calvinistic doctrine of predestination, however, troubled his soul. According to the seventeenth-century scholastic Reformed doctrines of predestination, humans are destined by God to act out the events of their lives. They exhibit neither control of their lives nor the ability to make free-will decisions. By the 1700s this doctrine was not only popular in Reformed circles, it was openly accepted in Prussian universities. Professor Christian Wolff, for example, espoused this theory at the University of Halle. Frederick William's problem was simple. He believed that Calvinistic predestination was disruptive to the order of the army. Frederick William reckoned that if a soldier believed he was destined to desert, he had no choice in the matter, so he would simply leave. Therefore, the king discouraged teaching predestination, even though he accepted it as gospel. In addition, he frequently favored Lutheran chaplains over Reformed ministers for his army because the Lutherans not only had a different doctrine of predestination but their attitudes toward authority provided better discipline for the army.[59]

The king possessed an accountant's eye for detail. Lacking imagination, the Frederick William spent hours examining the most minuscule elements of any governmental operation in Prussia. Untiring of details, he centralized all civil service, judicial, and military functions under his direct command. Every official, judge, and officer could be held accountable for any action, expense, or decision. No matter was too small. The king constantly demanded reports and frequently wrote instructions on the margins of the reports, sending them back for in-depth explanations. His keen eye for detail

---

[59]Ergang, 32–33.

and his ceaseless work allowed him to oversee virtually every function in the kingdom.[60] He frequently beat offending bureaucrats with a club or cane when their performance did not meet his standards.

All details focused on fiscal conservatism. Frederick William was a king who loved to collect his ducats. The reason for his keen interest in money was not altogether different from that of his grandfather, the Great Elector. Frederick William wanted to build the army. It was not enough, however, simply to build an army to protect the state. He believed that a prince could rule only when his edicts were backed by military force. Since Frederick William wanted to convert the Prussian army into one of the mightiest military juggernauts in Europe, the only way to finance the undertaking was to gather funds. The focus of Frederick William's accounting tendencies and of his centralized autocracy was to increase the base from which he could build the army.[61]

One of the most informal, unusual, and intriguing aspects of the king's government was the inner circle of military officers who gathered around him every night after dinner. Frederick William referred to this assembly as the "Tobacco Parliament." Each evening senior army officers, whom Frederick considered part of his family, would join him outside the royal residence. Coarse bowls of tobacco were placed on rough wooden tables in front of each officer. The king distributed clay pipes and liters of beer in a barracks-room atmosphere. Although he made many state decisions in the smoke-filled den, Frederick William used these occasions for practical jokes and soldierly camaraderie. He frequently called on his most trusted generals, such as his close friend Leopold of Anhalt-Dessau, to provide advice on everything from tactics to foreign policy. The Tobacco Parliament, albeit unofficial, became Frederick William's greatest source of relaxation and statecraft. It symbolized his special relationship with his soldiers. Officers from foreign domains would often marvel at the status of to their Prussian counterparts. Each Prussian officer had access to the king. Frederick William was reluctant to engage in a social life outside of the military.[62]

---

[60]Ibid., 161–86.

[61]The most comprehensive summary of Frederick William's reforms written in English is Reinhold August Dorwort, *The Administrative Reforms of Frederick William I of Prussia* (Cambridge: Harvard University Press, 1953).

[62]Ergang, 44–61.

The daily routine at court was a far cry from the pomp and circumstance that marked the reign of Frederick I. The king rose early and worked hard. He also simplified the royal robes. In Frederick William's court, the mark of nobility was the plain dark blue coat of the Prussian officer. The king wore a simple coat, meticulously groomed, and devoid of rank. He considered himself to be the first officer of state. Like most Prussian officers, he carried a cane frequently attached to his hand by a leather thong, which he did not use to enhance his gait. Rather, he used the cane to enforce discipline among the common ranks of soldiers. Frederick William used it on his servants, his bureaucrats, and only occasionally on his beloved soldiers. Unfortunately, he used it frequently on his youngest son.

Military Reforms

The man closest to the heart of the king was Leopold of Anhalt-Dessau.[63] The Dessau clan produced a number of officers for the Hohenzollerns, and Leopold earned the nickname "The Old Dessauer" as his sons joined the ranks of the Prussian officer corps. Leopold was a tailor-made field marshal for the court of Frederick William. Gruff and utilitarian, the Old Dessauer was interested in only one thing: the competence of his soldiers. Having endeared himself to Frederick William in the War of the Spanish Succession, Leopold gained immediate access to the throne as soon as the crown prince became king. The king wanted military reform, a large army, and efficiency. Nothing could have excited the Old Dessauer more. He set about a program that would bring the Prussian military competence to a new high.[64]

Leopold was not an intellectual. In fact, it was rumored that he forbade Moritz, his young son, from reading and writing so that he would become a better officer. Regardless, Leopold was responsible for three major technical innovations in Frederick William's army.[65] The first dealt with the rate of

---

[63]See Herbert Tuttle, *History of Prussia, Volume I, 1134–1740* (New York: AMS Press, 1971; orig. 1844), 390. Unfortunately for those who do not read German, a thorough summary provided by Karl Linnebach, *König Friedrich Wilhelm I und Fürst Leopold zu Anhalt-Dessau* (Berlin: B. Behr's Verlag, 1907). It has not been translated.

[64]Russell F. Weigley, *The Age of Battles: The Quest for Decisive Warfare from Breitenfeld to Waterloo* (Bloomington, Ind.: Indiana University Press, 1991), 128–29.

[65]For a summary of all military reforms see Ergang, 62–83.

musket fire from the Prussian infantrymen. During the War of the Spanish Succession, Leopold was often frustrated by the inability of troops to maintain continuous fire in combat. Prussian drill had improved their loading and firing sequence, but their weapons could not match the speed of the soldiers. The primary weakness in loading was the ramrod. Wooden ramrods frequently broke when jammed down the barrel in the fast-loading panic of combat. Leopold spoke to one of his friends about the problem and asked him to develop a prototype ramrod made of iron. With a wooden ramrod as a model, the friend produced the desired result. Leopold field-tested the new ramrod and was delighted. Iron ramrods held up under extensive and rugged use. The Old Dessauer was able to increase Prussian firing effectiveness so that a platoon could fire a volley three times a minute. Such a standard was then unheard of on the continent.

In order to assure firing efficiency, Leopold introduced another innovation. At Blenheim, the Prussian infantry line had proven to be the difference between life and death. Leopold figured that troops could form lines more quickly, maintain bayonet walls better, and deliver rapid-fire more accurately if they marched in step. He introduced the concept of in-step marching for platoons, companies, and, finally, battalions. With the king's suggestion, he imported a lock-kneed "goose step" from the Russian army to maintain the line during wheeling maneuvers. The concept of marching proved so successful that soon armies all over Europe were mimicking the Prussians. Standard marching, turning, and firing maneuvers increased the efficiency of an already deadly military force.

The concept of drill caught the king's eye. Frederick William, ever mindful of meticulous details, noted the increased efficiency in the marching drill. Conferring with Leopold, the king soon produced the first drill manual. This was followed by standard rules and regulations for the Prussian army. The drill manual was a well-kept secret and every officer received a copy. Most officer education centered on memorizing and implementing the drill manual. When an officer left the Prussian service, the king required that he surrender his manual. While the nobles of other countries were quick to serve in the army of their choice, Frederick William encouraged Prussian nationals to remain at home. He also actively recruited Prussian nobles for his officer corps and punished with death any revelation of the drill manual.[66] The first manual was published in 1714. Frederick William's concept of drill was decades ahead of the French and the Austrians, so he was probably correct to regard it as a strategic secret.

Leopold introduced one other reform that the king enthusiastically endorsed. The armies of the War of the Spanish Succession were gaily arrayed in a variety of costumes. Once again, Leopold concluded that uniformity would enhance the efficiency of troops in battle. Therefore, he proposed that the Prussian army adopt a single uniform. Many of the officers rejected the idea because regiments were known for and took pride in their regimental uniforms. The king endorsed the Old Dessauer on this point. Soon Prussian artillery and infantry units were assigned a regulation blue uniform. Facings, caps, and lace differed, but Prussian blue became the army standard. Psychologically, the king endorsed the concept for the officer corps when he himself donned the royal coat of Prussian blue. Only the cavalry maintained its standard regimental uniforms, but the king even encouraged their standardization. As usual, Frederick William took a microscopic view of the Prussian uniform, even regulating the amount of cloth that could be used in sewing a jacket. When compared to British or French coats, the king proudly figured that the sparse Prussian jackets resulted in a "free" coat with every fifth product. He also ensured that the minimum number of buttons were used.

Leopold's ideas were important to Frederick William, but the king frequently acted on his own. One of his most impressive actions was the tax system reforms that resulted in more incentives for Junkers to join the officer corps. Under the reign of Frederick I, the officer corps in Prussia gradually received nobles for a variety of economic reasons. Frederick William endorsed this policy wholeheartedly and expanded it. As the first officer of the state, he saw the sons of noblemen as the natural candidates for service to their king. He created two policies that assisted him in this endeavor. First, by assuming the rights of taxation and military conscription

---

[66]For an analysis of commands see Harald Kleinschmidt, "Studien zum Quellenwert der deutschsprachigen Exerzierreglements vornehmlich des 18. Jahrhunderts (III)," *Zeitschrift für Heereskunde* 51 (1987): 162–64. Kleinschmidt also has two articles on uniforms and facings in the same volume. For English translations of regulations see William Faucitt, *Regulations for the Prussian Infantry* (New York: Greenwood Press, 1968; orig. 1759), and William Faucitt, *Regulations for the Prussian Cavalry* (New York: Greenwood Press, 1968; orig. 1757).

over the estates of the Junkers, Frederick William promised the sons of Junkers careers as military officers in exchange for their feudal rights. Second, Frederick William further enhanced the lofty status of individual officers.

Officers in Prussia were treated unlike officers of any other country. Frederick William viewed young commanders as his children, those learning the profession of arms. He regarded senior officers as brothers. They were entitled to treat the king as a social equal provided that they remembered he was a superior officer. Commanders enjoyed an open door to the royal throne room. While a diplomat or a civil service minister might be kept waiting for an audience with the king, an officer could simply walk up to Frederick William and make his request. Frederick William's rules and regulations stipulated different codes of behavior for enlisted men and officers, and he made it clear that Prussian officers were a distinct class. Frederick William succeeded in solidifying Junker loyalty to the crown, a policy that would have far-reaching results in the next century.[67]

The Prussian system of civil administration and taxation served military needs. The country was divided into military districts. In turn, each military district was divided into cantons. The canton was based on the number of hearths in each area. Each district was required to provide a certain number of soldiers based on the number of fireplaces in its canton. Villages could avoid having local men conscripted into the army by offering to pay for recruits from other territories. Eventually, however, Frederick William's manpower needs became so great that conscription in Prussian territories became the norm.[68]

As the army grew during Frederick William's reign, it was mobilized each spring for military exercises. These increased the efficiency of the army and gave officers command experience, although—to cut costs— soldiers were seldom allowed to load and fire their weapons. There was, however, a major weakness in this system of maneuvers. Frederick William's love of precision and his desire to see each trooper in the proper place required that the exercises take place on parade grounds. Thus, the

---

[67]See Michael Howard, *War in European History* (Oxford: Oxford University Press, 1976), 69.

[68]See comments by Hew Strachan, *European Armies and the Conduct of War* (London: George Allen and Unwin, 1983), 9. For an analysis see William O. Shanahan, *Prussian Military Reforms: 1786–1813* (New York: AMS Press, Inc., 1966), 35–44.

army did not practice in combat terrain. As a result, Frederick William's cavalry and infantry marched well on parade, but the forests and valleys of war broke the meticulously organized Prussian formations. The king solved the problem by limiting his participation in war.

The greatest improvement in Frederick William's military force was the infantry. Blue-coated, marching in step, and armed with a non-breakable iron ramrod, the infantry could endlessly go through the motions of firing, reloading, priming, cocking, and firing again. Although training was limited to the parade ground, infantry efficiency increased drastically. Prussian regiments drew envious attention from neighboring kingdoms, and the infantry was second to none.

The tie that bound the Prussian army together was not a patriotic devotion to duty. The Junkers were loyal to the king and their honor, but the enlisted men held no such loyalty. Often conscripted, the enlisted men wanted to be anywhere but in the Prussian army. Discipline held them together. Rigid and merciless, the Prussian army functioned on a system that provided harsh physical punishment for the slightest infractions. Sergeants carried wooden spears, called spontoons, which were used both to beat the soldiers and to encourage them to stand in line. Officers walked with canes, and they were frequently used against a slouching soldier or an unfortunate lad who happened to turn left when the command for turning right was given. Generally, new recruits were not beaten right away. In fact, Frederick William said it took a soldier five years to learn how to drill. But mistakes were not tolerated among veterans. Men were beaten, noses were broken, and teeth were knocked out, all in the name of discipline. Hardly a day went by when a Prussian soldier was not broken by an officer's cane. Leopold and the other senior officers approved. Discipline, they believed, was the hallmark of Prussian success.

For all the improvements in the infantry, the other branches of service enjoyed no appreciable change. To be fair, the artillery was not emphasized during this period, but other European nations did employ cavalry frequently. Unfortunately, neither Leopold nor Frederick William understood the function of cavalry. Further, with his frugality, Frederick William would not allow horses to be used in realistic exercises. The accountant could not be reconciled with the militarist. Realistic training for the cavalry cost too much. Therefore, the cavalry remained unskilled, and Frederick William's

favorite charger was the country plow horse. His idea of equestrian efficiency was to place big slow men on equally slow horses.

The weakness in the cavalry also tied into a personal quirk of Frederick
William. Although he firmly believed that he followed Christian edicts to
the letter and he prided himself on being a man of high morals—with his
self-confessed weakness for drinking during the Tobacco Parliament—
Frederick William admitted that his one vice centered on a strange com-
modity. He loved tall soldiers and claimed that they were a moral weakness.
He spared no expense in recruiting "long fellows" and surrounded his court
at Potsdam with his own giant grenadier regiment. The affinity for tall soldiers made him the butt of many jokes among European royalty, but it had a
more devastating effect on the cavalry.

Frederick William's idea of a proper cavalryman was an extremely tall
soldier on an equally large horse. Successful cavalrymen know that the advantage of a horse soldier is speed and shock action. The purpose of cavalry
is to move quickly and to strike with speed and daring. Leopold and Frederick William did not understand this. Plow horses, draft horses, and any
horse that had an extremely large girth was suitable for Frederick William's
cavalry. A giant trooper mounted on the back of such an animal brought
warmth to the heart of the king. While they were impressive on the parade
ground, these cavalrymen were no match for the other cavalry units of
Europe. Most foreign units could ride circles around the Prussian cavalry.[69]
One observer commented that the cavalry was composed of giants on ele-
phants.[70]

The fixation with tall soldiers also affected foreign policy. As Frederick
William recruited Potsdam guards, a regiment of "long fellows" all over six
feet, he began raiding the treasury to recruit tall soldiers. In addition, he alienated his neighbors by sending "recruiting parties" into their territories to
kidnap tall people for service in the Potsdam giants. When kidnapping
proved inadequate, tall men were frequently invited to the Prussian court,
only to be thrown into jail until they signed enlistment papers for the
Potsdam guards. Peter the Great saw this as an opportunity to secure an alliance with Frederick William. As he traveled through Berlin, the tsar noticed

---

[69]Herbert Rosinski, *The German Army* (Washington, D.C.: The Infantry Journal,
1944), 16.

[70]Ergang, 67.

the king's affinity for tall soldiers. In the 1720s he began sending approximately 100 giant Russians to Frederick William every year. The first time it happened, a tearful Frederick William looked at the tall soldiers and proclaimed everlasting friendship with Peter and the Russian people. Peter did not understand it, but he emptied his jails annually with a contingent of giant prisoners for his newly found ally.

Not all of Frederick William's reforms were centered on the army. The king focused attention on Prussia's criminal justice system, its civil service bureaucracy, and forms of private enterprise. Yet, Frederick William was cast in the mode of his grandfather, the Great Elector. While he sought to bring efficiency to the courts, to civil service, and to business ventures, the underlying purpose of his reforms was to create a state that could support his vast military organization. Frederick William behaved true to his name. He was the soldier king.[71]

In this vein, Frederick William's correspondence with his confidant and friend, the Old Dessauer, is rather enlightening. Frederick William constantly indicated his concern with military affairs to Leopold. The state may be supported by ideas and people of letters, Frederick William intimated, but no nation listened to its sovereign unless the throne was backed by military force. Leopold's concern with military efficiency, the structure of regiments, and military proficiency blended well with Frederick William's ideas. Leopold moved into Frederick William's Tobacco Parliament and inner circle. The prince and king made every move with the army in mind.[72]

Father and Son

Discipline was the tie that held Frederick William's army together, and, unfortunately for the heir to the throne, the king approached parental duties in the same fashion.[73] He was a crude, abusive father. A mystic might argue

---

[71]Walter Hubatsch, *Frederick the Great of Prussia: Absolutism and Administration* (London: Thames and Hudson Ltd., 1975), 26–33. See also Dorwort, Chapter IV, note 7.

[72]Ergang summarizes, 142–211. See also Linnebach, 25–31.

[73]See colorful comments by Thomas Carlyle, *History of Friedrich II of Prussia Called the Great*, John Clive (ed.) (Chicago: University of Chicago Press, 1969; orig. 1900), 121–32. Clive presents an edited, abridged version of an eight-volume history. Written in the "heroic style" of nineteenth-century historians, Carlyle's literary talent illustrates the emotional and physical pain of Frederick William's heir. A popular mod-

that the rift between father and son was an evil Hohenzollern curse. Regardless, several historians and biographers of Frederick II, the crown prince and Frederick William's oldest surviving son, have pointed to the abuse the young lad suffered at the hands of the king. It is beyond the limited purpose here to deal in detail with Crown Prince Frederick's childhood, but it is necessary to mention this problem to help to explain the course of Prussian military history. Even the most neutral observer would be forced to conclude that Frederick was an abused child. He learned that hate, fear, and physical pain were normative expressions of the human experience, and he took these beliefs to the throne of Prussia.

When Frederick William thought about the ideal child, he often looked to the Anhalt-Dessau clan. Leopold had sons who aspired to be military officers. Leopold's oldest son, the Hereditary Prince Leopold, was in a regimental uniform almost as soon as he could walk. His brothers followed him in the tradition. The Old Dessauer's boys were concerned with hunting, drinking, fighting, and the command of military personnel. Nothing could have pleased the Old Dessauer more and, ironically, Frederick William wished that he had been blessed with the same offspring. When the young Frederick came into the world in 1713 he was not the king's heir, and his rugged, sturdy, and outgoing older brother promised to be the type of king that Frederick William wanted to leave on the throne in Berlin. But Frederick's older brother died of a childhood illness, and young Frederick ranked next in the line of ascension. Small, frail, and extremely articulate, he was the antithesis of everything the king considered manly.

Young Frederick was a child of brilliance. In daily contact with a French nanny, the boy became fluent in both French and German at an early age, developing an affinity for the liberal arts. To counter these tendencies, Frederick William prescribed an education for the young prince, an education dominated by the king's methodical style. Crown Prince Frederick was to arise at a precise time in the morning, spend fifteen minutes in prayer, fifteen minutes attending to personal needs, and then begin his first series of lessons before breakfast. The routine for the day continued in such a precise manner until the prince retired at night. Frederick William's purpose was to

---

ern portrait is rendered by Nancy Mitford, *Frederick the Great* (New York: Harper and Row, Publishers, 1970), 33–45.

create a soldier king to follow in his steps. Ironically, he helped to create a genius.

Unlike his gruff, unsophisticated father, young Frederick loved to read and enjoyed the novelty of new ideas. Music, art, philosophy, and the spirit of the Enlightenment formed the world of the young prince. As he grew older, he collected books on every subject and became well-known for his wit and ability to converse intelligently on a variety of matters. The dark secret inside the house of Hohenzollern conveyed another side of Frederick's childhood. His father was incensed with the crown prince's affinity toward liberal education, and he literally tried to beat discipline into the future king.[74]

The concern here is on the Prussian army. Poor Crown Prince Frederick came into that army under the hands of a horrid father. It is indeed strange that a king who was reluctant to employ his army because of concern for Christian values would treat a child with such intense physical abuse. The king regularly caned Frederick, and it was not unusual to see the small boy running from the king when he accompanied him on rounds through the Prussian towns. As Frederick came into adolescence, he began to challenge his father, which only resulted in more abusive wrath. His only consolation was a sister, Wilhelmina, who continually tried to protect him and his mother who shed tears over her son's perpetual beatings. Sympathy was not enough. Young Frederick learned the meaning of distrust and pain.[75]

When Frederick turned seventeen, he felt that he could take no more. Quite unrealistically, he planned to desert the Prussian army with two close friends. It was the fantasy of childhood, a dream of Enlightenment freedom. During this new age of reason, Frederick believed there were places where a future king could escape tyranny and await a time until he could reign with justice and truth. The crown prince had no understanding that political reality would accommodate neither desertion nor such a dream, but he was seventeen. In 1730 he attempted to escape while visiting the Rhenish provinces on an inspection tour with his father. The three young men were uncovered before the desertion plot hatched. One of the boys escaped, but Frederick and his closest friend were captured and court-martialed. Frederick's un-

---

[74]See summary by Ergang, 212–38.

[75]See Ernest Lauisse, *The Youth of Frederick the Great* (trans. by Mary Bushnell Colemen) (New York: AMS, 1972; orig. 1892).

lucky friend—the son of a general and the grandson of a field marshal—was executed, and only the intervention of General Kurt von Schwerin, the head of the court-martial board, and the Old Dessauer saved the crown prince.

After the attempted desertion, it took years for father and son to be reconciled, and it was a hideous process. At one point, Frederick William considered sentencing his son to death, and—even though Schwerin's court-martial board had recommended leniency—the king favored the idea. Frederick spent several years on fortress duty in Küstrin and only returned to the king's grace under the promise that he would devote all of his efforts to military matters. Although Prince Frederick wrote to his friend, Voltaire, that he would change the direction of Prussia when he assumed the throne, the prince assumed his military duties with cynical vigor. The genius he had shown in art and literature was transformed to the drill field and strategy manuals. He soon commanded his own regiment at Küstrin and earned his way back into his father's grace.

During Frederick William's single limited military venture, the War of the Polish Succession (1733–1738), the king sent a small military contingent to support the emperor in the Rhine Valley. Serving under the old warhorse, Eugene of Savoy, Crown Prince Frederick accompanied the army, and Eugene took notice of the young man's actions. Although the Prussians were involved only in a few minor skirmishes along the Rhine, Eugene was impressed with Frederick's ability to comprehend the most complicated military matters as well as his coolness under fire. A few years later, a group of Prussian generals complained to Eugene about their probable future when Frederick William grew gravely ill. Their careers promised to be grim when the crown prince came to power. Eugene interrupted their lamentations. Recalling his Rhenish experiences with Frederick, the Prince of Savoy told the officers that they would be surprised. Eugene's words were prophetic.

In May 1740, Frederick William's life came to an early end. Young Frederick came to Potsdam to visit his ailing father. In a very touching, emotional scene, the prince, in military uniform, managed a painful reconciliation with his father. With that, the king passed on, but not before planning a frugal state funeral, detailed to an infinite degree. The Enlightenment philosophers rejoiced when the crown prince became Frederick II, King in Prussia, mistakenly believing that new king would usher in an age of wisdom and peace. They would have been far wiser had they put down their philosophy books and looked at the king through the eyes of Eugene of Savoy. The entire world would be surprised by young Frederick's actions.

Chapter 5

The First Silesian Wars

Psychologists frequently argue that the actions people take are based on the result of past experiences. If this is true, Frederick William I accomplished more than the construction of a massive army. When Frederick II came to the throne in 1740, the 27-year-old king brought the legacy of his youth. His experiences were centered in constant physical abuse and mistrust. Even the most basic social unit—the family—was the source of violence and pain. If Frederick William I had failed to use the army out of a sense of morality, Frederick II would have no such reservations. Compassion, human love, and concern for humanistic values were part of the new king's professed worldview, but his first priority was war. Frederick II trusted few people and had few reservations about the use of power.

The Beginning of the First Silesian Campaign

Reconciliation between father and son was only superficial. The political situation in 1740 was tailor-made for a Machiavellian prince. England and France were locked in a struggle for global domination. Austria was weakened by perpetual martial struggle, and to make matters worse, Emperor Charles VI, left this world with no male heir. The patriarchal European monarchs viewed his daughter, the young Maria Theresa, as a weak opponent in the struggle for continental domination, and no monarch was in a position to exploit this weakness as much as the new Prussian king. Unfettered from the moral bridle of his father's reluctance to wage war, Frederick II immediately prepared to strike the Austrian Archduchess.

Yet, King Frederick II was, as Robert Asprey so aptly dubbed him, a magnificent enigma.[76] The king's first actions seemed to indicate that he

---

[76]Robert B. Asprey, *Frederick the Great: The Magnificent Enigma* (New York: Ticknor and Fields, 1986).

was a true man of the Enlightenment. He revised criminal law, abolishing medieval criminal codes. He called senior officers together and chastised them for the harsh discipline in the army. He revised the military regulations and forbade his officers and sergeants from beating common soldiers. On the surface, Berlin promised to be a haven for the arts, sciences, and literature. These changes were short-lived and superficial, however. Frederick's real reforms came as he prepared the Prussian army for war.[77]

One of Frederick's first actions was to revise the army's command structure. He alienated the Old Dessauer, seeking new blood for leadership. He found it in the person of Kurt Christoph von Schwerin, an experienced combat commander who had entered the Prussian service years earlier from Sweden. Schwerin embodied many of the traits of an Enlightenment man, a man of *Bildung* or character. He favored education in officer ranks and viewed discipline only in functional terms. The purpose of military discipline, Schwerin believed, was not to instill blind obedience but to ensure performance in combat. One of Frederick's first military actions was immediately to promote Schwerin to the rank of field marshal. Although the Old Dessauer complained bitterly, Schwerin gained the ear of the king.[78]

The struggle between the Old Dessauer and Schwerin became intense.[79] Their differing approaches to the army caused them genuinely to detest one another. The Old Dessauer had no use for anything save blind obedience, and he missed the informal relationship he had enjoyed with Frederick William. Schwerin was more reflective, with a broad appreciation of the finer things in life. Whereas Leopold always favored endless drill, Schwerin sought to build an understanding of military action among junior officers. The Old Dessauer's alleged abandonment of his younger son's education appalled Schwerin. Yet, it would be Schwerin who handed down more death sentences for insubordination, and Leopold was certainly more popular with the rank and file. Frederick did not intervene in the struggle. He was quite

---

[77]Ludwig Reiners, *Frederick the Great: A Biography* (trans. by Lawrence P. R. Wilson) (New York: G. P. Putnam's Sons, 1960), 1–5 and 89–105.

[78]Asprey, 83.

[79]See Pierre Gaxotte, *Frederick the Great* (trans. by R. A. Bell) (New Haven, Conn.: Yale University Press, 1942), 200–1.

content to let his two field marshals wallow in animosity toward each other.[80]

The question of Austria was another matter. In October 1740, Frederick began plans to attack Maria Theresa. He turned to Schwerin and excluded the disgruntled Leopold. Frederick's target was the rich northern Austrian province of Silesia. Directly south of Brandenburg, Silesia lay between the rolling hills of Poland and the mountains of Moravia and Bohemia. It had abundant minerals and rich farmland. The Hapsburgs claimed the province, but they had not set foot in it for over a century. The Hohenzollerns had an old claim to some of the land. While it was an insufficient diplomatic excuse to seize Silesia, Frederick was willing to use it as a pretext for invasion. All the European powers coveted the province, but it was closest to Prussia, and Prussia had a large army. Frederick believed ownership of Silesia would be decided only by the force of arms.[81]

The Austrian Archduchess had many problems when she assumed the throne in Vienna, and few people believed that she would be able to control the empire, despite her father's attempt to have her rule guaranteed by the so-called Pragmatic Sanction.[82] With her attention diverted, Frederick reckoned that Maria Theresa would be unable to hold the province of Silesia, and Frederick intended to take it. He sent Schwerin to prepare for the invasion, while ordering his foreign minister, Heinrich von Podewils, to lay the diplomatic ground work.[83] Both men advised caution, but Frederick was young and hungry for glory. With a corpulent war chest inherited from his father, Frederick secretly began buying grain stores to support a late autumn campaign. When the army's fall maneuvers ended, Frederick conveniently failed to demobilize the troops. All Berlin speculated about the king's plans, but Frederick kept his secret well. Only the top echelon knew that he was about to do the unthinkable. Suddenly, to Europe's complete surprise, the Prussian army entered Silesia five days before the winter solstice. No one

---

[80]Christopher Duffy, *The Army of Frederick the Great* (New York: Hippocrene Books, Inc., 1974), 157–59.

[81]Herbert Tuttle, *History of Prussia, Volume II, Under Frederic the Great* (New York: AMS, 1971; orig. 1888), 66–73.

[82]See C. T. Atkinson, *A History of Germany: 1715–1815* (New York: Barnes and Noble, Inc., 1908), 105–13.

[83]Tuttle, vol. II, 78–79.

could have predicted that Frederick would spend the next twenty-three years of his reign paying for this decision.

The First Silesian Campaign

Frederick's opportunity had come because the Emperor Charles VI died on October 20 leaving only his daughter, Maria Theresa, to assume the throne. Convinced that Maria Theresa would be the pawn of European powers, Frederick ordered Heinrich von Podewils to draft a statement making the Hohenzollerns' historic claim to Silesia. Diplomatically, the statement carried little weight, and Podewils protested. Frederick turned a deaf ear to his foreign minister, uninterested in Podewils's ethics. Silesia was ripe for the picking, and Frederick had no intention of being thwarted by diplomatic niceties. He continued planning for war.[84]

The invasion plan was fairly simple and governed by the geography of Silesia. The northern edge of the province bordered Brandenburg. Silesia was bisected by the Oder River, and Frederick planned the invasion route along its course. Christopher Duffy points to the logic of this decision. First, the Oder was an open channel for supply and communication, in effect, the eighteenth-century equivalent to a modern super highway. The river would serve as Frederick's supply line. Second, most of the critical fortresses controlling Silesia sat on the Oder. The first major fortress was Glogau, a town located a few miles south of the Brandenburg border, housing a small garrison of Austrian troops. Further up the Oder sat the free city of Breslau, the provisional capital of Silesia. Control of Breslau was critical to Frederick's campaign. Two other fortresses, Ohlau and Brieg, lay to the south of Breslau. Frederick's plan was quickly to isolate or capture the fortresses and then move to the west, closing the mountain passes from Moravia and Bohemia.[85]

Lower Silesia represented the most attractive economic area for Prussia, but upper Silesia was the strategic key. Geographically, upper Silesia con-

---

[84]One of the best modern summaries of the campaign and subsequent wars is found in Christopher Duffy, *Frederick the Great: A Military Life* (London: Routledge, 1988). Duffy and Asprey provide the best modern English summaries of Frederick's Silesian wars. Duffy's description (*Frederick*, 25-75) provided the most of the basis for the remaining segment of this chapter.

[85]Ibid., 159.

tained the only mountain passes through which the Austrians could respond to the invasion. Frederick and Schwerin believed that if they could seize the foothills in a winter campaign, they would effectively isolate the province from the Austrian army. Maria Theresa's army could not come through the passes in the winter. Frederick also knew that if this plan were successful, the Austrian troops inside Silesia would be cut off from rescue.

While the Old Dessauer fussed and fumed over his exclusion from the war plans, Frederick began moving troops to Crossen, the last major Prussian town north of the border crossing.[86] Although Frederick believed winter campaigns were wasteful and fairly ridiculous, he felt that his first military venture in the winter was justified. The Austrians, he believed, could not reinforce the area until spring, and if he moved in the winter, the Austrians would not be able to mount any type of effective resistance. A winter campaign would also completely surprise the Austrians. So, on December 16, 1740, Frederick crossed the border with 27,000 Prussian troops. It was indeed a surprising and unprecedented action. Germany would not launch another major winter offensive until December 16, 1944.

Students unfamiliar with military activities frequently do not understand why Frederick would enter Silesia with 27,000 troops when he had 84,000 men at his disposal. A short discussion of concentration and maneuver will prove helpful in explaining the reason. It is difficult to place the majority of troops in the field at any point in a military operation, and it is impossible to place the entire strength of an army in a combat setting. Military historians may study the movement of troops, but generals study logistics. The majority of military actions at almost any point in history are aimed at supporting troops in the field. In 1740, Frederick had an army of 84,000. To take nearly a third of that army into a winter campaign was a substantial endeavor.

In addition, Prussian geography dictated that a large portion of the troops remain in Brandenburg. The Hohenzollern domains were stronger than they had been a century earlier, but they were still scattered and they remained on the military crossroads of Europe. Prussia was, as Voltaire observed, a kingdom of borders. Frederick knew that every border demanded protection. To make matters worse, the three countries with larger armies than Prussia—Russia, France, and Austria—sat on Frederick's borders. If

---

[86]Tuttle, vol II, 58–59, and Asprey, 163.

he weakened a border, Frederick would invite the same type of invasion from his potential enemies that he was conducting in Silesia. Frederick also wanted an economized operation. Like any good commander, he wanted to use only enough forces to ensure victory, keeping the remaining units in reserve.

Weather was also a factor. Frederick could have used more troops in Silesia. Indeed, in the coming years he would operate with thousands more soldiers. Yet, in December 1740 the king needed a fast-moving, rather limited force. His purpose was to overpower Austrian garrison troops and gain winter quarters. Until he had those quarters, he had to bring supplies through a line also suffering from the effects of the winter weather. With 27,000 troops, even if Austria could manage a response, Frederick would be favorably matched with any opposition. Frederick based the number of Prussian troops for the operation on a logical decision.

The initial advance was remarkably successful. The smartly dressed Prussian troops moved over roads, taking the local Austrians detachments completely by surprise. The Silesians greeted Frederick, and he treated them as his subjects. To exploit the area and its resources, Frederick genuinely wished to treat the Silesians well. He had no desire for a local revolt. Within two days, he had advanced over twenty miles. Unfortunately for the Prussians, the weather began to turn bad on December 18.

Frederick's advance slowed as the local dirt roads turned into snow-covered mud. To make matters worse, the garrison at Glogau, even though overwhelmingly outnumbered, refused to surrender. Despite these obstacles, Frederick knew he had to continue to advance. He divided his army, improvising two corps. "Second Corps" was placed under the command of the hereditary Prince Leopold of Anhalt-Dessau, the eldest son of the Old Dessauer. Frederick and Schwerin continued down the Oder with the majority of troops in a unit Frederick called "First Corps." On December 31, Frederick arrived at the gates of Breslau. He was greeted by a delegation from the city, and the city fathers asked to be treated as neutrals. Frederick agreed and he celebrated the new year with a triumphant ceremonial entry into the town. Within a few months he would renounce his promise and take Breslau without a shot.

Frederick continued the advance. He divided the First Corps into two wings. Schwerin moved to the left along the Bohemian foothills. Although he was unable to control the mountain passes, he cut off the majority of

Austrian troops. Neisse, a small fortress close to the Moravian foothills, refused to surrender and became a problem, but Schwerin's operation was generally successful. On the Oder, Frederick took the minor fortress of Ohlou and, with the exception of Glogau and Neisse, Silesia fell completely by the end of January. It had been a daring operation, but the initial campaign was a military success. The only Austrians who could resist in Silesia were completely isolated. Frederick's plan had worked.[87]

Schwerin took command of Silesia in mid-winter, and Frederick returned to Berlin. Glogau fell to the hereditary Prince Leopold on March 9, clearing a logistical lane to all Prussian troops. It appeared to be an overwhelming victory, but the Austrians were unwilling to hand over Silesia without a fight. Realizing their perilous situation, the troops isolated in Silesia began raiding Prussian positions, and these raids demonstrated the weakness of Frederick's army. While his infantry was unparalleled, the king's cavalry was horrid. The Prussian horses was no match for the Austrians. The Prussians held the advantage in large-scale movements, but the Austrians outmaneuvered the Prussians in small battles. Frederick returned to the Oder in the early spring without totally understanding this situation. It was not the last surprise Maria Theresa had in store for the Prussian king.

In early April snow still covered much of Silesia. More importantly, the mountain passes were deemed impassable. Schwerin assured the king that no army could move through the passes until May. The field marshal encouraged Frederick to spread his cavalry in a large screen and to keep most of the infantry dispersed. The campaign season, Schwerin said, would begin in a few weeks. For now, the Prussian army was safe.

Maria Theresa, who was developing an intense personal hatred of Frederick, had no intention of waiting for the campaign season. Realizing that her enemy had grown complacent, she dispatched a small army to move into her contested province. In April, Field Marshal Count Wilhelm von Neipperg moved into Silesia with a force numbering 16,000. Neipperg's infantry striking power was limited—he commanded about 10,000 infantrymen—but the cavalry was seasoned. In Silesia, the cavalry began rounding up the scattered Austrian units, and Neipperg was able to relieve the makeshift siege around the fortress of Neisse. Schwerin and Frederick were

---

[87]Tuttle, vol II, 77–89.

caught completely off guard. They scrambled to assemble their scattered
forces in the face of a concentrated Austrian field army. [88]

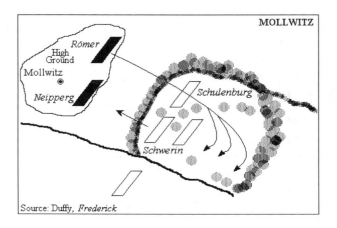

Source: Duffy, *Frederick*

On the snowy morning of April 10, 1741, Frederick concluded that he
must find the Austrian field force and destroy it. Otherwise the Austrians
would send more troops when the spring snows melted. Frederick did not
know the exact location of Neipperg, but he knew that the Austrian was in
the general area of Brieg. The king gathered 21,000 troops along the Oder
River shortly before dawn and moved north from Brieg, hoping to find
Neipperg. Schwerin assembled the troops in five columns as they trudged
through the snow. Frederick's scouts reported that Neipperg seemed to be
assembling a few miles to the west, and, around mid-morning, the columns
turned in that direction. By 10 A.M. the initial reports were confirmed.
Neipperg was in the vicinity of Mollwitz. Frederick moved the army
through a wooded area in the direction of the Austrian army, and, calling
Schwerin to his side, the young king began to plan an attack while the army
was still moving.[89]

---

[88]Reiners, 97–100.

[89]See Duffy, *Frederick*, 29–33, and *Army*, 160–61. For details on units, see the German language source: Curt Jany, *Geschichte der königlich Preussischen Armee bis zum
Jahre 1807, volume II, Die Armee Friedrichs des Grossen: 1740 bis 1763* (Berlin: Verlag

Frederick hoped to disperse the Prussians into two standard battle lines, a front line to engage the troops at Mollwitz and a second line to support it. He conventionally planned to put cavalry units on each flank. In nervous excitement, Frederick ordered the men to form battle lines long before they drew within range of the Austrians. The result was mass confusion. The infantry, still in the woods, dropped their packs in assigned supply areas and moved to form lines. The trees would not cooperate. The first line became so hopelessly confused that it failed to accommodate all the units. Several hundred soldiers were pushed to the rear with no place to go. The second line, under the command of the hereditary Prince Leopold, formed too far to the rear of the first line to offer much support. This was partially due to the stranded units who could not make it into the first line and the interference of the trees. The artillerymen cursed as they tried to move the cannon through the woods. Every soldier knew that this wooded assembly point was far different from a parade ground.

If things were bad for the infantry, the cavalry fared far worse. When the Prussians finally formed their battle lines, the cavalry moved slowly through the woods to take positions on the northern and southern flanks. The horses on the southern flank were deployed in the proper formation when the line began to move, but Frederick had over two miles to go. Some of the soldiers noticed that a small creek separated them from the infantry. As the troopers rode forward, the creek became wider. After a few hundred yards, a formidable stream separated the southern cavalry from the battle line. By the time the Austrians were in sight, the left wing could do nothing to help.

The right wing of cavalry fared somewhat better in the initial deployment. Although the infantry line was too compact, the cavalry advanced in good order and eventually came in sight of the Austrians. The right wing of cavalry was commanded by an old friend and teacher of the king, General Adolph von der Schulenburg. Although the Prussians had deployed far too early, Schulenburg kept his troopers under control, screening the cursing infantrymen as they fought their way through the snowy woods. It was not the drill field, but it worked. By mid-day the tired Prussians emerged from the woods facing Mollwitz. Although the front rank was too narrow,

von Karl Siegismund, 1928), 32–48. (Although an ultrapatriot, Jany writes with a quartermaster's concern for unit details.)

Schulenburg kept the northern flank secure, while the artillery pieces were unlimbered and set toward the Austrians. Neipperg was unprepared. Although he held the high ground, the appearance of the Prussians unnerved his command. The Prussian commanders were pleased. Schulenburg called his cavalry officers for a conference while the Prussian artillerymen fired a salvo of solid shot against the scrambling Austrians. As the smoke drifted by through the trees, Schwerin tried to find the king and to place the first rank in better order.

General Römer, the Austrian cavalry commander, was not aware of the confusion in the Prussian ranks, but he was cognizant that Neipperg needed time to prepare. Without waiting for orders, Römer launched an assault toward Schulenburg's troops, and the result was an absolute disaster for the Prussians. Schulenburg, still in conference with his officers, was completely surprised by the sudden Austrian onslaught. As he tried to gather his troops to meet the Austrians, the white-coated imperial troops smashed into the Prussians while they were sitting still. Schulenburg's command collapsed. Prussian troopers were unable to stop the flood of Austrian horses, and they were swept away by the momentum of the Austrian charge. The result was absolute chaos. Frederick was almost captured, but he made his way to the infantry line. Poor Schulenburg received a nasty facial wound, then he was killed trying to rally his forces. To the infantrymen looking from their ranks, the horrific scene appeared to be murder-on-horseback. Although there was no order, the king was frightened out of his wits. The common soldiers and their king believed that Römer was about to ride down the entire army.

Schwerin, the combat veteran, had seen this all before. He, too, was frightened by the Austrian cavalry, but panic in the musket ranks frightened him more. The Austrians rode around the entire Prussian flank, looking for a gap in the infantry line. The Austrian troopers knew that if they could find one break in the bayonet wall, they could cut the Prussians to pieces. They surged around the flank and galloped into the woods. What Schwerin did not know was that the Austrians were just as confused as the Prussians. Römer's men had been so surprised by the lack of resistance that they simply kept riding through the woods. Several dozen lost their lives in individual fighting, including General Römer, and the attack lost its impetus. Yet, neither Schwerin nor his infantrymen knew the situation. To them it seemed as if the Austrians would break through at any moment. The panicked infantrymen began discharging their muskets into the mass of horses, hitting both friend and foe. This not only further confused the cavalry fight, it scared Leopold's second rank to death. Believing they were receiving fire from the Austrians, Leopold's men fired a volley forward through the trees, a volley that smashed into the backs of the soldiers they were supposed to be supporting. Knowing the signs of a rout, Schwerin lost all interest in the cavalry battle. As his troops fired at each other, the field marshal believed that the day was lost. He began to search for his king.

Frederick was not in a comfortable frame of mind when Schwerin found him. The king had never seen a battle before, and he was trying to control his fear. Relived at Schwerin's appearance, he turned to his experienced marshal for advice. Schwerin told him that the undisciplined volleys caused him the most concern. It demonstrated that the army had lost confidence in itself and that it would soon abandon its position. The field marshal advised the king to leave the field to avoid capture. Frederick gathered some important papers and joined a small group of men in a cavalry escort. On a gray horse, which Frederick would cherish for the rest of its life, Frederick rode back toward the Oder at the gallop.[90] It was the only time the king ever left a field when the outcome was still in question.

For his part, Schwerin began to take immediate defensive actions. He sent a caustic message to Leopold, asking the hereditary prince to please refrain from firing into the first rank. He then examined the position. The Prussians had quite unintentionally achieved a major tactical advantage. The infantry units that had been forced out of the first rank at the beginning of the march now found themselves in a line between the first and second ranks. They were facing the Austrian cavalry in a perfect position to protect the Prussian right flank. The men were in a perfect U-formation, effectively blocking any Austrian attempt to break the Prussian infantry. With only the front still in question, Schwerin restored order, placing the front line in an oblique order; that is, in a line slanted away from the Austrian infantry positioned up the hill near Mollwitz. Schwerin was content to let Neipperg make the next move, and the Austrian marshal picked up the gauntlet.

Around 2:00 in the afternoon, Neipperg began a series of infantry assaults down the slopes from Mollwitz. The Austrians began shooting it out with Schwerin's oblique line, and that turned out to be a fatal mistake. The

---

[90]Gerhard Ritter, *Frederick the Great: A Historical Profile* (trans. by Peter Paret) (Berkeley, Calif.: University of California Press, 1970), 83–84.

parade ground, which had provided no help for the inexperienced Prussians that morning, had prepared them for an exchange of musketry. With step-by-step precision the Prussians fired and reloaded using the endless drill and the iron ramrods championed by the Old Dessauer. The Austrians, already weak in infantry, soon fell on the snow-covered slope. Thomas Carlyle provides one of the most romantic descriptions of the encounter between the two lines: "Three volleys for every two and iron ramrods against wooden ones."[91] The Austrians could not withstand the withering Prussian fire, and it turned the tide of the battle. After the fifth assault, Neipperg decided he could take no more. As darkness fell, the Austrians began a withdrawal from the village, causing Schwerin to probe their position carefully. After receiving no response, Schwerin moved forward and took Mollwitz. The field marshal then sent dispatch riders to find the king to inform him that Prussia had won its first victory.[92]

Frederick was quick to grasp the lessons of Mollwitz.[93] He had deployed his forces too early and incorrectly. The performance of the cavalry had been disastrous. The left wing had taken itself out of the battle and the right wing had been destroyed without resistance. Frederick realized that he needed to take corrective measures to deal with these deficiencies, and he spent most of the summer of 1741 doing so. Infantry formations needed to be geared to battlefield, not parade ground conditions. He encouraged commanders to practice deployment in less than ideal circumstances. The cavalry was another matter, and Frederick searched for men who would infuse the cavalry with a fighting spirit. Shock action did not come from giants on plow horses. He needed aggressive hell-bent-for-leather adventurers who would seek the initiative. He also never completely forgave Schwerin for advising him to leave the field. The king was always quick to blame others for any personal failure.[94]

---

[91]See the abridged version Thomas Carlyle, *History of Frederick II of Prussia Called Frederick the Great,* John Clive (ed.) (Chicago: University of Chicago Press, 1969), 373.

[92]Ergang quotes a popular satirical poem describing Frederick's flight. Robert Ergang, *The Potsdam Führer: Frederick William I, Father of Prussian Militarism* (New York: Octagon Books, 1972; orig. 1941), 252.

[93]Jay Luvaas, *Frederick the Great on the Art of War* (New York: Free Press, 1966), 12.

[94]Asprey, 206.

Ever the diplomat, Frederick took advantage of other opportunities. As the army trained, he approached the French and signed a secret treaty. He hoped that a Franco-Bavarian army would move down the Danube into Austria. At worst, they would force Maria Theresa to move her entire army away from Silesia, and at best, Vienna would fall. In the meantime, he continued to march and countermarch against the Austrians. Wilhelm von Neipperg, not willing to risk another defeat, wanted to avoid bringing the matter to a head. Throughout the summer, both sides engaged in several skirmishes, but both the Austrians and the Prussians avoided a major battle. By fall, however, Frederick's plans seemed to have gone awry. The French invaded Bohemia, not Austria, and Frederick became worried about their eventual plans for Germany. Since the French would not do his bidding, Frederick decided to take play another diplomatic card. On October 9, Frederick moved toward Neipperg. This time, his purpose was not to fight.

To the Austrians, the Franco-Bavarian invasion of Bohemia was just as disastrous as an invasion of Austria itself. With her enemies in Bohemia, Maria Theresa feared a general invasion, and she summoned Neipperg for service against the French. Vienna was ready to write off Silesia to save the homeland, and the archduchess needed Neipperg's army in Bohemia. This gave Frederick a diplomatic opportunity. Suggesting that Prussia and Austria could find peace, Neipperg asked Frederick to sign a secret peace treaty. The king was interested, especially since the French had disappointed him. On October 9, 1741, Frederick and Neipperg signed the secret Convention of Kleinschnellendorf. According to the treaty, the Prussians were to stay in the field, maintaining the illusion that they were fighting the Austrians. To help with the deception, the Austrians would pretend to fight the Prussians in the Silesian foothills. In reality, the convention allowed the Austrians to leave and move toward the battlefields of Bohemia. Frederick had no qualms about the arrangement, and, since the French had failed to invade Austria, he felt little loyalty toward them. Although Maria Theresa and Frederick did not believe the other would honor the treaty, Frederick became the ruler of Silesia and the Austrians secretly recognized it.[95]

Maria Theresa had about as much intention of honoring the Convention of Kleinschnellendorf as Frederick. She had taken Frederick's invasion of Silesia personally, and she wanted it back. Although she needed Neipperg's

---

[95]Ritter, 85–86 and Tuttle, vol II, 124–25.

troops, she intended to abide by the treaty only until the French were driven from Bohemia. Frederick was of the same understanding. He would follow the secret convention only as long as the French remained in Bohemia. His army "assaulted" Neisse and Glatz and conducted "aggressive patrols" in search of Neipperg's army. This pleased the French observers attached to Frederick's army. They assumed the king was still at war. Frederick knew better, but he also knew that he would soon be fighting Austria again. In January 1742, with the Convention of Kleinschnellendorf only a few weeks old, Frederick decided to plan an invasion of Bohemia. He had reason to do so. Maria Theresa's army had nearly driven the French from the area, and the Austrian archduchess had announced the terms of the secret treaty to Frederick's allies. Enraged by the deception, the French had demanded action. Frederick chose to give it to them.

The spring of 1742 found Frederick's army deep in Austrian territory. The invasion was a mistake from the beginning. Unable to bring the Austrian army to grips, Frederick moved into Moravia, extending his lines to the breaking point. This would not be the last time he would commit such an error while on the offensive. In early May Frederick received indications that the Austrians were massing to his rear. He reacted slowly, retreating to Bohemia. Seemingly oblivious to the Austrian threat, Frederick began gathering his forces at a leisurely pace. He eventually saw his mistake, and by May 10 he became alarmed. The substantial Austrian force was headed in his direction, causing Frederick to hastened the pace of his actions. With his army divided into two infantry groups, Frederick marched the groups to an assembly point near the Bohemian village of Chotusitz. Apparently, the Austrians were prepared to wage battle.[96]

As the Austrians moved in a determined manner toward Frederick, the king began his second battle as frantically as he had begun his first. On May 17, a large Austrian force assembled under Charles of Lorraine outside Chotusitz. It was equally matched with the Prussians, and Charles had managed to surprise Frederick. While the Austrians were formed just south of the village and ready to advance, Frederick's troops were spread over a four-mile front. The hereditary Prince Leopold held a small command on a hill bordering Chotusitz, but Frederick had the main infantry body four miles to the west. If Charles attacked, Leopold was in danger of being iso-

---

[96]Duffy, *Frederick*, 41–45, and *Army*, 162. See also Jany, vol II, 66–74.

lated. Frederick's only possible advantage was his unproved cavalry. General Henrik von Buddenbrock sat to Leopold's left in front of a small lake with six regiments of cavalry. This force could not be easily brushed aside, if it performed well. General Hans von Ziethen was ahead of the Prussian army, on the hereditary prince's right. He turned to rejoin the main force with another respectably sized body of cavalry. For his part, Frederick began marching as fast as possible to reinforce Leopold at the exposed position.

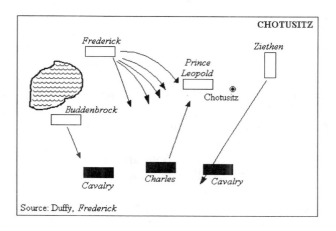

Source: Duffy, *Frederick*

Frederick hastily issued orders for the cavalry to join the battle before an Austrian assault on Leopold's position could begin. In the meantime, he briefed his infantry commanders, asking them to march their regiments forward as soon as possible to engage the Austrians along the axis of their advance as soon as they came in contact. Historians would later call this an "oblique action." Frederick's idea seemed to be to reinforce Leopold by striking the Austrian flank rather than moving to Chotusitz. Regardless, the orders for cavalry action did not arrive in time, but the local commanders took issues into their own hands. Ziethen launched a devastating assault toward the Austrian right wing of cavalry. The charge was extremely successful, so successful, in fact, that the Prussians continued chasing the Austrians after they drove them from the field. Returning retribution for Mollwitz, Ziethen's enthusiastic followers failed to check their advance. They continued charging, moving far beyond the rear of Charles's infantry formations. Many of the Prussian cavalrymen were later captured, including a promising young officer in the Rochow Cuirassiers, Friedrich Wilhelm von Seydlitz. On Leopold's other side, matters went from bad to worse. If Ziethen was too aggressive, Buddenbrock's right wing was exactly the opposite.

General Buddenbrock, mindful of the gap between Frederick's approaching infantry and Leopold's exposed command, launched an assault on the Austrians. He met the Austrian cavalry with deadly effectiveness, but in his fear for Leopold's position, Buddenbrock reassembled his cavalry after destroying the first few Austrian ranks. In other words, he retreated from the midst of battle just as he was winning. The experienced Austrian cavalry took advantage of the situation. Not willing to allow Buddenbrock the opportunity to threaten an infantry advance, they counterattacked. The result was a bitterly contested stalemate. The Austrian cavalry could not advance, but Buddenbrock could not move to protect Leopold. As Ziethen rode into the horizon and Buddenbrock sat trapped against the lake, Leopold saw the white-coated Austrian infantry marching toward Chotusitz. Unless Frederick arrived with the infantry, he would be destroyed.

The overall infantry strength of both armies was relatively even, but Leopold was outnumbered about four to one. He could not expect to hold his exposed position without help. As soon as the Austrians came within range, Leopold's regiments poured disciplined volleys into the Austrian ranks. Iron ramrods and rapid fire were not enough. The Austrians dislodged Leopold from the top of the hill and forced him to retreat. Enraged, Leopold set fire to the village and moved down the hill, noticing that the Austrians had failed to reform their ranks. Although outnumbered, Leopold decided to take the initiative back. Reprimanding his men for failing to hold their positions, the hereditary prince appealed to their sense of Protestant duty in the face of a Catholic army. His men responded. With the command, "In the name of Jesus, march!" Leopold led his small force back up the bluff in a bayonet charge. Surprised, the disorganized Austrians were forced to conduct their own retreat.

Leopold's actions were bold, but another factor caught Charles's attention. As he prepared to attack Leopold for a second time, the first segments of Frederick's army began to arrive, placing themselves along Leopold's flank. Charles's cavalry was unable to respond as each Prussian unit began to fire musket volleys. Buddenbrock's stalemate now hampered Charles. He turned to face the arriving Prussians, deciding that his only option was to stand and shoot it out with the Prussian infantry. It was a fateful decision. Not only were the Prussian muskets more effective, but they gradually increased in number as each new Prussian regiment fell in place to the right of the existing line. Within two hours, the entire Prussian infantry was firing into Charles's exposed troops. By noon, Charles had had enough. The Austrians retreated and Frederick had his second victory, although he appalled French observers when he stopped to bury his dead and allowed the Austrians to slip away. The French wondered about Frederick's loyalty to the alliance.[97]

The Battle of Chotusitz brought the first Silesian War to an end. Concerned with France, Austria wanted nothing more to do with the Prussians. They concluded a separate peace with Frederick on July 28 at Breslau over France's vehement protest. The Austrians formally ceded Silesia to Prussia, but Frederick was still not convinced that peace had indeed come. He continued to reinforce the army, to raise new regiments, and to focus on the weakness of the cavalry. At Chotusitz, the cavalry had exhibited a new fighting spirit. He now needed to reinforce that spirit with an intelligent commander, a man with both a fighting personality and the ability to control a field battle. Frederick did not have such a commander in 1742, but his attention was increasingly drawn to Friedrich Wilhelm von Seydlitz, who had recently reentered the Prussian service after being exchanged for Austrian prisoners. In the meantime, Frederick leaned heavily on the wiry Ziethen.

The Second Silesian War: 1744–1745

Frederick and Maria Theresa probably understood the diplomatic niceties of the Peace of Breslau. Like Kleinschnellendorf, it was a temporary solution until the situation with the French and Bavarians stabilized for the Austrians. Throughout 1743 Frederick trained his army and watched Austria with a cautious eye. When Maria Theresa's troops threatened complete victory, Frederick went back into action. In August 1744 he launched an invasion of Bohemia. He had no desire to face Maria Theresa on his own.

Frederick moved into Bohemia with an army larger than his father could have imagined. It included three columns totaling 80,000 men. He

---

[97]Basil H. Liddell Hart, *Strategy* (New York: Praeger, 1967), 106.

took the main body through Saxony—a feat Helmuth von Moltke would repeat 122 years later. The Old Dessauer's son, the hereditary Prince Leopold, brought 15,000 through Lusatia on the northern Prussian-Bohemian border. Schwerin brought the remaining 16,000 from Silesia. The object of the assault was Prague. Frederick was wise to use such a massive force. He suspected that the Austrians were better prepared than they had been in 1740. That assumption was correct.

Prince Charles of Lorraine had not lost favor with Maria Theresa. He was doubly ensconced in the Hapsburg family as both the brother of Maria Theresa's husband and the husband of her sister. Although he was defeated at Chotusitz, he had been rather successful against the Bavarians and the French. Upon Frederick's invasion, he beat a hasty retreat from western operations to Bohemia with the main bulk of the Austrian army. For her part, Maria Theresa had appealed to an Austrian sense of patriotism and assembled a new army in front of Vienna. Reinforced with these new garrisons, the approaches to Vienna were covered by the Austrian General Otto von Traun. The Saxons, angered over Frederick's invasion through Lusatia retreated south and hoped to join either von Traun or Charles. All in all, Frederick needed every soldier in his 80,000-man invasion force, for his enemies presented a formidable contingent.[98]

Frederick still held the strategic advantage. His columns fell on Prague and captured the city in early September. Rather than move to Vienna, Frederick decided to reduce the fortresses in southern Bohemia. Although it appeared to be a logical step, it turned out to be a disaster. His momentum slowed, allowing Charles the time to gather the dispersed Austrian forces. In hindsight, Frederick should have marched to Vienna, but by October it was too late. The momentum shifted as Charles conducted localized offensives against the Prussians.

Frederick realized the gravity of the situation. In hostile territory, he felt that his only option was to retreat. He moved his army north to the Elbe River and sought winter headquarters. The Austrians continued to advance. Reluctantly, Frederick pulled his troops out of winter quarters and withdrew his garrisons from Prague. Harassed by Austrian light cavalry, his army trudged through the snowy passes into Silesia in late autumn. Frederick had lost several thousand men, fought no major battles, and gained nothing save

---

[98]Tuttle, vol II, 223–57.

a major Austrian army on his doorstep. Always ungracious about accepting defeat, Frederick blamed Kurt von Schwerin for the disaster. The king failed to acknowledge his talent for overextending his strategic lines.

Frederick waited for the Austrians to move in the spring, while Maria Theresa concluded a treaty with England, Saxony, and Holland, forming the Quadruple Alliance against Prussia, Bavaria, and France. Frederick's concern, however, was more immediate. His forces in Silesia were continually harassed by Austrian cavalry units that had moved through the mountain passes as soon as the weather permitted. He hastily concentrated his forces, now numbering about 60,000, at the village of Frankenstein, only to find that Charles had crossed the border with 80,000. After assessing the situation, Frederick retreated north in an effort to lure Charles away from the hills.[99]

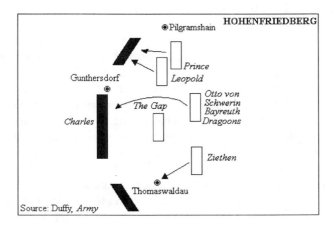

By late May 1745 Frederick was in northern Silesia, still close to the hills that bordered Saxony. He stopped his army on June 1 near the village of Striegau. Charles, assuming Frederick to be in complete retreat, moved to go on to Breslau. His troops were overconfident and security was lax. Charles's columns spread for a distance of several miles. Unbeknownst to the Prince of Lorraine, Frederick had been personally reconnoitering the

---

[99]Ibid., 268–73.

area of Charles's advance. On the evening of June 3, the king saw that the Austrians were moving for a night march from their camp near Hohenfriedberg.[100]

In the days before general staff systems, planning and preparation were left to commanding officers. In the Prussian army, Frederick alone handled these tasks. At his headquarters on the evening of June 3, he began a detailed preparation for the movement of his army by night.[101] Working diligently through the evening, Frederick planned a silent march. He hoped to deploy secretly along the side of the Austro-Saxon force and to strike it before battle formations could be developed. At 9:00 P.M. the Prussian army moved out, marching through the wetlands of a small stream called the Striegauer. Shortly after midnight, Frederick stopped and summoned his generals. He told them that, before daylight broke, he wanted to take the Prussian army across the Striegauer, form a battle line, and move toward the Austro-Saxon force. The army rested for an hour and at 2:30 A.M. on June 4, the movement across the stream began.

In most battles, plans change as soon as the first units move, and Frederick's maneuvers across the Striegauer proved no exception. Frederick did not know that Charles had deployed a cavalry screen along his eastern flank, and Frederick's leading cavalry elements ran into Charles's cavalry while it was still dark. A fierce engagement began to rage, and some Saxon infantry regiments were pulled into the fray. The Prussian cavaliers, ready to avenge themselves for past mistakes, were spurred on by adrenaline. They slaughtered the hapless Saxons without mercy after driving the Austrian cavalry away. Although the main Austro-Saxon infantry body remained intact, it was marching north—in the direction of Frederick's position the night before—about a mile from the cavalry battle on the Striegauer. Charles did not move them to assist the units embroiled in the nocturnal battle. The main Saxon units continued to march north toward Pilgrimsham, while the main Austrian body followed south of the village of Gunthersdorf. The position of Charles's forces could not have pleased Frederick more.

Frederick's infantry elements began to approach the Austro-Saxon line shortly after daybreak. They passed by the engaged cavalry units, and the main body of infantry under the hereditary Prince Leopold struck the north-

---

[100]Duffy, *Frederick*, 59–66, and *Army*, 163–65. See also Jany, vol II, 128–45.
[101]Tuttle, vol II, 304–7.

ern Saxon line. Surprised and outnumbered, the Saxons were in no position to hold. Leopold's infantry fire carried the day, inflicting heavy casualties on the Saxons. By 7:00 A.M. the Saxons were finished, and Charles's main Austrian infantry units were in no position to help. They were forced to turn to protect themselves from a possible attack between Gunthersdorf and Thomaswaldau. Frederick had anticipated just such a predicament as he placed a second body of Prussians to face Charles. It was a marvelous piece of tactical ingenuity. Leopold attacked the weakest point of the Austro-Saxon line with the strongest elements of the Prussian infantry. At the same time, Frederick moved enough troops to the south of Leopold to prevent an attack by Charles. Consequently, Charles could only sit and defend while his northern flank was decimated. Leopold continued to move forward in the vicinity of Gunthersdorf.[102]

It was a brilliant plan perfectly executed. Ironically, the perfect execution caused things to fall apart. Leopold's attack was successful, but action was confused. As the infantrymen went forward, a perilous gap began to appear in the Prussian line. The Prussian northern flank moved beyond the village of Gunthersdorf, taking the bulk of the infantry striking power with it. Charles looked at the situation and saw an opportunity. His main infantry body to the south of Gunthersdorf was now deployed and ready for battle. Outnumbering the troops Frederick had assembled to protect Leopold's flank, Charles decided to move forward and split the Prussian army by filling the gap caused by Leopold's advance.

The Prussian cavalry should have offered some protection for Frederick's southern flank. While the growing gap between Leopold's forces in the north and the southern Prussian units invited an assault, Prussian cavalry units continued to cross the Striegauer. Realizing that any Prussian cavalry reserves might pose a threat to an Austrian counter strike, Charles launched his own cavalry toward Thomaswaldau. He hoped to seize the village, outflank the Prussian infantry, and occupy the Prussian cavalry reserves. The surprise move did not work. Before the Austrians could secure the southern Prussian flank, Ziethen arrived with two regiments. He drove the Austrians back. While this protected the southern Prussian flank, it did nothing to

---

[102]See Asprey, 322. (Asprey divides the battle of Hohenfriedberg into two separate encounters. He states that the first battle was against the Saxons and that Frederick was forced to fight a second battle against Charles.)

bridge the gap by Gunthersdorf. Charles, disappointed not to have gained Thomaswaldau, moved his infantry forward to the gap. It appeared that the Prussians had no more cavalry to protect their exposed infantry.

At the same time, Colonel Otto von Schwerin, the brother of Field Marshal Kurt von Schwerin, was attempting to join the battle. Evidence suggests that his regiment, the Bayreuth Dragoons, may have been completely lost. He managed to cross the Striegauer with his elite cavalry regiment, but he had not been able to find the fighting. Smoke and confusion reigned, and he heard the familiar crack of the infantry volleys, but the Austrians were nowhere to be found. He rode forward, in the direction of Gunthersdorf, searching for a place for his regiment. No one is quite sure what happened next. Some argue that Frederick, aware of the gap at Gunthersdorf, ordered the Bayreuthers forward to close the gap. Others believe that Schwerin simply rode forward and found a place for his regiment. Regardless of the orders, the Bayreuth Dragoons soon found themselves in the annals of military legend.

As Otto von Schwerin's regiment rode forward, the colonel saw an alarming gap in the Prussian line, a gap under assault by an advancing Austrian infantry force. Seeing the Austrian skirmishers in front of them, Schwerin ordered the Bayreuth Dragoons to charge about 8:15 A.M. It became one of the most celebrated military actions in the Silesian Wars. The Bayreuthers took a volley from the skirmishers and smashed into them. The main body of Austrian infantry, which marched directly behind the skirmishers, was unable to bring its muskets to bear. If they fired, they would mow down their comrades, but if they failed to fire, the Bayreuthers would ride into their own ranks. Unfortunately, there was no time to debate, and the Austrians were too stunned to act. The impetus of the cavalry charge carried the dragoons into the main ranks of Austrian infantry. The Bayreuthers slashed the infantry to ribbons.

For the next twenty minutes, a hellish hush fell over the battlefield. Muskets and cannon were silenced as witnesses heard the sound of steel crashing into steel. The Bayreuth Dragoons indiscriminately cut the Austrian infantry into pieces, inflicting thousands of casualties. The extent of the slaughter was evident from the trophies that Schwerin brought back to his king. In the eighteenth century an infantry unit depended on its standard or colors. This unit flag showed the men where their comrades were located. Since maintenance of the infantry line was frequently a matter of life and death, men rallied to their unit's standard. It symbolized order in the midst of battle, and it was not unusual for men to give their lives to keep their unit colors from falling into enemy hands. When Otto von Schwerin rode back to the Prussian lines, the Bayreuth Dragoons carried sixty-seven Austrian infantry standards. They had destroyed the main body of the Austrian infantry at the cost of ninety-four men. At 9:00 A.M. Charles left the field with his surviving forces.

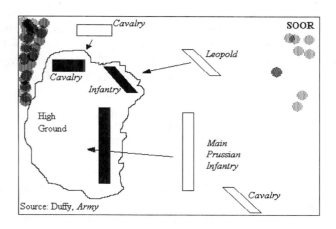

Many historians consider Hohenfriedberg Frederick's greatest victory.[103] The title *"der Grosse"* or "the Great"—so often inappropriately used by royalty—stuck with him after the victory. Indeed, he had planned and executed a difficult maneuver. This was certainly the work of tactical brilliance that might atone for his autumnal blunder in Bohemia. Frederick's aggressive actions, as well as those of his local commanders, carried the day. Despite the tactical plan, much of the credit was due to the murderous aggressiveness of the Prussian cavalry and much was due to simple luck. Regardless, Charles fled toward Bohemia.[104]

---

[103]As an example see William F. Reddaway, *Frederick the Great and the Rise of Prussia*, (New York: G. P. Putnam's Sons, 1911), 146–48.

[104]Asprey, 324–29. (Asprey argues that Frederick failed to exploit the victory and that its strategic significance was lost.)

Frederick pursued Charles with about 30,000 men. Once inside Bohemia, Frederick found himself in a cat and mouse game of maneuver. Charles, slowly reinforcing his army, soon became too strong for Frederick to fight. By September, the Prussians began to retreat again toward Silesia. Christopher Duffy states that Frederick seemed to exhibit an uncanny ability perpetually to be surprised by the Austrians, and he indeed is correct. On the morning of September 30, 1745, near Soor, Frederick learned that Charles had cut off his retreat.[105]

The Austrians held the high ground to Frederick's rear. Frederick was forced to wheel his army and launch a series of infantry assaults. No other army in the world could have accomplished this feat, but the well-drilled Prussians moved through the Bohemian countryside. The maneuver had a surprising effect. The Austrians did not believe Frederick would waste infantry by attacking the high ground. However, Frederick's movement gave him a numerical advantage on the Austrian left. His disciplined formations took the high ground despite withering Austrian infantry fire. Charles was again forced to withdraw. With his retreat line reopened, Frederick moved back into Silesia.[106]

Still the Austrians were not finished. Believing that the House of Hapsburg would be best served with an offensive action, Charles planned to move the theater of operations. As he collected reinforcements in Bohemia, he planned to invade Brandenburg directly from Saxony. The Austrians began moving into Saxony in October. With the main bulk of his army committed to Silesia on the wrong side of the mountain passes, Frederick planned a two-stage defensive. The Old Dessauer, who had remained in Brandenburg training the army, was to gather an army in Magdeburg and move into northern Saxony to block Charles. At the same time, Frederick hoped to move his Silesian forces into Saxony to drive Charles back. Ziethen helped by launching a series of surprise raids on Charles' supply bases in November. Frederick thought this would end the issue, but he was wrong.[107]

Ziethen stopped Charles but not the main Austrian army. Under the command of Marshal Rutowski the Austrians moved toward Dresden.

---

[105]Duffy, *Frederick,* 68–72, and Jany, vol II, 146–51.
[106]Asprey, 338–39.
[107]Ibid., 342.

Withdrawing from his encounters with Ziethen's cavalry, Charles sent as many units as possible to join Rutowski. Rather than retreating back into Bohemia as Frederick had expected he would do, Charles wheeled the remaining part of his army to the west, effectively blocking any reinforcement that Frederick could provide to the Old Dessauer. On December 13, the leading elements of Charles's army joined Rutowski near Dresden. If Rutowski could get past the Old Dessauer, he could open the road to Brandenburg. With 31,000 men the Austro-Saxon force moved out of the area of Dresden on December 14. Rutowski's army prepared for battle near the village of Kesselsdorf.

The Old Dessauer was prepared for Rutowski.[108] With 25,000 troops, he had laboriously moved down the Elbe toward Dresden. He lined up for battle, attacking the Austro-Saxons head-on while they sat in a prepared defensive position. Just before the assault, Leopold rode out in front of them to offer a prayer. "If you do not wish to aid me," the Old Dessauer asked God, "please wait passively on the sidelines for the outcome of the battle and do not help the rogues on the other side."[109] With that, the old Prussian field marshal moved his troops forward. The next day Frederick learned that Leopold had won a decisive victory over Rutowski at Kesselsdorf. It appeared that the Austrians could do no more.

On Christmas Day, 1745, Maria Theresa's representative met Frederick to sign the Treaty of Dresden. Other than the aborted Bohemian campaign, the Austrian army had been defeated in every major battle. In addition, Maria Theresa faced other problems. Her English allies had deserted her to deal with an uprising in Scotland. On the continent the French were gaining ground and threatening once again. Maria Theresa felt she had no choice. Reluctantly she recognized Frederick's claim to Silesia and asked for peace.[110] Frederick acquiesced.

---

[108]Ibid., 346.

[109]Philip Haythornwaite and Brian Fosten, *Frederick the Great's Army 2 Infantry* (London: Osprey, 1991), 10.

[110]Liddell Hart, 106.

Chapter 6

The Third Silesian War: 1756–1763

In his definitive military biography of Frederick the Great, Christopher Duffy says the Seven Years' War was elevated to the status of heroic mythological narrative in Prussian military circles.[111] General Curt Jany's view of the war serves as an example. Jany states that the leaders were like heroes of old, cast in the mode of a Greek epic, and he patriotically praises all those who fought in the war.[112] Duffy's analysis, which is more realistic, points out that Frederick the Great was not so quick to immortalize the Seven Years' War. If you had seen the cold, hunger, and the pain closely, the king concluded, you might be inclined to revise your romantic views. Frederick understood the truth. Prussia had nearly been destroyed in the Third Silesian War. Those who marched into its many battles—and managed to survive them—hardly saw it as an adventure.

Austria Isolates Prussia

In Europe, the period of history between 1756 and 1763 is known as the Seven Years' War, but in Germany it is known as the Third Silesian War—for good reason. Even though fighting began in the American Ohio River Valley, the European phase of the conflict originated with Frederick's invasion of Silesia in 1740. Sixteen years had not blunted Maria Theresa's desire to regain Silesia, and in the inter-war years she found a champion to accomplish the task, Count Wentzell von Kaunitz.[113] As Ambassador to

---

[111]Christopher Duffy, *Frederick the Great: A Military Life* (London: Routledge, 1985), 242.

[112]Curt Jany, *Geschichte der königlich Preussischen Armee bis zum Jahre 1807, Volume II, Die Armee Friedrichs des Grossen: 1740–1763* (Berlin: Verlag von Karl Siegismund, 1928), 342.

[113]See analysis by Gerhard Ritter, *Frederick the Great: A Historical Profile* (trans. by Peter Paret) (Berkeley, Calif.: University of California Press, 1970), 95–107.

France, Kaunitz won both the attention of Maria Theresa and the admiration of the French. She recalled him to Vienna in the early 1750s and elevated him to Foreign Minister. Maria Theresa hoped that he would be the man who could destroy the Hohenzollern dynasty.

Kaunitz believed that Austria could regain Silesia through the diplomatic isolation of Prussia. While the Austrian army trained, the new foreign minister sought alliances to support military maneuvering. His first move focused on shoring the traditional bonds with Austria's old friend, Russia. Promising to partition Prussia and give the tsar the choicest land, Kaunitz cemented warm relations between Vienna and St. Petersburg. Oslo came next. The Swedes were interested in the return of Pomerania, and Kaunitz was happy to oblige—for the price of a Swedish alliance. West Germany was critical to Kaunitz's scheme, so the minister asked the German princes to reaffirm their loyalty to the Hapsburgs. With the exception of Hesse, Hanover, and Brunswick, most of the minor German provinces pledged their allegiance to Vienna, even agreeing to send their troops to an imperial army—the *Reichsarmee*—if Frederick drew his sword against Austria. But the greatest prize in Kaunitz's policy of "Reversement" was yet to come. Kaunitz wanted to forge a pact with Austria's deadly arch enemy. The minister wanted to become allies with France.[114]

As long as France was a Prussian ally, there was little Austria could do to thwart Frederick's power. Kaunitz's objective was to reverse this position, not only by getting the French to leave Prussia but also by accomplishing the absurd. The foreign minister proposed military comradeship between Paris and Vienna. At first, the idea seemed preposterous, drawing rage then laughter from both capitols. Yet Kaunitz, enduring the initial hostility, employed all his diplomatic skills. Reminding the French that their concern was Britain, not Austria, Kaunitz told Parisian diplomats that the Prussian army would never march to the Ohio Valley to fight for France, but it might cross the Rhine to march on Paris. When fighting on the Ohio erupted in 1754, French diplomats began sifting through the logic of Kaunitz's arguments. The Austrian minister responded by telling the French court that neither Austria nor the Holy Roman Empire had any overseas imperial interests and that it was inconceivable for France and Austria to clash in the

---

[114]G. P. Gooch, *Frederick the Great: The Ruler, the Writer, the Man* (Hamden, Conn.: Archon Books, 1947), 30.

Americas. The two countries were, Kaunitz concluded, "continental allies." As the British began placing troops in Hanover, Kaunitz also reminded French diplomats that a *Reichsarmee* could defend France on German soil. The French slowly accepted Kaunitz's position, and the once preposterous notion gave birth to a strange, ahistorical new friendship—a friendship consummated by a military alliance declared formally on May 17, 1756. Kaunitz had achieved the impossible. Frederick stood alone.

Prussian Military Preparations

As in the interlude between the First and Second Silesian Wars, Frederick never really believed Maria Theresa would abandon Silesia. Therefore, he kept his army intact, increasing its size and expanding conscription. Maintaining adequate artillery units—and even some middle-class officers who knew enough calculus to aim the guns—Frederick matched his cannon to the standards of the age. The king's main efforts, however, focused on the infantry and the cavalry, and he raised new regiments, drilling them to Prussian standards. Cavalry regiments were mounted on horses bred and trained exclusively for war. The king continued to seek aggressive commanders who could take advantage of the disciplined cavalry and infuse it with a fighting spirit. Despite the improvement in the horse, Frederick and his generals still believed that the next war would be decided by the ability of infantrymen to deliver volleys. Therefore, Frederick's training always focused on infantry drill under battlefield conditions.

The leadership of the Prussian army was excellent.[115] Kurt von Schwerin, unfairly blamed for the failed Bohemian campaign of the Second Silesian War, earned his way back into the king's good graces by increasing the coordination between infantry and cavalry. The aging field marshal was supported by an able set of field commanders. A Scotsman, Field Marshal James Keith, became a close personal friend of the king. Although his mastery of German was always suspect, Keith inspired his troops, and his tactical competence combined with a love of learning to make him the king's personal favorite. The general built a house in Potsdam near Frederick's Sans Souci, and the two men passed many hours in lively conversa-

---

[115]For a full discussion of Prussian leadership see, Hubert C. Johnson, *Frederick the Great and His Officials* (New Haven, Conn.: Yale University Press, 1975).

tion.[116] Hans Joachim von Ziethen proved to be a horrid peacetime soldier, but his cavalry was second to none. Loved by his troops—they dubbed him the "Hussar King"—the wiry, hard-drinking Ziethen helped to instill a fighting spirit that dominated the cavalry.[117] Lieutenant General Hans Karl von Winterfeldt was an able commanding officer and an excellent administrator. Like Keith, Winterfeldt became one of the few members of Frederick's inner circle.[118] In military terms, Winterfeldt joined Schwerin as one of the few commanders entrusted with the army's secrets. [119]

The House of Hohenzollern and its noble cousins also provided the Prussian army with military expertise. Both of Frederick's brothers came into the profession of arms, and, although Prince August Wilhelm fell from Frederick's grace, Prince Heinrich became one of the most accomplished soldiers of the age. Other royals, Duke August Wilhelm von Bevern and Ferdinand of Brunswick, also participated in Frederick's entourage. Prince Ferdinand, a disciple of Schwerin, gained a rare appointment from Frederick—an independent field command. Married to Frederick's favorite sister, Ferdinand commanded a combined group of German, British, and Prussian forces in western Germany during the Seven Years' War.

Frederick also had several other competent field commanders, including: Field Marshal Hans von Lehwaldt, and Generals Christoph von Dohna, Johann von Hulsen, and Heinrich von Manteuffel. The Old Dessauer passed away in 1747, but his children remained in the Prussian service. Prince Moritz, the Old Dessauer's youngest son, was not the brightest of the family, but his loyalty to Frederick was unwavering. Despite the rumors about his inability to read and write, Moritz performed all duties to his utmost ability, including the instructions he received in written form. The hereditary Prince Leopold rose to the rank of field marshal and fought with the reputation he earned in the first two Silesian wars. The remaining male

---

[116]Robert B. Asprey, *Frederick the Great: The Magnificent Enigma* (New York: Ticknor and Fields, 1986), 394.

[117]Ibid., 209.

[118]Ibid., 394.

[119]See also comments on Keith, Schwerin, Winterfeldt, and the Anhalt-Dessaus in Christopher Duffy, *The Army of Frederick the Great* (New York: Hippocrene Books, Inc., 1974), 165–66.

members of the Anhalt-Dessau clan served Frederick with a variety command ranks.

Despite the talent that surrounded him, Frederick always retained control. He was his own war minister, field commander, and chief of staff, and he abandoned these roles only when forced by the situation. Frederick preferred to manage his battles directly rather than to trust his marshals with the authority for independent action. Schwerin was privileged with inside information and Winterfeldt was forever the king's confidante, but Frederick relied on the majority of his commanders to do one thing. They were to accept his orders and move their troops according to his specifications. The king rarely forgave insubordination or failure, and if a commander faltered in battle, his career was finished. When whole units failed, the result was more drastic. Officers from the offending regiment could expect neither promotion nor decoration, and Frederick's attitude frustrated the career aspirations of many men who had to accept the blame for his failures. Independent actions were rare and accepted only when dictated by circumstance. When given the chance, Frederick commanded alone.

One general did create an unusual relationship with the king—a relationship that often bordered on disrespect and insubordination. Frederick rewarded this general with early promotion, his personal admiration, and an independent command. At the beginning of the Seven Years' War, Friedrich Wilhelm von Seydlitz, the young officer who caught Frederick's attention after Chotusitz, led the Rochow Cuirassiers. The thirty-six-year-old colonel came into the Prussian cavalry through military socialization. As a young boy, he had been orphaned and placed in the court of a west German nobleman. The youngster learned to drink hard, gamble, and fight before he reached puberty. Strikingly handsome, he was almost devoid of sexual morality. He came to epitomize the devil-may-care spirit Frederick so desperately needed for his cavalry. Seydlitz had one other skill. From childhood on, he was one of the finest horsemen in Europe.[120]

---

[120]There are no biographies of Seydlitz published in English. Duffy's *Frederick* does an outstanding job of weaving Seydlitz in and out of Frederick's campaigns. Duffy also recommends Buxbaum's *Friedrich Wilhelm Freiherr von Seydlitz* in his bibliography. F. N. Maude, *The Jena Campaign,* 1806 (London: Swan Sonnenschein & Co., 1909), 31–52, describes Seydlitz's impact on European cavalry.

Seydlitz entered a hussar regiment as soon as the army would take him. As his frame filled during his teenage years, he transferred from the hussars to the dragoons. His continued growth and aggressive attitude finally won him a commission in the cuirassiers. Seydlitz was one of the few officers who had actually served in all three cavalry branches. He fought with distinction in the first two Silesian Wars and, in the peace after 1745, rose to command the Rochow Cuirassiers. His regiment gave him an outlet for his aggressive personality. Rather than satisfying himself with parade-ground maneuvers, Seydlitz taught his cuirassiers to move at full pace through any terrain, preparing the regiment for combat conditions. Realizing that shock action was the hallmark of distinction for the cavalry, Seydlitz taught his men to fight from the saddle. Under his direction they continued to fight when dismounted, and he even required his troopers to teach their horses to fight. Horses in Seydlitz's regiment bit, reared up to strike fallen riders, and jumped up to kick riders off their horses with their hind legs. It was not unusual for the Rochow Cuirassiers to suffer casualties in peacetime operations. Seydlitz was a fighting commander.

Despite his affection for drink, gambling, and women, Seydlitz became the epitome of a gentleman officer. Observers noticed that he seemed to be poured into his tight-fitting dress uniform. He was gracious toward his enemies, despite his ferocity in battle, and he exhibited a kindness that allowed him to admire both enemies and the men under his command. All of this made Seydlitz an extremely popular figure, not only in the Prussian army, but in the armies of his enemies. Undoubtedly, he became one of the finest cavalry commanders in Prussian history. In the Seven Years' War cavalrymen all over Europe modeled themselves after Seydlitz's tactics and behavior.[121]

Frederick's enemies did not have comparable commanders. The Russian army was highly nationalistic and motivated, but it was far from the best force in Europe. It did not deserve the absolute contempt that Frederick exhibited toward it, but even the most neutral observer would agree that the army was mismanaged. It was designed for frontier warfare, and officers were continually forced to keep an eye on political affairs in St. Petersburg. The French army was outmoded. The class unrest that would sweep France

---

[121]See R. D. Pengel and G. R. Hurt, *Prussian Cavalry, Seven Years War* (Birmingham, England: 1981).

into the revolution had yet to appear, and incompetent officers were given commands because of their relationship with the court in Paris. Despite their formidable reputation—and the unabashed admiration of Frederick the Great—the French were no match for the Prussians in 1756. In the north the Swedes had better leadership, but they lacked numbers. Austria was another matter. After the War of the Austrian Succession, Maria Theresa worked hard to improve her army. Troops underwent rigorous training, the equipment was modernized, and Austrian infantry formations improved their rate of fire. Officers received special instructions in tactics specifically designed to counter the Prussians. With increased discipline and efficiency, the Austrian army of 1756 was a far cry from the force that had taken the field at Mollwitz.

Strategy and Tactics in the Seven Years' War

The Prussian strategy in the Third Silesian War resulted from its diplomatic isolation.[122] As Kaunitz increased the allies' power, Prussia was surrounded by his enemies. Frederick found himself defending four separate fronts. He could launch no grand offensives, and he would be forced to keep his enemies from uniting. The king divided the army into three major commands. A western group was to watch France and the German states sympathetic to the empire. A second command was established in the east to defend against the Swedes from the north and the Russians from Poland, while a third army group assembled in the south.

The strategy was simple. Since he was surrounded by hostile powers, Frederick's three armies formed a "defensive perimeter." Frederick's goal was to fight a strategic defensive, to keep enemy forces from uniting inside the confines of his perimeter.[123] Tactically, he had a different objective. If any army breached his perimeter, the king intended to attack and lay it waste. With military forces concentrated in the east, south, and north, the main power in Frederick's strategic defensive was in a fourth army under

---

[122]Gooch, 34–35. See also Nancy Mitford, *Frederick the Great* (New York: Harper and Row, 1970), 220–26.

[123]Basil H. Liddell Hart, *Strategy* (New York: Praeger, 1967), 108–9.

his personal command. This "roving group" would respond to any crisis by joining the local command to attack an invader.[124]

This plan was always subject to modification. During the first three years of the war, Frederick aggressively moved toward any breech on the perimeter and, at times, even took the war beyond his planned limits. After 1760 he had suffered too many casualties to continue this policy. Unable to keep his enemies apart, the king attempted to outmaneuver them or to strike them in some indirect fashion. Even with these modifications, Frederick still hoped to destroy enemy formations before they entered his defensive perimeter. In the closing years of the war, as Frederick's supplies dwindled, he became more and more cautious, yet he continually attempted to seize the initiative whenever possible. [125]

Two geographical factors dominated the defensive plan. First, Prussia's many isolated provinces prevented Frederick from defending the entire country, so the western states and East Prussia were placed on the sacrificial block to keep Brandenburg intact. Second, the southern flank, the critical juncture of the entire perimeter, was directly threatened by a large Austrian army. Frederick decided to occupy Saxony and to drive the Austrians further south to keep them from linking with the French or the Russians. The plan was sound. The French, supplemented by the *Reichsarmee,* were hesitant to leave west Germany, and the Russians, who intended to join the Austrians, were separated by distance. Frederick sought control of Saxony to keep the allies separated. The small country became part of his defensive perimeter.

The Opening Campaign: 1756

Frederick the Great was never one to wait for the unfolding drama of history. Concerned with his precarious position, the king saw no need to delay until his enemies gathered for battle. Even though he was diplomatically surrounded, militarily he held the initiative.[126] The Russians were far

---

[124]Walter Goerlitz, *History of the German General Staff: 1657–1945* (trans. by Brian Battershaw) (New York: Frederick A. Praeger, 1962), 6.

[125]Martin Van Creveld, *Command in War* (Cambridge: Harvard University Press, 1985), 54–57.

[126]Walther Hubatsch, *Frederick the Great of Prussia: Absolutism and Administration* (London: Thames and Hudson, 1975), 112.

away, and the Swedes had yet to cross the Baltic. The French army was large, but it was on the other side of the Rhine. Frederick saw this military situation as an opportunity to strike first. Taking Schwerin and Winterfeldt into his complete confidence, the king invaded Saxony on August 29, 1756. The Saxons begged for Austrian assistance, while their small army, numbering less than 20,000, retreated to the south, hoping to find help from Vienna. Dresden, Saxony's capital, fell on September 10. By the end of September, 14,000 Saxon soldiers were surrounded in Pirna, a fortress along the Elbe River. Unless they were relieved by the Austrians, the entire Saxon army would be forced to capitulate.

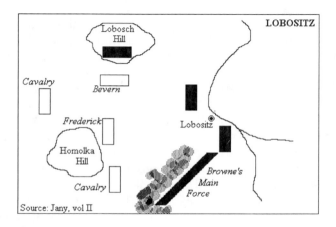

The Austrians took little time in responding. When word of the siege at Pirna reached Vienna, Maria Theresa dispatched Field Marshal Maximillian von Browne to Bohemia with an army of 50,000 soldiers. Frederick got wind of the Austrian move, and his bother-in-law, Ferdinand of Brunswick, crossed the border into Bohemia to keep an eye on things. Ferdinand watched and saw that the Austrians intended to relieve Pirna. He sent word to Frederick. Forewarned by Ferdinand's observations, Frederick left the siege at Pirna, moving up the Elbe River in search of Browne. On the

morning of October 1, the king found him near a bend in the Elbe at Lobositz.[127]

Neither Frederick nor Browne knew where the other was, and a thick
autumnal fog made their search even more difficult. Frederick, who had
taken a position to the west of Lobositz, believed that Browne was somewhere to his front, but that was only a guess. Neither Frederick nor his
commanders could see through the mist. The king's friend, James Keith,
summarized the attitude of the Prussian leadership. As Keith peered into the
early morning fog, he reported that he could see nothing. The Prussian army
was blind. Knowing that the fog would eventually clear, Frederick decided
to deploy anyway. His troops sat before a hill to the north and a wooded
area to the south, and the king guessed that Browne was directly in front of
him in Lobositz. As the fog lifted, the king realized that his estimation was
only partially correct. The Austrians were indeed in Lobositz, but they also
held the hill on the northern Prussian flank. What Frederick could not see,
however, was that Browne had gathered his main strength on the Prussian
south, behind the wooded area. The path to Lobositz was flanked by two
Austrian forces.

Browne had selected his position with care. Not anxious to engage Frederick, he had cleverly arranged his army in front of Lobositz to deceive the
Prussians. The northern Austrian flank was covered by regular infantry
units screened by light troops, but the troops in Lobositz were nothing more
than an advanced guard supported by the main battery of Austrian artillery.
The artillery, in turn, was supported by cavalry flanking Lobositz. To any
observer in Frederick's position, the main Austrian strength appeared to be
centered in Lobositz, but that was purely a misconception. The forward
cavalry unit was supported by thousands of Austrian horsemen, just out of
Frederick's sight. This force was flanked by 34,000 infantrymen, completely hidden from the Prussians even after the fog lifted. As Frederick
prepared to attack Lobositz, he was neither aware of the reinforced infantry
to his north nor the tremendous Austrian powerhouse sitting in the woods to
his south. [128]

---

[127]Duffy, *Frederick,* 102–8, and *Army,* 167. See also Jany, vol. II, 362–74.

[128]Herbert Tuttle, *History of Prussia, Volume IV, Under Frederic the Great: 1756–
1757* (New York: AMS Press, 1971; orig. 1896), 25–26.

Frederick attacked as soon as the fog cleared, ordering the Duke of Bevern to clear the hill on the northern flank with a few infantry battalions. The king believed that this would open the path to Lobositz, but Bevern soon discovered Browne's surprise. The northern flank was not as weak as it seemed. When the light infantry was cleared, Bevern ran head long into two formations of Austrian regulars. The Austrians, who had spent several years learning to fight in the Prussian style, were a far cry from the troops Frederick had faced at Mollwitz and Chotusitz. Bevern's attack ground to a halt under the thunder of enemy muskets. As his casualties mounted, he called for help.

While Frederick's infantry and cavalry stood watching Bevern's assault, the Austrian artillery suddenly fired a salvo of solid shot at the Prussian line, followed by a steady cannonade of explosive shells. The Prussians had never experienced a bombardment of this magnitude, and it shook their nerves. Deciding to respond, Frederick launched several regiments of cavalry toward Lobositz, hoping to divert the Austrian artillery. Even the bravado of a cavalry charge did not prepare the Prussian troopers for what they found next. As they neared Lobositz, Browne's second surprise came into view. The Prussians were not attacking a few regiments in view before Lobositz, they were riding into thousands of enemy horsemen. The Austrians lowered their sabers and galloped forward. The king, who could not see the counterattack, was infuriated when his cavalry turned and ran before reaching the Austrian lines.

Within minutes, the troopers from Frederick's diversionary assault came riding past the Prussian formations, including the main body of Prussian cavalry. Sitting behind the infantry and suffering from the Austrian bombardment, Frederick's cavalrymen watched their comrades in disgust. The cavalry officers believed that a small Austrian unit had driven the attack back, and they wanted vengeance. Without waiting for orders, some of the units surged forward, and that spark ignited the tinder. The whole body of Prussian cavalry suddenly launched a spontaneous counterattack, not knowing that the entire Austrian cavalry was moving toward them. Disappearing in the smoke of the Elbe valley, the Prussian cavalry first slowed in muddy field, then smacked into the Austrian horse. The whole area became a mess of confused troopers and countless hand-to-hand struggles. The Austrians sounded recall, leaving the Prussian cavalry to suffer the effects of direct fire artillery fire while they tried to extricate themselves from the mud. As the survivors rode back, they told the king of a large Austrian infantry body hidden in the woods south of the battlefield. Frederick was in trouble.

Fortunately for the Prussians, Browne did not seize the initiative. With the Prussian cavalry removed from the picture, Browne was satisfied to let his own cavalry roam in front of the Prussians and continue the artillery bombardment. He kept the Austrian infantry in its positions, giving the initiative back to Frederick. If the cavalry had failed the Prussians, Frederick vowed that the infantry would not. Shortly after noon, he ordered Bevern to dislodge the troops on the northern flank by bayonet. The order came at a good time. Bevern's troops had virtually depleted their ammunition by exchanging volleys with the Austrian regulars, and he either had to move forward or retreat. Bevern ordered his troops forward, and shortly after noon the hill on the northern flank was under Prussian control.

As Bevern's final assault was getting under way, Frederick ordered Ferdinand of Brunswick to take the village of Lobositz. The Prussians had numerical superiority over the Austrians in the area and, despite the improvement of Austrian infantry, the Prussians still fired faster than their Austrian counterparts. Ferdinand's troops forced the Austrian cavalry to retreat, and his men disappeared in a cloud of smoke. Browne, leaving his main body of infantry intact in the woods, made no effort to reinforce the village. This gave Ferdinand an advantage in Lobositz, and the Austrians soon gave way to his attack. The Prussians entered the smoking village to find the Austrians running for their lives. By mid-afternoon Frederick completely controlled the ground from Lobositz to the north.

Browne had to make a decision. With his infantry intact, he was in a position to fight, but his objective was to relieve the siege at Pirna. He did not want to seek a decisive engagement on the Elbe. Since his northern flank was gone, Browne chose to move, retreating up the Elbe in good order.[129] As with many battles, the outcome of Lobositz was initially unclear. Frederick had taken the high ground to the north as well as the village, but the main body of Austrian infantry had marched out of harm's way. Both sides lost about 3,000 men. As the days passed, however, the situation became more focused. On October 14, Pirna surrendered to Frederick. With that, Browne withdrew into Bohemia, and Frederick went into winter quar-

---

[129]Asprey, 434–38.

ters after conscripting the entire Pirna garrison into the Prussian army.[130] (The uninspired Saxons deserted at every opportunity.) All in all, the end of 1756 had been kind to the Prussians. As the snows fell, it was clear that Frederick had been the victor at Lobositz.

Success, Failure, and Success

When the campaign season of 1757 opened, Frederick's strategic defensive perimeter seemed to be intact. The king had 50,000 men near Swedish Pomerania and an additional 40,000 men in Hanover, under the British Duke of Cumberland, protecting him from France. When reports came that the Austrians were gathering their forces in northern Bohemia, Frederick decided to destroy them before they could move north. In April, the Prussians launched a four-pronged invasion of Bohemia. Prince Moritz of Anhalt-Dessau and Frederick moved two groups from Saxony, while Bevern and Schwerin moved two other groups from Silesia. Frederick's plan was to have all four bodies join together as they marched south.[131] The tactic worked, catching Browne completely by surprise. He hastily reacted by concentrating his forces at Prague. On the fifth of May, the four wings of the Prussian invasion joined together just north of the city with a total strength of 65,000 soldiers.[132]

---

[130]Tuttle, vol. IV, 32–34.
[131]Duffy, *Frederick*, 115–21, and *Army*, 169–70. See also Jany, vol. II, 394–406.
[132]Asprey, 443–46.

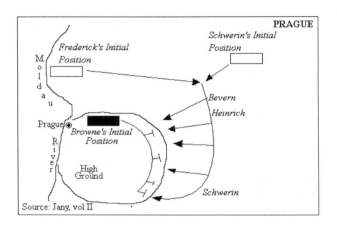

Source: Jany, vol II

The Austrians gathered nearly 70,000 troops in Prague in an excellent defensive position. They concentrated on the high ground just to the east of the city and south of the Moldau River, leaving only their front exposed. Unknown to the Austrians, however, Frederick had no intention of launching a frontal assault. On May 6, Frederick moved from the Moldau valley, veering from the Austrian front. Instead of moving forward as the Austrians expected him to do, Frederick moved to the east, completely crossing the Austrian front. Not to be outdone, the surprised Browne ordered the Austrian army to mirror Frederick's movement, and the Austrians slowly turned to face the east. They still held the high ground, but the Moldau now guarded their left flank and a small marsh covered their front. Unluckily, their line had not remained intact during the difficult maneuver. Frederick had not expected the Austrians to react. Since they had merely turned their front, the king decided not to attack, but Schwerin convinced Frederick that the Austrian defensive line had been weakened. Under his field marshal's urgings, the reluctant king ordered his units forward, directly into the new enemy front.

Seeing the Prussians preparing for an assault, Browne ordered his cavalry forward. He hoped to keep his enemy in the swampy ground to the east of Prague, but General Hans von Ziethen saw the ploy. As the Prussian infantry prepared to march up the hill in the direction of Prague, Ziethen moved toward the new Austrian flank. Before the Austrian cavalry could take any effective action, Ziethen ordered all of his cavalry units forward, scattering the Austrian cavalry into the hills. Quickly forming above the Prussians, the Austrian infantry stood without the benefit of cavalry. It appeared that Schwerin had been right.

Despite the initial appearances, the fates were not kind to Schwerin that day. The Prussian plan looked good on paper, but it did not match the actual condition of the ground. Reconnaissance in the 1700s was primitive compared to modern military operations, and the Prussians failed to notice what Browne accidentally discovered. The area in front of the Austrian position was not merely marshy, it was a virtual swamp. As soon as the infamous blue-coated lines marched toward the Austrians, the Prussian troops found themselves knee-deep in mud below Browne's new position. The foot soldiers became mired in the muddy mess. The Austrians, now partially formed, responded by pouring musket fire into the faltering Prussian troops. Hans Karl von Winterfeldt, leading his regiment through the muck, was one of the first casualties. Falling wounded to the ground with a ball through the back of his neck, he was joined by hundreds of Prussians as Austrian musket balls slammed home. The Austrian fire was effective.[133]

The swampy area became a killing ground. Each minute, more and more Prussian infantrymen fell to the ground with ghastly wounds. Their marching pace slowed, then ground to a halt. Unable either to extricate themselves from the mud or advance up the slopes to Prague, individual units began unshouldering their muskets. If volleys were good enough for the Austrians, perhaps they would be good enough for the Prussians. Unhappily, though, the Prussians were forced to fire up hill toward a partially hidden enemy. Their fire was not effective, and their casualties multiplied. Men often cannot fight under such conditions. With the prospect of being slaughtered on the marshy ground increasing, a few units began to falter. At first one or two men dropped their muskets and ran, then the panic spread to whole platoons. The sight of a few platoons running turned entire companies, and eventually three regiments simply turned and ran. The agony in the mud had come to an end.

Field Marshal Kurt Christoph von Schwerin watched the unfolding scene from a hill behind the Prussian lines. His face was grim. Three whole

---

[133]Tuttle, vol. IV, 86.

regiments—6,000 men—were running in the face of the Austrian musketry, and Schwerin began to fear that the entire army might be destroyed, if the Austrians launched a successful counterattack. As he searched for a solution, one of the regimental standards caught his eye. It belonged to the 24th Infantry Regiment. Despite his concern for the whole battle, this action drew Schwerin's complete attention as if it were a microcosm of the entire encounter. Although he was a field marshal of the Prussian army, second only to the king, Schwerin was also the *Chef* of the 24th Infantry Regiment—and his colors were fleeing from the Austrians.

In order to understand Schwerin's feelings, it is necessary to consider briefly two of the more important offices in the Prussian army. Each company was commanded by a main captain called a *Hauptmann*. Even though a company captain handled the day-to-day affairs of the unit, the *Hauptmann* generally held the main captain's position for life, even when he was promoted to higher ranks. The company belonged to a man, and regimental leadership worked in the same fashion. Like a company *Hauptmann,* the regimental colonel, the *Chef*, maintained his position no matter what command he held. Many field commanders even wore their regimental uniforms when they were promoted to general. Appointed directly by the king, colonels had great affection for their regiments, and their units returned the affection. Schwerin may have been a field marshal for the king, and probably the most important officer in the Prussian army, but before all else he was *Chef* of the 24th Infantry Regiment—a regiment running from the Austrians. It was more than the field marshal could bear.

Leaving his safe position, Schwerin galloped down into the muddy bog until he reached the fleeing men of the 24th. Snatching the regimental colors from the guidon bearer's hands, the field marshal drew his sword and faced his soldiers. With the regimental standard in his left hand and his sword in his right, Schwerin asked if anyone wanted to continue running. Ashamed, the soldiers of the 24th rallied to their *Chef.* If he could brave the Austrian line, so could they. Nodding approval, Schwerin turned his horse around to face the Austrians with a cry of, "Here, my children!" He led the attack, his muddy soldiers following him in perfect step. Somewhere on the other side of the line, an Austrian artillery crew saw what was happening, and they fired a blast of grapeshot in Schwerin's direction. Balls entered Schwerin's head, chest, and stomach as the *Chef* tumbled backwards from his horse. Many Prussian soldiers believed it to be the most gallant death ever recorded for a Prussian general, but gallant deaths are for the safety of history books. At the instant it occurred, the 24th Infantry Regiment ran back down the hill.

Schwerin's sacrifice was not in vain. Even with Winterfeldt wounded and Schwerin dead on the field, Frederick's movement across the Austrian front had caused more problems for the Austrians than the Prussians imagined. The main body of Prussian infantry was falling back in front of the Austrian musketry, but the Austrians had failed to transfer their front smoothly. Browne was in a strong position above the marshes—at the exact spot where Schwerin lay dead—but his left flank was scattered. The men had not turned properly, and there were gaps in the Austrian line—gaps spotted by commanders of Prussian reserves. Bevern sent a regiment into the point where the Austrians had turned south and discovered there was nothing in front of it. Other Prussian units, including one led by the king's brother Prince Heinrich, also discovered the weakness. Heinrich's regiment forded a small river and made particularly good headway. As the slaughter continued to the south, local commanders in the north penetrated the Austrian lines. By noon they were on the high ground outside of Prague, and the Austrian flank was in jeopardy.[134] When Browne was informed of the Prussian advance, he feared a complete rout.

With their northern flank collapsing, the Austrians continued to move closer and closer to Prague throughout the afternoon. They fought well, but their position of strength had been turned into a position of weakness. In the early evening, Field Marshal Browne fell, mortally wounded. By night, several thousand Austrian soldiers moved inside the city to set up a defensive position, while others fled across the Moldau. The battle had been terrible, both sides losing about 14,000 men. The Austrians still held Prague, but Frederick had his second victory. When the fighting subsided, a shaken king found his brother, still soaked from head to foot. He informed Heinrich of the slaughter in the south and Schwerin's death. The two men embraced one another in a shower of tears.

---

[134]Chester V. Easum, *Prince Henry of Prussia: Brother of Frederick the Great* (Westport, Conn.: Greenwood Press, 1971; orig. 1942), 41–44.

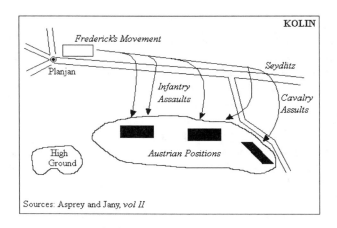

Sources: Asprey and Jany, *vol II*

All the Prussian casualties were in vain, for the victory at Prague was hollow. Frederick could not take the city due to the relative Austrian strength inside the walls. To make matters worse, a large Austrian force, under the command of Field Marshal Leopold von Daun, was on its way to relieve the city. In June Frederick was forced to abandon siege operations and meet the challenge posed by Daun. Frederick left Prague, marching nearly sixty miles to the west. He located the Austrian force near the village of Kölin on June 18.[135] It was an extremely hot day, and the Prussians were exhausted from marching. Frederick had 34,000 men under his command compared to Daun's 54,000. A less offensive-minded commander might have decided to await an Austrian move, but such passivity did not suit the Prussian king.

Some of Frederick's commanders did not react to the idea of a tactical offensive with enthusiasm when they saw Daun's battle line. The Prussians were aligned along a road to the north of the Austrian position, and Daun held the high ground.[136] Agreeing that a frontal assault would be suicidal, Frederick decided to use the same tactic that had been successful at Prague.

[135]Duffy, *Frederick,* 124–31, and *Army,* 171–73. See also Jany, vol. II, 407–18.
[136]Jay Luvaas, *Frederick the Great on the Art of War* (New York: Free Press, 1966), 216–23.

His intention was to cross the Austrian front and launch an attack on its right flank, where he hoped to achieve local numerical superiority. He would leave part of his advancing army along the center of the Austrian force, "refusing" his own right flank. The exhausted troops began the maneuver about 1 P.M. The tactic might have worked, but the cavalry turned too soon. Ziethen led an assault, only to receive a wound that removed him from command. Like Winterfeldt he would return to fight on another day, but the main body of the Prussian cavalry withdrew. Thinking that Ziethen had moved on the Austrian flank, the infantry moved into the area vacated by the cavalry.

Rather than striking the flank, the local commanders began assembling their lines before they had completely crossed the Austrian front. The result was horrific. Instead of marching into the Austrian right flank, they marched headlong into the center of the Austrian line. The Austrians replied with direct canister fire. The Prussian infantrymen were hampered by two factors. Not only were they directly in front of a superior force, but the vegetation and the smoke hid most of the Austrian soldiers from Prussian view. The result was that the Prussian soldiers were standing on a hillside fighting a numerically superior army that they could not see. In addition, individual commanders lost contact with one another. The brave soldiers accepted the Austrian fire without retreating, but they could not advance. Casualties were heavy.[137]

With the infantry being chopped to pieces, the remaining cavalry units assembled on the roadway to the rear of the Prussian infantry. The new commander, Major General Christian von Krosigk, temporarily replaced Ziethen, and he conferred with his deputy, Colonel Friedrich Wilhelm von Seydlitz. Both men agreed that the infantry would be murdered unless the Austrian line were broken. After obtaining the king's permission, Krosigk and Seydlitz decided that a second cavalry charge on the Austrians could do the job, if it were delivered on the Austrian flank. The Prussian cavalry moved further to the right and launched an assault that completely surprised Daun. By five o'clock the Prussians opened a hole in the Austrian line. All this came to naught, however, as the exhausted infantry was unable to disengage and move down the road. The lack of infantry support allowed Daun to send his own reserves to the area, driving Seydlitz away and killing

---

[137]Tuttle, vol. IV, 96.

Krosigk.[138] Frederick, roaming up and down the line, was unable to assist Seydlitz. By six o'clock the Prussian line began to break, and some units retreated on their own initiative. Infuriated by his inability to crack the Austrians, Frederick attempted to repeat the gesture Schwerin had used at Prague. He grabbed a regimental standard, drew his sword, and screamed out, "Rogues, do you want to live forever?" He then moved forward as if he would attack Daun all by himself. Even this did not inspire the infantry. Frederick's soldiers pulled him back and convinced him that the day was lost. Frederick abandoned his gesture as Daun moved Austrian cavalry around the Prussian flank. The Prussian offensive spirit was gone, and the king was forced to defend his line. Seydlitz set up a rear guard while Frederick withdrew.

Having suffered 12,000 casualties—approximately one third of the total Prussian force—Frederick was lucky to escape with his remaining forces. In a manner that would come to typify Austrian behavior, Daun failed to exploit his victory. He suffered 8,000 casualties and was in no position to follow. By fortune and sheer tenacity, the king had never known defeat until June 18, 1757, but the Battle of Kölin was Frederick's first major failure.[139] The Prussian army had been soundly defeated, and the battle had been mismanaged. The only bright point of the day came in the action of Seydlitz. The king noted his role and commended him for breaking the Austrian line with the second cavalry charge. Seydlitz was promoted to major general and given the *Pour l' merit*, one of Prussia's highest military decorations.

The situation did not improve. Frederick's other brother, Prince August Wilhelm, found himself outfoxed by Daun and forced to retreat a few days after the Battle of Kölin. Frederick was so angered that he dismissed his brother in front of the prince's own troops. August Wilhelm left the Prussian camp in tears and died a few months later of a brain tumor.[140] Frederick was forced to abandon his siege of Prague and retreat to Saxony. The Bohemian invasion had cost nearly 30,000 casualties and had come to nothing.

Kölin turned out to be a strategic defeat, and in July the news became worse. The allies finally began to move in conjunction with one another. The French, under Field Marshal Louis de Estrées, moved into northern

---

[138]Asprey, 457.
[139]Ibid., 459–60.
[140]Tuttle, vol. IV, 104.

Germany with over 100,000 men. Another French force joined the men of the *Reichsarmee* to form a united Franco-German army under Charles, Duke of Soubise. In the south, the Austrians reinforced their small army, sending the combat veteran Prince Charles of Lorraine to command it. A Russian army under Field Marshal Stepan Apraksian invaded East Prussia to add to Frederick's woes. Finally, the allies rubbed salt in the Prussian wounds as 16,000 Swedes crossed the Baltic to encamp in Pomerania. Prussia's enemies were closing.

June ended with two complete disasters. In the west Field Marshal de Estrées had outmaneuvered the British Duke of Cumberland at Hastenbeck. On July 26 de Estrées attacked Cumberland, and the Anglo-German army was forced to abandon Hanover completely. The Prussian contingent of Cumberland's army retreated to Magdeburg. The east brought news of yet another disaster. The Russian Field Marshal Apraksian defeated Prussian general Hans von Lehwaldt at Gross-Jägersdorf. This defeat opened the door to Berlin. Frederick rushed to make defensive arrangements, but it seemed that collapse was eminent.[141]

Fortune saved Frederick. Even though his western and eastern flanks were virtually destroyed, the Russian army failed to exploit the advantage, and the French spent August carefully monitoring Austrian and Russian activities. This delay gave Frederick the time he needed to prepare a defense. Leaving Bevern in charge of Bohemia, Frederick moved west to meet the French, but fortune was not altogether kind. After arriving in the west, the king received word of a personal tragedy. The king's confidant, Hans Karl von Winterfeldt, had been killed in Silesia. The badly shaken king continued in his duties. Frederick, who could count his cronies on only one hand, found the number of close friends dwindling.

If Frederick suffered from personal tragedies, the war took a turn for the better. Soubise's *Reichsarmee* proved nearly as lethargic as the Russians, and De Estrées's invasion of Hanover came to naught. Paris replaced De Estrées with the Duke of Richelieu and urged Soubise to take offensive actions. Neither move had much of an impact. Richelieu, like his predecessor, failed to advance, while Soubise and his deputy, Austrian General Joseph von Saxe-Hildburghausen, merely moved 60,000 men to Magdeburg. When

---

[141]Ibid., 101–10, and Jany, vol. II, 426–35.

the broken-hearted Frederick moved toward Soubise, the French general re-treated. Frederick took his troops back to Bohemia to face the Austrians.

October 1757 brought a series of strategic marches and counter marches. The Russians continued to sit, while the Austrians decided to evacuate Bohemia to move into Silesia. Prince Charles of Lorraine reckoned that Silesia presented more of a threat to Frederick than Bohemia, and when Frederick failed to follow him, the Austrian commander sent a raiding party to Berlin. Frederick dispatched Moritz of Anhalt-Dessau to deal with the situation, only to receive word that Soubise was moving forward again, in the direction of Charles of Lorraine. If Soubise managed to reach Bohemia, Charles could return from Silesia. Frederick could not allow the two allies to unite, so in late October he again marched toward Soubise with a small Prussian force. Frederick's intended to destroy the *Reichsarmee*.

By October 30, Soubise's 60,000-man force retreated from Frederick's 20,000-man army, leading the king to believe that Soubise would not fight until he joined the Austrians. Seydlitz, who had been given command of the cavalry, kept contact with the *Reichsarmee*, but it was also clear to him that Soubise wanted to avoid battle. Soubise's attitude frustrated Frederick. Continued pressure from Charles in Silesia and the threat of the Russians in the east did not afford Frederick the luxury of lingering in the west. He or-dered Prince Ferdinand of Brunswick to prevent Richelieu from joining Soubise and searched for strategy that would bring the *Reichsarmee* to bat-tle. In the end, Frederick wanted battle more than the French commander wanted to avoid it. Hastily returning from his defense of Berlin, Moritz joined Frederick on November 3. In addition, one of Frederick's few re-maining close friends, the Scotsman Field Marshal James Keith, brought a small contingent to the Prussian force, bringing Frederick's total strength to just over 21,000 men. As Frederick set up camp near the village of Ross-bach on November 4, he noticed that Soubise's forces were in the area. He was determined to bring them to battle, provided that battle would be on Prussian conditions. The next morning, however, seemed to bring anything but Frederick's terms.

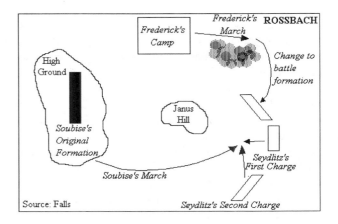

Source: Falls

As the Prussian soldiers rose from sleep on November 5, scouts reported that the Franco-German force was assembling outside the main camp on a hill overlooking Rossbach.[142] A hasty reconnaissance confirmed the initial reports. Soubise was organized on the high ground just to the west of the Prussian camp. Frederick conferred with his officers. He wanted a decisive battle, but he did not want to launch a frontal assault against an enemy holding a hill, especially when he was outnumbered nearly three to one. Frederick decided to do the next best thing. Placing his troops in column formations, he retreated from the Rossbach camp and marched to the east, planning to turn and strike the French on their flank. Soubise, who had never witnessed Frederick's tactics, had no inkling of Frederick's intentions. When he saw the Prussians move away in columns, he assumed they were retreating.

If Soubise had understood Frederick, he would have been more cautious.[143] In continental terms, when a commander formed columns, it meant that he was marching with no desire to fight. Soubise did not know that Frederick employed column formations to enhance mobility during battles. Unlike any other army in Europe, the Prussians could form columns, march

---

[142]Duffy, *Frederick,* 140–45, and *Army,* 175–76. See also Jany, vol. II, 436–44.
[143]Luvaas, 223–31.

into battle, and form battle lines while they were still moving. Frederick had the advantage of an army drilled to near perfection. In an age where the battle line meant the difference between survival and death, only the Prussian troops could change formations while marching. As the Prussian army trudged out of Soubise's sight—apparently running to fight another day— the poor French commander had no idea that Frederick was moving to strike the Franco-German flank.

If ignorance is bliss, Soubise had reached Nirvana. He had, he believed, faced the dreaded Prussian king and achieved a tactical victory by maneuver—victory without a shot! Watching the Prussians in apparent retreat, the relieved French commander ordered his entire army to stand down. Within two hours—two stationary hours—their battle line was reconfigured into a column formation, and, at long last, the French and their German allies were ready to march. They left the hill in early morning, marching away in columns. Cavalry massed in the front while the infantry lined up behind them. It was not a fighting formation, but Soubise had no intention of fighting—he planned to march away. His blissful ignorance would have evaporated had he known that the Prussians were turning south just a few miles to the front of his columns.[144]

One can only guess at the horror that overcame the leading elements of the French cavalry as they rounded an area known as Janus Hill about 9 A.M. The *Reichsarmee* had been marching for over an hour into a plain along the side of the hill, completely unaware that each step brought them closer to the Prussian army. Since they were in a column formation, most of the men could not see what lingered to their front. The leading ranks, of course, could see, and they tried to take action. Frederick had not retreated. Instead, he was in front of Soubise, forming a battle line as he moved. Sitting beside the line, in a straw-yellow regimental uniform, was the form of Friedrich Wilhelm von Seydlitz. As French cavalrymen scrambled to their colors, Seydlitz turned to face his command. Throwing his clay pipe into the air—his proverbial sign for an all-out charge—Seydlitz galloped toward the astonished French cavalry. Thousands of Prussian troopers spurred their horses in response.[145]

---

[144]Tuttle, vol. IV, 127.
[145]Asprey, 471–72.

The initial cavalry battle took about five minutes. In column formation there was little the French could do to meet Seydlitz's murderous onslaught. The Franco-German infantry following the cavalry had little or no idea of what was happening. To their surprise, the French cavalry simply galloped past them as if the were running for dear life. They were. Next, the astonished infantrymen saw not French but Prussian cavalry swarm around their entire position. The stupefied Franco-German soldiers were goaded by panicked shouts as officers and sergeants attempted to bring the long column into some type of battle formation. It worked. Seydlitz could not break the infantry column, and he was forced to ride away. Perhaps, the common soldiers may have thought, their impromptu action had saved the day. If they believed this, they were disillusioned when the first Prussian artillery shells began to bounce through their tightly packed ranks. Even the lowliest private began to realize the situation. The *Reichsarmee*—in marching formation—was facing an army arrayed for battle. Disaster loomed.[146]

If Soubise had been surprised by the cavalry action, Seydlitz knew exactly what the Prussians should do. Commanders of lesser ability would have continued pursuing the French cavalry as the Prussians had done at Chotusitz, but Seydlitz had a more strategic target, the Franco-German infantry. Instead of allowing his troopers to engage in a fruitless pursuit of the French cavalry, Seydlitz ordered his forces to halt and he reassembled them on the southern flank of the battlefield. He hoped to have the opportunity to strike again, and his target was the undefeated infantry.

Frederick was keenly aware of the gift that had been placed in his hands. As the tightly packed mob scrambled in front of him, Frederick's disciplined battle line delivered volleys, causing the king to become so excited that his own enlisted soldiers had to pull him back when he wandered in front of the firing line. After fifteen minutes, the musket and artillery fire had done its worst. Looking at the carnage, an enthusiastic Seydlitz sent word to his sovereign, telling him that the cavalry was assembled for another assault. The elated king ordered his troops to cease fire and gave Seydlitz permission to charge. Ordering a chaplain forward, Seydlitz commanded his men to bow for a blessing. As the chaplain prayed, a horsefly startled an animal in the front rank, causing Seydlitz to interrupt the prayer

---

[146]C. T. Atkinson, *A History of Germany: 1715–1815* (New York: Barnes & Noble, 1969; orig. 1908), 221–22.

with a steady string of stalwart profanity. When the flustered minister delivered his final "amen," the Prussian troopers looked up to see their general several yards in front of them, already riding hell-bent-for-leather toward the *Reichsarmee*. No one could resist such inspiration. The men kicked their horses to action, and the ground resounded with the sound of pounding hoofs and the battle cries of blood-thirsty warriors. Within minutes the survivors of the *Reichsarmee* understood the meaning of defeat.

The Prussian victory at Rossbach was decisive in the west.[147] Ferdinand of Brunswick would continue operations in support of the main Prussian force, but no French army would ever again muster a direct threat against Frederick. Tactically, the Prussians suffered 548 casualties at Rossbach, while inflicting over 5,000. Strategically, Britain immediately warmed relations with Berlin, agreeing to reinforce Ferdinand and to pay a subsidy to the Prussians. Seydlitz, one of the Prussian casualties, was wounded with a prognosis for a full recovery. He stayed at Rossbach to marry a seventeen-year-old girl he had briefly courted before the battle. A surprised Frederick promoted his hard-charging warrior to lieutenant general of cavalry. At some point in the midst of the victory celebrations, Seydlitz contracted syphilis.

Frederick had no time to savor victory. Even as Soubise left the western theater, the Austrians, under Charles of Lorraine, moved into central Silesia with nearly 80,000 troops.[148] The local Prussian garrison was completely outmatched and forced to retreat. Frederick faced two choices. He could either let Charles remain in the Prussian defensive perimeter throughout the winter, or he could attempt to drive the Austrians south. Neither option was completely palatable. Under no circumstances could the king tolerate the presence of the Austrians, yet his army appeared too weak to attack. Undaunted by the numbers, Frederick took a small group of soldiers from the Rossbach campaign and marched east. By early December he joined the local Prussian troops, giving him a total of 30,000 men. Since the Austrians had 80,000 troops in a defensive position, the king assumed that a Prussian attack would surprise Charles of Lorraine.

---

[147]See comments by Gerhard Ritter, *Frederick the Great: A Historical Profile* (trans. by Peter Paret) (Berkeley, Calif.: University of California Press, 1970), 115.

[148]Duffy, *Frederick,* 148–54, and *Army,* 177–79. See also Jany, vol. II, 453–58.

The morning of December 5, 1757, was frigid as the Prussian army rose early, hoping to find the Austrians. Nearly freezing and outnumbered, the Prussians had no reason to be in good form, but their morale was excellent. Their king, unlike any other king in Europe, was beside them. He had joined the forced march and slept out in the open during the pre-winter nights. The royal bottom had been bared during latrine breaks, and the unwashed king shook with the same cold that engulfed the lowliest private. He had not acted as the sovereign. During the march, Frederick had taken to addressing the troops unofficially while accepting their informal comradeship. Some of the men even addressed the king as "Fritz." Before seeking this decisive encounter with Charles, Frederick had spoken to a number of the men in an emotional appeal. Unlike the previous battles, Frederick told them, this time the army was fighting for its homeland.[149] Everyone else had deserted, the king said, only true Prussians remained. The message took hold. On December 5, Frederick's small army was ready to do battle.

As the sun came up, the advance Prussian guard located Charles's main body on high ground to the east of the Prussian march. It was not a pleasant sight. Charles was arrayed for battle, stretching a series of defensive lines from a wooded area south along a hill facing the Prussians. His position was fortified, and the northern section of Charles's flank was anchored by a large grove of trees. The southern flank turned at the village of Leuthen, and it was supported by reserve troops. Under normal circumstances an attack would have been out of the question. Frederick, however, felt that he held two trump cards. The Austrians had nailed themselves to a defensive position along a ridge, leaving the king free to maneuver, and Frederick knew the area like the back of his hand.

Frederick called the senior officers together and explained the plan of attack. The entire army would move forward to the Austrian north and begin an attack. Once the attack commenced, the real action would take place. With the exception of a small corps, the entire Prussian army would disengage and march back to the west. He hoped the Austrians would assume that this was a retreat. Frederick's actual purpose would be to march the troops through a series of low hills—unseen by the Austrians—and attack them on the southern flank in the late afternoon. It was a daring plan, but it

---

[149]Duffy, *Frederick,* 146–47, summarizes the speech, and Luvaas, 231–35, prints a text Frederick wrote from memory.

could work, if the Austrians took the bait. Shortly after 8 A.M., the first Prussian soldiers engaged the Austrians on the northern flank.

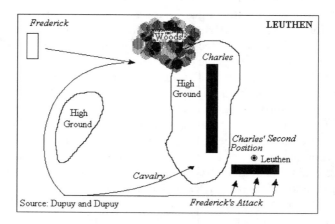

Source: Dupuy and Dupuy

Everything went according to plan.[150] Charles assumed that Frederick had fallen on the strongest point of the Austrian line. As Prussian shells fell on the fortified Austrians, Charles bolstered his strength by transferring reserves from the southern flank. When the Prussians came forward, the Austrians easily drove them back. Charles ordered his troops to step up their action and watched intently. The next hour brought the desired result. The small Prussian army seemed to be no match for the Austrians. Leaving a rear guard, Frederick looked as if he were retreating. Charles had no idea what the king was actually doing.

The maneuver was very complicated and probably could not have been accomplished by any other army. Frederick took his troops out of a battle and assembled them in columns while they were marching. From the Austrian vantage, the Prussians simply seemed to march off. The ground appeared flat, but Frederick knew better. Having conducted several peacetime exercises in the area, the king moved his troops into the low hills and they disappeared. Quickly, the Prussians wheeled south, marching for about four miles, and by late afternoon the king reappeared on the southern flank. As

---

[150]Luvaas, 235–40.

the Prussians surged forward, Charles believed that a second Prussian army had entered the area.

Frederick was under no illusions. He knew that his attack had to be successful. Leaving his horse, he moved among the men of the front rank and explained the critical situation to them. They had to be successful, the king said. Bolstered by the king's plea, the troops went forward. For the next two hours, until darkness, they battled around the area of Leuthen. Although they did not break through initially, Charles was not able to respond. His troops were committed in the north, and he could not shift them in time to meet Frederick. The Austrian infantry arrived in piecemeal fashion, only to be ground up by Prussian infantry volleys and cannon shot. When the Prussian cavalry rode forward to blunt an Austrian cavalry attack, the fate of the Austrians was sealed. Just after darkness, the Austrian line broke. The Prussians swarmed into the Austrian rear taking thousands of prisoners. As snow fell in the darkness, the Austrian army collapsed.[151]

Historians refer to Leuthen as one of Frederick's greatest victories— some claiming it to be greater than Hohenfriedberg.[152] To be sure, Frederick struck an entrenched army that outnumbered him three to one, and he annihilated it. Charles lost nearly one half of his army, whether killed, captured, or wounded. Austrian soldiers littered the ground, along side of the Prussian dead and wounded. Yet, in the midst of the bloody victory there was one moment of humanity that reached far beyond the miserable glory of Rossbach and Leuthen. It happened somewhere along the front, gradually spreading from soldier to soldier. Each heartbeat increased the momentum of the event and brought more and more men—men who had survived the carnage—to the realization that they were still alive. As a gentle snow covered the wounded and the dying, the entire Prussian army gradually began singing. Within minutes the field resounded to the unaccompanied strains of "Now Thank We All Our God," sung by the survivors of a hellish day while they moved through the darkness.[153] Perhaps, it was an affirmation that

---

[151]Asprey, 475–81.

[152]For example, see Pierre Gaxotte, *Frederick the Great* (trans. by R. A. Bell) (New Haven, Conn.: Yale University Press, 1942), 329–30; Atkinson, 227; and Tuttle, vol. IV, 148–49.

[153]See artwork in Werner Knopp, *In Remembrance of a King: Frederick II of Prussia, 1712–1786* (Bonn, Inter Nationes, 1986), 38.

there was more to life than the slaughter of a nameless enemy, or perhaps they were simply thankful to be alive. No none knows the answer. When the Prussians moved into winter quarters, they did not know that the worst was yet to come.

Chapter 7

The Third Silesian War Continues

The concept of the strategic defensive had worked well for Frederick in 1757, despite the problems of summer. Kölin represented a turn of fortune, but the army was able to recover. When the Russians failed to exploit their success at Gross-Jägersdorf, Frederick snatched victory from his enemies at Rossbach and Leuthen. Moreover, the victory at Rossbach resulted in a favorable alliance with Great Britain, and all-important British fiscal support began flowing into Frederick's coffers. The year even seemed to end on a favorable note, yet there was a deeper truth. Although France and Austria had been decisively defeated, they had not been driven from the field. Prussia was still ringed with enemies.

1758 and the Tactical Offensive

As 1758 opened Frederick sought to solidify the Prussian defensive perimeter.[154] Believing that the west was in the safe hands of Ferdinand of Brunswick, Frederick wanted to offset Russian advances in the east. To do so he needed to neutralize the Austrians, and this entailed a different strategy from 1756. In 1758 Frederick simply wanted to drive the Austrians as far from his southern flank as possible. A second Bohemian campaign—like the campaign of 1756—was pointless. Instead, Frederick moved south from Silesia into Moravia, hoping to divert the Austrian forces. By May 20 Frederick laid siege to Olmütz, one of the key fortresses in Moravia.[155]

Austrian Field Marshal Leopold von Daun was destined to become Frederick's greatest nemesis. As Frederick tried to divert the Austrians toward Moravia, Daun planned another maneuver. Refusing to confront Frederick

---

[154]See M. S. Anderson, *Eighteenth-Century Europe: 1713–1789* (London: Oxford University Press, 1968), 35–37.

[155]Robert B. Asprey, *Frederick the Great: The Magnificent Enigma* (New York: Ticknor and Fields, 1986), 486–509.

around Olmütz, Daun destroyed the Prussian supply lines and reinforcements sent from Silesia. With his supplies in shambles and his lines of communications threatened, Frederick was forced to abandon all hope of an offensive against the Austrians. In June he raised the siege at Olmütz, and in August he retreated to Silesia. Daun had managed to turn Frederick's attempted diversion into a Prussian defeat. The Russians had another nasty surprise for Frederick. As soon as the Prussian forces stopped retreating, a new Russian commander arrived in the east. He was said to favor offensive action.

While Frederick had focused his attention on the Austrians, General William Fermor had taken command of the Russian army in early 1758. He immediately entered East Prussia with 34,000 troops and took Königsberg in the middle of winter. Reassembling the Russian forces, he began a slow march toward Pomerania with nearly 66,000 troops. As Frederick busied himself at Olmütz, Fermor moved toward Brandenburg. General Christoph von Dohna, who commanded the Prussian troops along the eastern perimeter, was no match for Fermor, and Dohna was unable to stem the Russian advance. When Frederick pulled back into Silesia in August, he was greeted by frantic appeals from Dohna. Fermor was directly adjacent to Brandenburg along the Oder River. Frederick left Silesia with 26,000 troops, marching hastily toward Dohna.[156]

Although the large Russian army was camped only a few dozen miles from Frederick's capital, fortune smiled on Prussia. While Fermor was more competent than Apraksian, the Russian army remained lethargic. The Russians were threatening Brandenburg, but they failed to take full advantage of their opportunities. Rather than advancing rapidly, they moved slowly, giving Frederick time to react. While Fermor's presence was a threat to Berlin, his actions were not. He spent August sitting on the Oder, not moving deeper into Brandenburg. In addition, Frederick's enemies failed to act in concert. The Austrians made no move to support the Russians when Frederick marched to the eastern theater. Therefore, when the king joined Dohna on August 11, he had to face only a single foe. Had Daun moved to support Fermor, Frederick's story might have ended.

---

[156]F. W. Longman, *Frederick the Great and the Seven Years' War* (New York: Charles Scribner's Sons, 1926), 147.

If his allies lacked coordination and tenacity, Frederick's army embodied the opposite qualities. Frederick and Dohna united to form a joint command of 37,000 men. Despite the August heat, the Prussians moved forward at a rapid pace, covering more ground than any other European army was capable of doing. By August 22—although his troops were totally exhausted—Frederick located Fermor near the village of Zorndorf.[157] Fermor's Russians numbered nearly 50,000, and they held a strong defensive position. General James Keith, a veteran of the tsar's army, warned Frederick that the Russian soldiers were rested and capable of hard fighting. Frederick, who had only contempt for Russian commoners, was undeterred by his friend's advice. The king believed that his best course of action was to attack immediately.

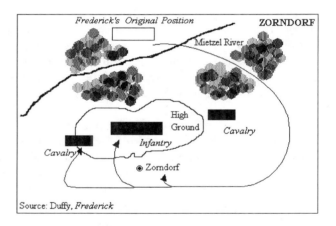

Fermor's Cossacks, spread throughout the countryside, were quick to bring word of Frederick's presence, and the Russian general watched closely as the Prussian king entered the area. The Russians had been laying

---

[157]Christopher Duffy, *Frederick the Great: A Military Life* (London: Routledge, 1985), 163–72, and *The Army of Frederick the Great* (New York: Hippocrene Books, Inc., 1974), 182–83. See also Curt Jany, *Geschichte der königlich Preussischen Armee bis zum Jahre 1807, Volume II, Die Armee Friedrichs des Grossen: 1740 bis 1763* (Berlin: Verlag von Karl Siegismund, 1928), 487–99. See commentary by Martin Van Creveld, *Command in War* (Cambridge, Mass.: Harvard University Press, 1985), 53.

siege to the fortress of Küstrin, but Fermor abandoned the enterprise as Frederick drew near. The king's reputation preceded his army, and Fermor wanted every soldier he could muster in the Russian ranks. When Frederick neared the village of Zorndorf, Fermor assumed a defensive posture. The Russians had more troops and more guns, but Fermor was not about to take the offense. Sitting on the high ground, he waited for Frederick to attack.

At 3 A.M. on August 25, Frederick began moving from the Prussian camp just to the north of the Russian positions. The plan was somewhat standard by now. Frederick had no intention of launching a direct assault. Rather than moving forward, the king sent his army on a long flanking maneuver, planning to strike Fermor from the south. The Prussian soldiers moved through a wooded area, marching far from their original positions. Somewhat to Frederick's dismay, the soldiers broke into song, singing a number of hymns. The king commented to Seydlitz that the men seemed to be afraid of dying. The Prussian soldiers had a right to be concerned, and had Frederick been able to observe the Russian army, he too might have uttered a song. Fermor was gradually moving the Russian front to meet Frederick's flanking movement. Maintaining his position on the hills around Zorndorf, Fermor kept his face toward the Prussian army. The Russians were hardly outflanked.

By mid-morning the Prussians were in position to the south of the Russian troop formations, but the king was shocked to see that the entire morning had been wasted. Although he hoped that he would strike the weakest part of the Russian line—the thin southern front—Frederick saw that Fermor had shifted his entire position. Instead of attacking the weakest point, Frederick stood directly across from the strongest point, and, without waiting for the king to move, Fermor took even more of the initiative away. He sent his Cossacks to burn the village of Zorndorf. The smoke covered his new position, further hiding the Russians from Frederick's army. Frederick had two options, either launch a frontal assault or retreat. He chose the first.

Frederick might have been forced into a frontal assault, but he chose to make it on his own terms. There was no need to line the troops in a single formation and move directly toward the Russians. The king decided to send the main body of Prussian troops, under the command of General Heinrich von Manteuffel, into a single point of the Russian line. Two other groups would support the attack, but their purpose was to support Manteuffel's flank and to keep the Russians from reinforcing the line at the point of the main attack. A fourth group of Prussians would move to the left of the supporting units, while refusing to engage the Russian soldiers. Seydlitz was to wait for the opportune moment and strike the Russian line as soon as Manteuffel's infantry had weakened it. The Prussians' version of a frontal assault was not as direct as one might expect.[158]

Although Frederick often expressed contempt for the artillery, he was willing to set aside his principles in the face of Fermor's strong position. The king assembled a large battery of cannons and began a two-hour bombardment. Unfortunately, the artillery did not have the desired effect, and as Manteuffel's troops marched toward the Russians, smoke and confusion overtook them. With Zorndorf burning, the Prussian cannonade added to the confusion by covering Manteuffel's path with a heavy layer of smoke. Manteuffel's men were out of contact as soon as they reached the smoke, and when supporting units began their maneuver to assist Manteuffel, they could not find him. As a result, Manteuffel moved directly into the Russians without any help, while Russian soldiers poured volleys of musket and cannon fire into his exposed ranks. Whole companies were blown away like chaff, and the ineffective infantry assault did nothing to weaken the Russian line. Frustrated, Frederick sent a message to the cavalry, telling Seydlitz to charge the unbroken Russian infantry.

The newly promoted lieutenant general of cavalry sat in front of several cavalry regiments, searching through the smoke for an opportunity to assault. A virtual legend throughout the army after his famous ride at Rossbach, Seydlitz was aware that his men were capable of breaking an enemy when the circumstances were correct. The solid line of Russian infantry presented no such opportunity. Seydlitz, always known for speaking his mind in the king's presence, was not impressed when Frederick's order to attack arrived. Aware that Manteuffel's disciplined rapid fire had failed to make an impression on the Russian line, Seydlitz knew that the time was hardly ripe for a cavalry charge. The Prussian horsemen would be gunned down before they reached the Russian ranks. Refusing to attack, Seydlitz sent his king a terse response. Tell his majesty, the general said, that at the end of this day my head will be at his disposal. While it is still attached to

---

[158]Asprey, 494–99.

my shoulders, however, I intend to use it in his good service.[159] He sent three regiments to cover Manteuffel's flank and kept the remaining cavalry in position. Frederick, irritated by the response, could do little as Manteuffel's attack fell apart.

General von Dohna had taken his orders to heart. Not the man to command an independent wing in Frederick's army, Dohna made sure that his troops were "denied." When Manteuffel turned toward the Russians, Dohna marched to the west. He continued this path, despite the noise of battle to his north, and by early afternoon he had nearly marched himself out of the battle. Frantic, Frederick rode from the field in an effort to locate his general. If Seydlitz would not attack, he reasoned, Dohna would. Yet, Dohna could only attack if the king could find him. Just as Frederick left to search for Dohna's wing, the Russians began to waver. Even though Manteuffel's men had suffered heavily, their disciplined volleys were taking a heavy toll, finally breaking Fermor's forward line into several isolated units. While the king looked for Dohna, Seydlitz looked through the smoke. At long last, Seydlitz thought, the time to strike had arrived. Without waiting for orders, he tossed his clay pipe into the air.

Zorndorf was not Rossbach. As the screaming Prussians rode toward the Russian infantry, the Russians closed their formations. Instead of turning and running—the action Seydlitz thought the frightened soldiers would take—the Russians formed a bayonet wall. The Prussian cavalry came to a dead stop. Realizing that his cavalry units would be destroyed in this position, Seydlitz ordered his buglers to blow recall, and the Prussians reassembled in a make-shift reserve area. He reformed the regiments and waited for another opportunity to strike. A lesser commander would have called it a day.

To the left of Seydlitz and Manteuffel, Frederick finally found Dohna and, in a rage, ordered the stunned general to move back toward the center. Dohna immediately changed directions and marched into the smoke, hoping to find the weakened Prussian center. Fermor was not willing to wait for Frederick to repair the damage. With Seydlitz repulsed, he felt that victory was near. Fermor launched his own cavalry assault, thinking this would break the Prussian infantry. Seydlitz had prepared for such a turn of events.

---

[159]Walter Goerlitz, *History of the German General Staff: 1657–1945* (trans. by Brian Battershaw) (New York: Praeger, 1962), 4.

Ever mindful of the need to keep his squadrons assembled, Seydlitz was in position to counter the Russian strike. The famous clay pipe was thrown into the air again, and the rugged cavalier led his troopers forward. The result was one of the most savage cavalry engagements in history. Prussian and Russian cavalry troopers grappled with swords, fists, and anything that could serve as a bludgeon. At some places the ground turned red with blood. In the end, Seydlitz's men rode back to their lines, but so did the Russians. Seydlitz had saved the Prussian center.

Dohna finally arrived on the main battle line late in the afternoon, and his presence proved decisive. Small units of Prussian soldiers began to advance, and the Russians could not hold them. Seydlitz moved the cavalry into the arena, even though the troopers were nearly spent. The center evolved into a hellish pushing match, with the Russians slowly giving way. Piles of bodies grew from the ground, but individual Russian soldiers refused to throw in the towel. By nightfall, however, the Russians quit firing and moved from their positions. The Prussians were in no condition to pursue. More than a quarter of Frederick's army lay in the gore of the hotly contested field.

As the sun rose on August 26, no one was quite sure what to make of the situation at Zorndorf.[160] Thousands lay dead on the ground and thousands more moaned beside them. Prussian and Russian looters slipped from their camps and silently slit the throats of friend and foe alike while robbing the dead and wounded. The looters made no attempt to distinguish age or uniform. Frederick and Fermor still sat facing one another, neither knowing if the other would accept defeat. The Russians, with more than 20,000 casualties, blinked first. In the early afternoon, as patrols began firing at one another, Fermor believed a second battle was evolving—a battle for which he had few supplies. With his lines threatened, Fermor pulled the Russian army from the field, leaving the thousands of decomposing Russian bodies to fester in the August heat. Frederick thanked Seydlitz for his wise insubordination of the previous day and soon became particularly joyful. The Russians were not such good soldiers after all, he said. Surveying the casualty-laden ground, Seydlitz pointed out that the "inferior" Russian soldiers were responsible for all of the Prussian deaths. Once again, Frederick

---

[160]The ferocity of the battle is mentioned in general historiography. See, for example, Arthur Hassall, *The Balance of Power: 1715–1789* (London, Rivingtons, 1925), 259.

tolerated Seydlitz's insubordination. He could afford to be magnanimous. He was the victor.

As usual Frederick could not linger to savor the results. No sooner had he defeated the Russians than he received word that the Austrians were moving against his brother Heinrich in Saxony. Frederick assembled 15,000 troops and moved to assist Heinrich in another series of forced marches. He arrived in northern Silesia in September and moved to Saxony to join Prince Heinrich, increasing the total Prussian force to 30,000 men. Frederick knew that Austrian Marshal Leopold von Daun was also in Saxony, and the king wanted to keep an eye on him. What Frederick did not know—and what he certainly could not have guessed—was that Daun had 90,000 troops. In October the Prussians moved into the village of Hochkirch, not knowing that they were outnumbered nearly three to one.

Field Marshal Daun had been carefully waiting for the opportune moment to strike. When Prussian reinforcements arrived in Silesia in September, Daun prudently avoided battle. Having learned to fear anything Prussian, Daun wanted to engage Frederick on Austrian terms. As Frederick camped in Saxony, Daun gathered his strength. If Daun seemed hesitant, his disguised actions bolstered Frederick's confidence. Since the Austrians refused to give battle, Frederick simply believed that he was outmaneuvering them. Daun knew better.

On the night of October 13, the Prussian army was camped around Hochkirch.[161] The troops were split into various contingents, and they were not arrayed for battle. Frederick's generals complained about the position. The Prussian camp was surrounded by a forest, and the troops were exposed, causing James Keith to say that the Austrian commanders should be hanged if they failed to attack.[162] Frederick laughed at his friend's comments, firmly believing that Daun would never take the initiative. The general did not share Frederick's confidence as he watched the flickering Austrian campfires a mere 600 yards away from the Prussian perimeter. It was not a comforting sight.

The Austrian attack completely surprised the Prussians. Realizing that Frederick had committed a tactical blunder, Daun struck the next morning. He moved the Austrian forces in three large columns through the pre-dawn

---

[161]Duffy, *Frederick*, 173–79, and *Army*, 184–86. See also Jany, vol. II, 500–8.
[162]Longman, 152.

hours, silently and effectively. Just before sunrise, the forward echelons of the Austrian army slipped over the Prussian barricades. Prussian soldiers were killed in their tents as they slept. Frederick awoke shortly after dawn to the sound of gunfire. He emerged from his tent to observe the action, only to conclude that he was under a light cavalry assault. Nothing could have been further from the truth. Daun had nearly surrounded the Prussian camp, and his troops had infiltrated its perimeter. Frederick's scattered forces were scrambling to assemble in any type of battle formation, but the haphazard manner in which they were encamped hindered their movement. Daun's soldiers moved into Hochkirch, setting fire to the village, and as the sun cleared the remaining darkness, the Prussians gradually realized the predicament. Frederick the Great was in a fight for his life.

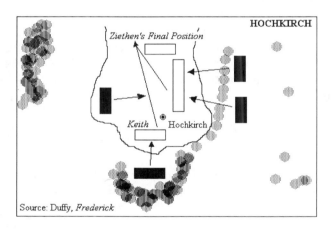

Keith was able to assemble an ad hoc formation on a hill just outside the burning village, while other units aligned themselves in Hochkirch's fiery streets. The alarm sounded throughout the camp, and other Prussian soldiers quickly moved to battle formations. It was a terrible scenario. The Prussians were surrounded on three sides, their commanders were not in contact with one another, and their southern flank—anchored by Keith and the burning village—was in immanent danger of collapsing. Junior Prussian officers formed their units so that platoons and companies could fire in two directions simultaneously. The troops inside Hochkirch were packed so

closely that men could not fall after they were wounded. Within minutes
Daun began moving his artillery forward for the final blow. It seemed Frederick was in his grasp.

By 8 A.M. the situation had become critical. The Prussians were offering resistance, but they could not support one another from their scattered
positions. Daun maximized the danger by skillfully employing the Austrian
artillery. His cannons sent salvos of grape shot directly into the Prussian
formations, killing dozens of men every minute and terrorizing hundreds
more. The Prussians might have turned in panic, but there was nowhere to
run. They could only return fire and wait to be gunned down.

While the entire position was in crisis, no one had it worse than James
Keith. With each Austrian volley reducing his ranks, Keith's men fired back
as rapidly as they could reload. Surveying the scene, Keith thought that his
only salvation was movement, and he could move in only one direction. Despite the withering fire, Keith informed his subordinates that he intended to
advance. The surviving soldiers in the Scotsman's tenuous position realized
that disaster was impending, and they followed their field marshal toward
the Austrians. Prior to his counterattack, Keith managed to send Frederick a
message. Tell his majesty, the field marshal said, that we will try to hold.
Keith added a farewell, stating that he did not expect to see the king again.
A few minutes later a musket ball struck Keith in the lower abdomen. The
wound condemned Keith to an agonizing death.

Inside the Hochkirch things went from bad to worse. The Prussians
were packed so tightly that they could not offer effective resistance while
merciless Austrian artillery tore through their ranks. The soldiers would
later refer to the position as *Blutgasse,* "blood alley." Things were not much
better outside the village as the battle threatened to become a confused rout.
Prince Moritz of Anhalt-Dessau fell prisoner to advancing Austrian troops,
while the queen's brother, Frederick of Brunswick, was beheaded by an
Austrian cannon ball. Even Frederick's horse was shot out from under him.
It appeared as if there were no salvation. Fortunately, local Prussian commanders took a different view. Realizing that they were being driven from
the field, they fought to open the northern flank and an avenue of escape.
Their actions caused Daun and Frederick to arrive at the same conclusion
by mid-morning. Annihilation or survival depended on Daun's ability to
encircle the Prussians completely. Accordingly, Daun tried to close the trap
while Frederick called for Hans von Ziethen.

The hussar king was a horrid peacetime soldier, often drinking too much and gambling his wealth away, but he was at home in the saddle. In addition, Ziethen inspired the soldiers under his command. With the army struggling for survival, Ziethen was Frederick's best choice for action. Ziethen formed all available cavalry units, virtually under the muzzles of his enemy's guns, and he charged north. His men swept past the Prussian infantry ranks into the closing Austrian formations. Stunned by the shockingness of the action, the Austrian infantry retreated, opening a small corridor to the north. Lieutenant Colonel Christoph von Saldern, a cavalry commander of the Seydlitz-mode, moved a combined cavalry and infantry unit into the gap. Ziethen, his task accomplished, sent word to Frederick. There was a small escape route to the north, and the army could utilize it, if the king hurried.

Hurry the king did. Daun was slow to react to Ziethen's cavalry action, but Frederick flooded the area with Prussian soldiers. While the Austrians had numerical superiority, they were concentrated in the south. Ziethen had attacked the weakest area of the Austrian encirclement, and when Frederick moved forward, he outnumbered the Austrians. The Prussians inflicted heavy casualties as they retreated north, and Daun, mindful of Frederick's mobility in combat, ordered his troops to halt. Although he could have moved to re-engage Frederick's retreating army, the Austrian marshal feared to fight a battle of maneuver against the legendary maneuverability of the Prussians. He satisfied himself with a victory and made no attempt to destroy the remnants of the Prussian army. Daun had a right to be content. One third of Frederick's men lay dead or bleeding on the ground around Hochkirch.[163]

The heavy Prussian casualties would have destroyed an army fighting under different rules of warfare. Daun, dismayed by 8,000 dead and wounded in the Austrian ranks, refused to pursue Frederick. A Napoleon, Moltke, or Patton would have behaved differently, but Daun was fighting by eighteenth-century standards. In the Seven Years' war, generals seldom delivered the coup d' grace to a defeated opponent. The Prussian losses were severe—9,500 men, including two field marshals, hundreds of officers, the queen's brother, and 101 guns—but Frederick's army was not destroyed. There was, however, another belated casualty. Moritz of Anhalt-Dessau had

---

[163]See Asprey 502 ff. for comments on the campaign.

been hit in the leg before falling into Daun's hands. The youngest Dessauer gained his parole and painfully made his way home. His leg wound healed but not before developing a cancerous tumor. He died on the family estate. Like Keith and Frederick of Brunswick, Moritz symbolized the fate of thousands of unknown common Prussian soldiers. They no longer cared whether Hochkirch was victory or defeat.[164]

The State Under Siege

Frederick went into a deep depression after Hochkirch. Aware of the magnitude of the defeat, the king was nearly shattered by a series of personal tragedies. He frequently broke into tears when remembering the death of his friend, James Keith, and his condition worsened when a messenger brought news of his sister Wilhelmina's death. His beloved sister—the girl who had tried to protect Frederick from his abusive father—was gone. A dejected Frederick sent for reinforcements and marched against Daun. Although Daun possessed superior strength, he was too cautious, and his caution gave the king a decided tactical advantage. The Prussian army marched around Daun, and by the end of November they threatened his supply lines. The Austrian marshal was forced to retreat to Pirna for the winter. As usual, neither the French nor the Russians took advantage of Frederick's position, and the army was able to move into winter quarters. Frederick had risen from the ashes like a phoenix.[165]

The spring and summer of 1759 were relatively quiet. The king was in no position to attack, and he spent most of his time trying to maneuver his enemies out of the Prussian defensive perimeter. The issue was different in the west where Ferdinand of Brunswick moved aggressively against the French and the *Reichsarmee*. By April he was advancing toward Frankfurt-on-the-Main. The French, tiring of the war on the continent and afraid of combat on their own soil, were quick to respond. They sent 60,000 troops, forcing Ferdinand to retreat throughout the summer. By August the French

---

[164]For a personal view of the battle see Henri de Catt, *Frederick the Great: The Memoirs of His Reader: 1758–1760* , *Volume II* (trans. by F. S. Flint) (London: Constable and Company, Ltd., 1916), 33–43. Both volumes contain firsthand accounts of Frederick's actions during the middle of the war.

[165]Asprey, 502–9.

established a strong defensive position in the area of Minden, but Ferdinand had had enough. He made plans to attack.

The French were ensconced in an elongated set of field works to Ferdinand's front outside of Minden.[166] Ferdinand was outnumbered—he commanded 45,000 men—yet he understood that victory rested on his ability to force the French from their strong position. Ferdinand developed a plan that left most of his troops in reserve. He hoped to attack the French right with a small contingent of troops and either force his enemies from their defensive works or get them to advance into the open field. Ferdinand attacked the French early in the morning of August 1, 1759. The relative safety of the defensive positions proved too much for Ferdinand's British and German soldiers. The French held their line and drove the attackers back. Then they made a dangerous mistake. In their enthusiasm, the French troops left their fortifications to continue counterattack. Ferdinand had drawn them into the open.

Ferdinand called on his Prussian troops to meet the French, and the soldiers marched forward to deliver disciplined volleys. The tactic worked, but the attack bogged down when the French soldiers retreated to their original positions. Try as they might, the Prussians could not break through the fortifications. By early afternoon the French had regained the advantage, and they began massing their cavalry to overrun the hapless Prussians. Ferdinand had to stop the French cavalry. If the Prussians broke, the prince knew that he would be forced to retreat. Since he had no cavalry to stop the French horse, he sent a British infantry brigade forward. In some of the most savage fighting of the day, the British attacked the French cavalry, and, even though they suffered heavy casualties, they continued to press the attack. By evening, the French abandoned the field after suffering 10,000 casualties of their own. Ferdinand sent word to Berlin and London: he was advancing toward the Rhine.

Frederick received word of Ferdinand's victory at Minden with relief. Occupied by a new Russian threat, the king did not have the forces to fight off another attack by the French or the *Reichsarmee*, and he needed the summer to rebuild his dwindling army. General Peter S. Soltikov, yet another commander from St. Petersburg, had taken control of the Russian

---

[166]See Reginald Arthur Savory, *His Britannic Majesty's Army in Germany during the Seven Years' War* (Oxford: Clarion Press, 1966).

forces in the east, and he was proving to be too much for the rigid Christoph von Dohna. Frederick replaced Dohna with General Johann von Wedell, and he ordered Wedell to attack Soltikov. It was a mistake.

In fairness to the Prussian general, Wedell had never held an independent command, and Frederick did little to prepare his generals for such activities. On July 23 Soltikov sat near the village of Kay on the Oder River.[167] Wedell tried to attack him in Frederick's style, but Wedell was no Frederick. As the Prussians approached the Russian lines, Soltikov brought his artillery forward. Frederick usually left his enemies guessing about the point of attack, but Soltikov had no uncertainty about Wedell. When the Prussians attacked, Soltikov fired on them with devastating effectiveness. The massed Prussian troops lost of 8,000 men, and Wedell was forced to retire. As Wedell retreated, Soltikov advanced, linking his forces with 35,000 Austrians—an army from Silesia under the Austrian General Gideon von Loudon. For the first time Frederick's enemies were about to act in concert.

Frederick was forced to react. Coming from the southern theater, he gathered as many troops as he could muster and marched toward Soltikov and Loudon. By August 9 he had reached the eastern front with 49,000 troops, and on the evening of August 11 exhausted Prussian troops formed outside the Austro-Russian camp near Kunersdorf.[168] The camp was arranged along a good defensive position of low sandy hills. Frederick decided to attack the area in typical fashion, hoping to move around the front and strike the Russians from the rear. At 2 A.M. the tired Prussian soldiers were aroused from sleep and sent into the flanking maneuver.

The topography was not especially conducive to the Prussian scheme of battle, and Frederick's movement required the entire army to march through a large wooded area. This proved to be his undoing. The Prussian lines, so well schooled in linear tactics, began to lose cohesiveness in the woods. As officers reassembled formations, they were forced to abandon the tree cover to reform in a clear area. Once exposed, they marched toward the Austro-Russian camp—much earlier than Frederick had intended. As they approached the Austro-Russian lines, something seemed wrong, and the closer they got, the clearer the mistake became. Instead of advancing on the weak

---

[167]Longman, 157 and 171.
[168]Duffy, *Frederick*, 183–90, and *Army*, 188–89. See also Jany, vol. II, 530–39.

rear line of the enemy force, the Prussians were marching straight into a
Russian artillery position.

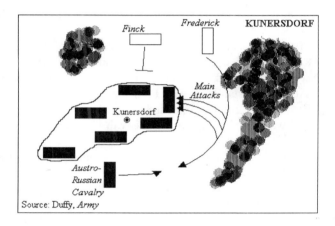

Source: Duffy, *Army*

The king was committed and could not be dissuaded from attacking.
Bringing his own artillery forward, Frederick began his attack at 11:30 A.M.
under a massive artillery bombardment. At 12:30 the infantry moved into
assault lines as the artillery began to fire at secondary targets. The over-
heated infantry marched forward just as the first Prussian cannonballs
landed behind the Russian strong point. Soltikov began shifting infantry to
the east in an effort to stem the Prussian assault, and the field was soon covered in black powder smoke.

   At first it seemed that Frederick's insistence on the offensive would pay
off, as Prussian troops took the Russian position with little difficulty. The
problem came when the infantry tried to roll up the flank. Soltikov had reinforced the position, and he could not be moved. Seydlitz came forward with
cavalry in an attempt to force the issue, but nothing came of his efforts. In
the process, the oft-wounded Seydlitz was hit in the hand and carried from
the field. The shaken Prussian troops were soon more concerned over their
own mounting casualties than with the mangled hand of their popular general. The attack threatened to become a repeat of Wedell's disaster at Kay,
and Prussian casualties mounted.

The attack continued throughout the afternoon, but Soltikov continually reinforced his position with fresh Russian troops. The situation evolved into a battle of attrition, which Frederick could ill afford. In addition, Loudon's Austrians waited at the rear for a chance to strike the Prussians. In desperation, Frederick ordered an all-out attack on the Russian position from three directions. At first the Russians wavered, but Austrian reinforcements stabilized Soltikov's position. Frederick moved the cavalry to attack the Austrian rear, only to have it receive a thunderous barrage of Russian canister before it reached the Austro-Russian line. The Prussian cavalry was forced to retreat, leaving Frederick's infantry exposed on the ground before Soltikov. Loudon decided it was time for the Austrian cavalry to make an appearance.

Loudon's Austrians were reinforced by a contingent of Russian cavalry. They seemed to appear from nowhere as they rode into the retreating Prussians. The Prussians turned and engaged them for more than an hour, but the Russians and the Austrians were fresh. By 6 P.M. the Prussian cavalry was galloping from the field. After steadfastly holding his defensive position all day, Soltikov at last ordered the beleaguered Russian infantry forward. They were eager to extract vengeance, and within minutes Frederick's infantry was in retreat. Fearing a Russian breakthrough, Frederick personally tried to rally his troops in a defensive position, only to have his horse—the third of the day—killed beneath him. Spilling into the dirt, the king rose and shook his fist at heaven, asking for some accursed bullet to strike him dead. He then mounted another horse and continued forming a defensive line.

Even the king's presence could not stop the Russian onslaught. The defensive position could not hold, and the Battle of Kunersdorf became a complete rout. Frederick was able to make a final stand about a mile from the field with 5,000 troops and fifty guns, but the scene turned to a nightmare when darkness fell. With the majority of the army destroyed, the few Prussian survivors trembled through the night as they listened to the screams on the field before them. The Russians had decided to take no prisoners, and Cossacks roamed among the wounded, murdering any Prussian soldiers they could find. The Russians killed or murdered nearly 20,000 Prussian men and destroyed most of the army's equipment. In addition,

Loudon and Soltikov were fifty miles from Berlin, and Frederick had little to stop them.[169]

The next few days were like an impending Greek tragedy. Gloom hung over the Prussian army as Frederick gathered every soldier within walking distance. By August 16 the king had about 33,000 men. He prepared for battle outside Berlin, fully expecting to lose his city, his lands, and his life. He did not expect Soltikov's next move. Coalition warfare is difficult because national rivalries continue even among the best of allies, and Soltikov's feelings for Loudon illustrate the point. Soltikov believed that his troops had done most of the fighting at Kunersdorf and that the Austrians had contributed very little. In addition, he had no intention of destroying a king or a royal state. He informed the Austrians that they should attack Berlin, if they wanted Frederick so badly, and he pulled the Russian army back across the Oder. The Austrians were astounded and so was Frederick. He called the event "the miracle of the house of Brandenburg" and immediately made plans to attack the Austrians.[170] Daun, who was marching north to help destroy Berlin, returned south. Loudon ran back to Silesia.

Frederick joined his brother Heinrich and maneuvered against Daun, safe for the time being from Soltikov.[171] Daun was forced to move further south, but the year did not end with a Prussian victory. Frederick wanted to attack Daun in Saxony. He ordered nearly 12,000 troops under the command of General Frederick von Finck to cut Daun's line of retreat. Unfortunately for Finck, Daun had also learned to act as a master of maneuver. The Austrians slipped away, completely eluding Frederick. Finck, obeying his king, moved to the Austrian rear, not knowing that he could not be supported. On November 21 Finck's relatively small command was surrounded in the village of Maxen by 42,000 Austrians. The Prussians fought hard, but they could not move. Daun managed to do what he had failed to accomplish at Hochkirch, close the trap. By the end of the day, Finck's detachment was

---

[169]Asprey, 514–22.

[170]Gerhard Ritter, *Frederick the Great: A Historical Profile* (trans. by Peter Paret) (Berkeley, Calif.: University of California Press, 1970), 122.

[171]Chester V. Easum, *Prince Henry of Prussia: Brother of Frederick the Great* (Westport, Conn.: Greenwood Press, 1971; orig. 1942), 115.

forced to surrender. It was a small defeat, but on the heels of Kunersdorf it left the Prussians severely strained.[172]

Fighting to Survive

For the first time in his military life, Frederick the Great did not seek a decisive tactical confrontation when the campaign season of 1760 bloomed into summer.[173] Prussia needed to recuperate from the losses, and the Prussian army was greatly outnumbered. Daun faced Frederick in Saxony, holding an advantage of 100,000 men to Frederick's 40,000. Soltikov occupied East Prussia with 50,000 Russians. Given the king's new-found aversion for battle, Frederick detailed 15,000 troops to screen the Russians, scattering them thinly over the eastern sector. A more immediate threat was Loudon in Silesia with 50,000 Austrians. Prince Heinrich faced Loudon with 34,000 men. Although Prince Ferdinand forced the inept French army to retreat from western Germany, Frederick was not in a position to launch an offensive.

Still, the Prussian king was not a man to wait for his enemies to come to him. He believed that he might be able to launch a limited offensive against the Austrians and the *Reichsarmee* in Saxony. Specifically, he hoped to drive the Austrians from Dresden while keeping contact with his brother in Silesia. The king ordered Prince Heinrich to keep watch on Loudon, while Frederick established an independent corps to keep open lines of communication between Saxony and Silesia. He selected Lieutenant General Henri-Auguste von Fouque to command the corps.

On the other side of the hill, Gideon von Loudon carefully analyzed Frederick's troop movements, and by mid-summer he believed he saw an opportunity. The Prussian army was divided between Saxony and Silesia, with only the relatively weak corps of Fouque to link them. Frederick, on the other hand, did not see this as a weakness, presumably believing that the Austrians would not attack. The king had learned no lesson from the disaster at Maxen. He blamed the loss of the corps on Finck's ineptitude, not realizing that isolated forces cannot stand against a concentrated enemy. Loudon had a more objective view. As Fouque's detached corps sat along

---

[172]Longman, 163.
[173]See Jany, vol. II, 550–63 for a summary of events in 1760.

the Silesian border, Loudon prepared for another Maxen. Attacking early in the morning on June 23, the Austrians struck Fouque with 30,000 troops at Landeshut. The Prussians established a defensive front on the high ground in front of a small river, but the Austrian strength was too great. Loudon overran Fouque, and Frederick lost another corps. The Prussians could not afford such losses.

Frederick rarely questioned his own generalship and blamed Fouque for the failure at Landeshut. He concentrated on the more immediate objective, Dresden. By late July he prepared to lay siege to the city, only to break away from the battle when he received news from Silesia. Loudon did not behave like the lethargic Russians. After his victory at Landeshut, he drove into the Prussian-controlled areas of Silesia, taking the fortress of Glatz and threatening Breslau. Another Austrian army under Francis Lacy guarded Loudon's rear. Frederick had to abandon Saxony to go to Silesia. As Frederick marched to join Prince Heinrich, Daun followed the king with another Austrian army. Once again, it appeared that the Austrians held the upper hand.

Daun began moving toward Frederick and soon linked with General Francis Lacy, giving the Austrians a total strength of 90,000 men. Loudon commanded another 37,000 soldiers. Although Frederick gathered a number of troops from Prince Heinrich's army, he could not match the Austrian numbers. He was forced to maneuver and to seek an avenue for retreat. As he moved, the Russians entered the fray again when Soltikov sent his advance guard to join the Austrians. Commanded by General Czernichev, the Russians brought an additional force of 30,000. Toward the end of July Frederick began a series of forced marches, which had now become standard in the Prussian army, in an effort to slip away from his enemies. To the king's surprise, Daun outmarched him. By August Frederick was in dire straits. His troops had suffered heavily through the end of July, and even though they had not fought a major engagement, many men had died from the forced marches. The king had a mere 30,000 men against a force five times this size. The Prussians had managed to win against tremendous odds before, but Frederick had always maintained the initiative. In August his army seemed to have little left, and the Prussians had never faced such tremendous odds. As Frederick sat encamped along the Katzbach creek in Silesia, it appeared that the game was over.

Daun intended to finish the Prussians once and for all by encircling them as he had planned to do at Hochkirch. He asked Czernichev to move from the north along the Katzbach, while ordering Loudon to move from the south to cut off Frederick's line of retreat. Daun incorporated Lacy's command with his own army and prepared to strike the Prussians directly. By August 14 Daun was ready to strike. He ordered Loudon to cross the Katzbach and move toward the Prussian camp. Neither of the Austrian commanders knew that Frederick had gotten wind of the plan from a deserter.[174]

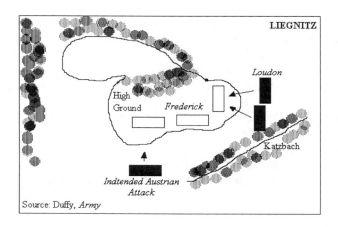

Source: Duffy, *Army*

Frederick realized that his enemies were closing in on him and their assault was imminent. He watched the Austrians and moved his camp to the outskirts of Liegnitz.[175] On the evening of the fourteenth he received word from a deserter that the Austrians were planning to strike the Prussian camp in the dark. The information matched Frederick's suspicions. He quietly moved most of his forces to a plateau overlooking Liegnitz and waited to see the Austrians move. In the early morning a scout rode into the main Prussian camp to inform the commanders that the Austrians were indeed moving. Frederick ordered his men forward in the darkness. Loudon's men,

---

[174]Easum, 140–41.
[175]Duffy, *Frederick*, 202–7, and *Army*, 193–94. See also Jany, vol. II, 564–78.

who had expected to march to the Prussian camp unhindered, were surprised when their cavalry screen encountered Prussian resistance. As Loudon ordered his infantry forward, the volleys of musket fire told him that he had met the main strength of the Prussian infantry. Frederick had not been surprised.

The Prussian infantry fought with zeal. One regiment—an assembly stripped of its regimental insignia after failure in a previous engagement—charged the Austrian cavalry with bayonets and routed it. The men were trying to raise their status in the king's eyes, and it worked. The king, who was rarely concerned with the welfare of individual regiments after they broke in combat, was moved. Apparently, the remainder of the army caught the same spirit. The Austrian attack at Liegnitz was intended to take the Prussians by complete surprise, but Loudon found himself locked in a fierce battle without knowing the location of his enemy. As the Prussian infantry slaughtered the on-coming white-coated masses, their morale increased.

Loudon had only one course of action. Fully committed to an attack, he had no opportunity to find Daun, and changing the plan was out of the question. As his men marched through the darkness into Prussian canister and muskets, he could only hope that Daun would strike with the remaining forces. Around 4 A.M. Frederick believed that his maneuver had isolated Loudon's troops. Gambling that Daun could not support Loudon in the darkness, the king ordered his infantry to counterattack. While Loudon was supposed to be supporting a coordinated attack against the Prussian camp—an attack bolstered by an almost unbeatable ratio of strength—he found that the Prussians had gained numerical equivalency on his front. Supporting Austrian units were somewhere in the vicinity, but they were of no help. In conjunction, the Austrians had been forced to attack an unseen enemy while the Prussians poured grape and ball into their confused ranks. When Frederick counterattacked, Loudon retreated to save his men. By 6 A.M., his army left the field with 3,000 casualties.[176]

Frederick could not comprehend the extent to which he had surprised the Austrians and destroyed their plans. He knew only that he had beaten one section of the massive army, and he was not quite sure where the other Austrians were. Realizing that he needed to retain the initiative, Frederick decided to engage in subterfuge. It was no secret that the Russians were less

---

[176]See Asprey, 534–38.

than eager to pursue the war. Their listless actions in Prussia and Soltikov's refusal to take Berlin were evidence that the hearts of the Russian commanders were not filled with anti-Prussian zeal. Frederick took advantage of the situation. With the sun revealing that Loudon had abandoned the field, Frederick sent a message to Czernichev, telling him that the entire Austrian army had been defeated. Czernichev, who did not want to face Frederick alone, retreated. When Daun arrived to deliver the death blow, he found 60,000 Austrians and Russians retreating from Liegnitz. Frederick was nowhere to be found.

The Prussians did not have the last laugh of the campaign season. Daun, ever reluctant to attack Frederick directly, looked for an indirect approach. Since Frederick could not abandon Silesia, Daun moved back to Saxony, joined the *Reichsarmee*, and hoped to force Frederick's hand. In October the Austrian marshal dispatched General Lacy to take Berlin. Czernichev sent 16,000 troops to help with the task, and on October 9 the city surrendered. To everyone's surprise, the Cossacks did not reduce the city and murder the population. The Austrians behaved differently. Holding Frederick personally responsible for the past three wars, the Austrian army wanted revenge. Sans Souci was subjected to military occupation, and the entire cadet corps from the Berlin officer school was carried off into captivity. The Saxons also wanted in on the fun. Angered by Frederick's earlier decision to bombard Dresden, they took particular delight in destroying a Hohenzollern hunting lodge outside Potsdam. Lacy's force left only after the citizens of Berlin paid a huge ransom. The Austrians did not need to stay. Complaints from his citizens forced Frederick to move north at a time when the Prussians could not afford to attack.

The most effective means of tackling the Austrians, Frederick reasoned, was to move the bulk of his army back to Saxony to face Daun. He began on October 11 while Berlin was still under occupation. Lacy retreated from the Prussian capital to join Daun, and Czernichev moved back to East Prussia, taking the Berlin cadet corps with him. Daun had been building strength south of Magdeburg, and, unbeknownst to Frederick, he had been ordered to stand and seek a decision against the Prussians. Knowing that the *Reichsarmee* would probably flee at the first sign of Frederick, Daun chose his ground carefully, selecting a fortified area near Torgau.[177] The eastern end

---

[177]Duffy, *Frederick*, 210–18, and *Army*, 194–96. See also Jany, vol. II, 579–94.

of the field was anchored by the Elbe River. The high ground contained a large plateau that could be attacked from the west and the north, but it stood protected on the south by a body of water known as the Great Pond. If an army placed itself on the plateau, it could be attacked only from the north or west.[178] Daun moved his troops to the plateau when he heard that Frederick was in the area.

Frederick arrived in Saxony in late October and began to search for Daun. While it may have been in the king's best interest to avoid battle, Frederick could not afford to leave Daun so close to the Prussian border, and he did not wish to allow the Austrians to live off Saxon supplies through the winter. The Prussians opened the campaign with a series of moves against the *Reichsarmee*, and the frightened French and German ran away, leaving Daun to his fate. Daun, sitting at Torgau, obeyed his orders and did not move. When Frederick reached the position on November 2, he decided to attack.

Frederick's plan was predictable. His army was positioned on Daun's southern flank, but Frederick intended to strike from the north. Hans von Ziethen was to play a critical role by conducting a supporting attack near the Great Pond. If Frederick could seize the high ground on the plateau, Ziethen's position would block Daun's best escape route. On the other hand, if Frederick's attack failed, the main effort would fall on Ziethen. The forces were nearly evenly matched, 48,500 Prussians to 53,000 Austrians, and, since many of the Prussian officers knew the ground, it appeared to be an attack worth risking.

Planning an attack and executing it are two different matters.[179] On the morning of November 3, things started horrendously. To begin, the weather refused to cooperate. Frederick's army moved off at 6 A.M. under a blackened sky and in the face of an icy German north wind. Things grew worse when a rainstorm pummeled the battlefield. By noon the Prussians were cold, wet, and miserable. To make matters worse, Frederick lost contact with Ziethen. As the king pondered the possible location of Ziethen's force, he heard musket fire to his south. Historians remain unclear as to why it happened, but they do know, as Frederick knew, that Ziethen had begun to

---

[178]Heinrich had prepared the position a few years earlier (Easum, 152–54).

[179]Jay Luvaas, *Frederick the Great on the Art of War* (New York: Free Press, 1966), 243–50.

attack far too early. He was not in position to support Frederick's main effort, and the intensity of the musket fire told the king that his hussar general was fully engaged. If Frederick wanted to attack, he would do so alone.

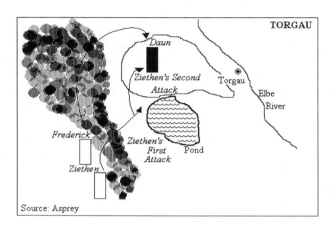

Source: Asprey

Frederick's columns were marching through a wooded area out of sight from the main Austrian positions. Frederick's first impulse was to march on, trusting Ziethen to handled the Austrians, but he knew this would divide the Prussian force. He left the woods to survey the Austrian position on his right. The Austrians appeared to be arrayed for battle at the top of a slowly rising hill, and their artillery was positioned in front of the infantry. The ground was soaked and it would be difficult to traverse. After considering the options, Frederick felt that he could either attack over the wet ground or retreat. He called for his officers and brought the troops out of the woods directly in front of the Austrian guns. At 2 P.M. the troops went forward with the elite Prussian Guards leading the way. The Austrians responded with a barrage.

Daun had not only chosen the ground well, he had fortified it with the proper amount of firepower. More than 270 cannons sat atop the high ground overlooking Frederick's position. As the Guards came forward from the trees, Daun's artillery opened fire with solid shot. The first volley went high, cutting the tops from the trees. The second salvo was lower, and the guards were driven to the ground in agony. The cold, heavy rain added to their confusion and misery. In a few minutes most of the Guards lay killed or wounded on the rain-sodden ground, and Frederick sent a second wave of infantry over their prone bodies. The famous Prussian marching precision was nowhere to be found, as men picked their way over muddy ground clogged with the bodies of dead and wounded comrades. The attack did not go well, and neither did the three attacks that followed it. Each effort terminated in the mud as the troops were subjected to a merciless greeting of Austrian canister.

After two hours the king himself became a casualty. Urging the men forward, Frederick was hit in the chest with a either a musket ball or a piece of grape shot. The impact sent him reeling to the ground in pain, but his lined vest had stopped the ball from penetrating his body. The shot was lodged in the lining of his clothing. Severely bruised, but sturdy, Frederick surveyed the field in front of the Austrian position as evening approached. The ground was riddled with the mass of Prussian dead and wounded. Several attacks over three and a half hours had resulted in no gain, and even Daun marveled at Frederick's willingness to sacrifice his infantry. Frederick reluctantly decided that nothing could be gained in the face of such a strong enemy position, and he called for a retreat. Daun, who had been wounded in the foot, prepared a message for the Hapsburg court while a physician cleaned his wound. At long last, the field marshal reported, the Austrians had thrashed Frederick the Great.

Ziethen was never one to grasp fully the logic of a situation, and he had no intention of admitting defeat. After fighting throughout most of the afternoon, the general wondered where his sovereign had gone. Since he did not seem to be supporting Frederick's attack, Ziethen pulled his men out of the battle at dusk. The Austrians assumed they had defeated him, but nothing could have been further from the truth. Ziethen immediately reassembled his troops—no small feat in the midst of battle—and marched north through the woods in an attempt to find Frederick. His forward echelons made contact with Frederick's main body shortly before dark. Both Frederick and Daun then suddenly heard—without warning—the tremendous thunder from thousands of muskets. Hans von Ziethen may not have been the most sophisticated of all Frederick's commanders, but he always attacked when ordered to do so. Neither Frederick nor Daun knew that Ziethen was attacking in force.

Frederick may have given up hope, but his local commanders did not. Spurred into action by the sound of fighting to the south, Frederick's officers ordered the main body of troops forward. They did not know what was happening, but they could see the Austrians were in disarray. They also did not know that the wounded Daun had been carried from the field. They simply saw an opportunity and decided to seek the initiative without the king's permission. Fortunately for them, it worked. A courier brought word to a dejected Frederick, asking him to come forward with all possible speed. Frederick arrived to find the local commanders talking of victory. Further investigation revealed that Ziethen was in the area and that elements of Frederick's own command were fighting for control of the plateau. Although both sides were in utter confusion, the king ordered reinforcements forward. The Prussians took the plateau around 9 P.M. in complete darkness. Torgau belonged to Frederick.

Victory often has an illusive quality. The Prussians had killed or wounded 4,000 Austrians, and they had captured 7,000 more, but their losses were staggering. Frederick's army had 16,000 men dead, dying, or wounded. In conjunction with this, the army was exhausted, and its leadership was depleted. Torgau was a victory for Prussia, but Prussia could not continue fighting indefinitely. In desperation, Frederick turned to the middle class for help. His officer ranks filled with merchants and craftsmen, including a Major Clausewitz who stuck a "von" in front of his last name in the hope of changing his social status. These men served Frederick with far more loyalty than he would ever serve them. The enlisted ranks were another matter. The strength of the army had been maintained by increasing conscription, but Frederick had nearly exhausted the pool of recruits. Driving an Austrian army from the plateau at Torgau could not change Frederick's growing weakness. To make matters worse, Daun set up winter quarters in Saxony doing exactly what Frederick had tried to prevent. The tactical victory at Torgau brought no strategic results.[180]

A Second Miracle of the House of Brandenburg

The year 1761 witnessed a complete reversal of roles for Frederick. Not only did he remain on the strategic defensive, but he was forced to abandon

---

[180]See Asprey, 540–44.

the tactical offensive as well. For the first time, Frederick did not seek a decisive engagement with his enemies. The Russians, under another new commander, marched into Silesia in the late spring. Frederick spent the entire summer trying to keep his army between Loudon's forces and those of the new Russian general, Alexander Buturlin. In his spare time, Frederick tried to increase the dwindling Prussian ranks. Since conscription was no longer effective, Frederick turned to another recruiting mechanism. Prussia had always maintained semi-mercenary troops known as "free battalions." Although they performed nowhere near the standard of regular Prussian troops, these indefinite volunteers served to augment the regular and more reliable line regiments. Frederick offered monetary rewards and gave his new officers a chance to command the free battalions, more than doubling the number of semi-mercenary troops in the Prussian army. He also began raising new irregular cavalry units that he called "Bosniaks." Neither Frederick William I nor Frederick would have reverted to such tactics in normal times, but 1761 was not normal. The army's numbers increased while its quality declined. [181]

By carefully marching, by blocking roads and passages, and by maintaining fortress garrisons, Frederick was able to keep the Austrian and Russian armies separated through most of the summer. Then on August 20 disaster struck. Buturlin—stretching his supply lines—managed to march the Russian army around Frederick's position while Loudon marched against it. The two allies met each other with a combined force of 130,000. Frederick, with only 55,000 men, was outnumbered again.[182]

Since the Prussians could not attack, Frederick decided to let his enemies come to him. In camp at Bunzelwitz in northern Silesia, Frederick set about fortifying his position.[183] The risks were great. If the defense failed, the Prussian army would be crushed, and if the defenses held, the losses could be staggering. Yet, Frederick had no other choice. If he attempted to retreat, he could be attacked in the open field. After grimly waiting for what he believed would be his last battle, Frederick was amazed to find that Bu-

---

[181]For comments on logistics see H. M. Little, "Thomas Powell and Army Supply, 1761–1766," *Journal of the Society for Army Historical Research*, 1987 (26): 92–104.

[182]Luvaas, 251–56.

[183]See Archer Jones, *The Art of War in the Western World* (Urbana, Ill.: University of Illinois Press, 1987), 304–5.

turlin suddenly abandoned offensive operations. Complaining that the Austrians were not providing enough logistical support, the Russian army marched away on August 26. Frederick had a reprieve.

The Prussians were in no position to attack, and neither was Loudon. Angered by the Russian departure, Loudon decided to move against the Prussian supply line. This resulted in another disaster for the Prussian army. On September 20 Loudon captured the main Prussian supply base in Silesia, the fortress of Schweidnitz. A shaken Frederick moved into winter quarters in December, only to find that a new British government was seeking peace. It eliminated the Prussian subsidy and talked of pulling the British units from Ferdinand's army.[184] Frederick also learned of an attempt against his life. Things had looked bleak in the past, but in December 1761 they looked hopeless. Little did Frederick suspect that he was about to be saved by the Russians.

Elizabeth, Tsarina of Russia, held feelings of hatred toward Frederick rivaled only by those of Maria Theresa. Even when some of her marshals were less than enthusiastic about warring with Frederick, the tsarina's hatred drove the Russian army. Elizabeth's son Peter, however, was cut from a different cloth. The future tsar was vain, grossly immature, and half-witted at best. Yet, Peter simply loved Prussia, even desiring a Prussian field commission as much as he lusted for the Russian throne. On January 5, 1762, Elizabeth died, and Peter III became tsar. The event evoked a second miracle for the house of Brandenburg.[185]

Frederick's bleak outlook turned to joy with the news of Elizabeth's death. He sought to open peaceful relations with the Romanovs as quickly as Peter proposed peace with the Hohenzollerns. Peter not only withdrew from the war, but he placed a Russian army corps at Frederick's disposal. Frederick knew a good thing when he saw it. He immediately inducted Peter into the Order of the Black Eagle, Prussia's highest military decoration, and within days he made Peter the *chef* of his own regiment. A few months later Peter was killed in a plot hatched by his wife Catherine. The new tsa-

---

[184]K. W. Schweizer, "The Bedford Motion and the House of Lords Debate, 5 February 1762," *Parliamentary History*, 1986 (5): 107–23. See also Little and W. O. Henderson, *Studies in the Economic Policy of Frederick the Great* (London: Frank Cass, 1963).

[185]Herbert Rosinski, *The German Army* (Washington, D.C.: The Infantry Journal, 1944), 20.

rina was hardly a dim-wit like her husband, and she had a more realistic assessment of Prussia. Yet, she was not eager to renew the war. Peter's brief reign gave Frederick the opportunity he needed to recover, and for a short period the Russians were no longer a threat.

With the Russians removed, Frederick turned to the Austrians. When the campaign season began in May 1762, he assembled the army in two main groups. Prince Heinrich commanded about 30,000 men in Saxony, while Frederick held the main force, about 70,000, in Silesia. Frederick knew that Prussia was almost exhausted, but he believed that the Austrians could not continue fighting forever. With the British and French making peace overtures, Frederick believed that the fate of Prussia would be decided by its geographical position during any future peace talks. Accordingly, the fortresses in Silesia became critically important. Whoever controlled the fortresses controlled Silesia, and control of Silesia during the peace process meant annexation at the end of hostilities. Of the fortresses in Silesia, the most important was the arsenal at Schweidnitz. Frederick had successfully laid siege to the fortress in 1758, but the Austrians had forced him away during the Liegnitz campaign. Frederick wanted it back because it served as the trump card in any peace settlement. The stage was set for the last round in Silesia.

Neither Daun nor Frederick was anxious for a set piece battle. Frederick moved through Silesia in early summer, hoping to outmaneuver the Austrians and force them to retreat. By July he reached Breslau, and his cavalry units raided into Bohemia. Daun stayed in western Silesia using the Oder River to cover his front and supplying his forces from Schweidnitz. Daun also understood the importance of possessing Silesia, and he was not about to abandon his strong position to deal with cavalry raids in Bohemia. Frederick might move further south, but Daun realized that the Prussians could not ignore the Austrian presence and maintain an offensive. If the Prussians wanted to take the position, they would have to move against it directly. Frederick came to the same conclusion in late July. If he wanted western Silesia, his Prussian army would have to take it by storm. With his old nemesis Czernichev now an ally, Frederick moved from the Oder to attack Daun.

On July 18 Frederick faced Daun at Burkersdorf in western Silesia.[186] Daun had fortified his position on high ground, and he hoped to hold the Prussians at bay. For the first time in several years of fighting, Daun faced superior Prussian numbers. Frederick planned to assault the position in a series of independent attacks designed to support one another. This was uncharacteristic of Frederick's style, but he was not eager to suffer casualties. There was also a second reason for Frederick's hesitation. The night before Frederick's planned attack General Czernichev told the Prussian king that Peter had been assassinated and that Catherine had ordered him to abandon the war. The news pained Czernichev because he had developed a new-found respect for his former enemy, but he had his orders. Frederick asked Czernichev at least to stand by with the assembled Russian corps—the Russian quickly agreed—and he decided that he must force Daun's hand before the Austrians heard the news. Frederick's army attacked the next morning.

The battle of Burkersdorf was unlike Frederick's other battles.[187] Both Curt Jany and Christopher Duffy note that Frederick's commanders fought with semi-autonomous units, much like commanders in modern war. The Austrian front combined two strong positions, one commanded by Daun and another by one of his subordinates. The Prussians attacked with five main groups and a massive artillery battery to support the main assault. The Austrians were driven from their positions, and Daun was forced to retreat to Moravia. The real outcome focused on Schweidnitz. With Daun retreating, Frederick was able to open a siege against the arsenal. It fell on October 9, leaving Silesia in Prussian hands. Frederick's third Silesian war had nearly come to an end.

The last Prussian battle in the Seven Years' War did not belong to Frederick. The *Reichsarmee* attempted to make one last stand against Prince Heinrich in Saxony. Acting in tandem, Heinrich and Seydlitz struck the *Reichsarmee* at Freiberg on October 29, 1762. It was a Prussian victory, and if the Prussians were exhausted, so were their enemies. With Britain and France seeking peace on the high seas and in America, the continental powers decided to seek accord. France had no interest in returning Silesia to Austria, and Russia was more concerned with the future of Poland. For its

---

[186]Duffy, *Frederick*, 236–40, and Jany, vol. II, 640–47.
[187]See Luvaas, 298–305.

part, Austria had thoroughly demonstrated that it could not retake Silesia on its own. Burkersdorf was simply the coup de grace, and Freiberg demonstrated the inability of the *Reichsarmee* to win concessions by force of arms. In January 1763 Austria and Prussia agreed to preliminary peace terms. The delegates settled for a stalemate, status quo ante bellum.

In the current definitive English version of Frederick's military life, Christopher Duffy provides a most moving synopsis of the Third Silesian War by pointing to Frederick's own analysis.[188] Frederick said that people who lionized the experience had not witnessed the hunger, the cold, and the suffering. They would change their attitude, the king said, if they had seen the war. He was correct. Statistics and tales of military glory do not account for the magnitude of grief that devastated countless families throughout Europe—families whose lives were forever ruined by the death of a parent or child. There was hardly a family in Prussia that had not lost someone to the war. One poor Junker woman lost twenty-one male relatives, including her husband, sons, and brothers. Thousands upon thousands of common soldiers filled nameless graves, followed by many members of the officer corps, including the general officers Schwerin, Prince Moritz, Winterfeldt, and Keith. Frederick concluded that Prussia had been saved only by sheer luck. Christopher Duffy adds that survival was also due, in no small measure, to one other factor—the military genius of Frederick the Great.

---

[188]Duffy, *Frederick*, 242–43.

Chapter 8

The Army in Decline

Frederick the Great won the Third Silesian War, but victory had a cruel consequence. The Prussian army was decimated. Thousands of soldiers—including a substantial portion of the elite officer corps—lay strewn throughout the fields of central Europe. The newly commissioned middle-class officers might have been able to lay the foundation for the future, but an ungrateful Frederick either dismissed them or relegated them to minor duties. He replaced middle-class officers with Junkers, and when he issued commissions after 1763, they went to the nobility. Unlike the members of the middle class, most of the Junkers were true believers in the social caste system. Under their leadership, over the next forty years, the army evolved into a rigid, unyielding bureaucracy. When the transformation was complete, the army was shrouded in the tactics of the past and governed by officers more intent on preserving their social status than fighting a modern war. On the other side of the Rhine, quite the opposite was happening. Stung by defeat, the French middle class was prepared to revolutionize the methods for waging war. Nothing could have been worse for Prussia.

Frederick's Remaining Years

After French troops occupied Berlin in 1806, Napoleon took a group of officers to Frederick the Great's grave in Potsdam. Hats off, gentlemen, the emperor said, if he were still alive, we should not be here.[189] While this is great material for the Frederickian legend, the conclusion is not altogether correct. Napoleon was standing in Potsdam because the Prussian army had been unable to match him in the field. The Prussians had beaten the French several times in the Seven Years' War, and the French had taken the lesson to heart. They paid special attention to their defeat at Rossbach and the

---

[189]Nancy Mitford, *Frederick the Great* (New York: Harper and Row, 1970), 291.

training of Seydlitz's cavalry. While Frederick and his successors rested on their laurels, social revolution stirred the middle class in France. When the French bourgeoisie tore power from the hands of the royalty, they developed a system of combat free of class restrictions and biases. The French progressed while the Prussians declined. Ironically, this process began during the reign of Frederick the Great.

Although the king is frequently listed among the "Enlightened Despots" of the eighteenth century, Frederick's apparent love of war turned many people away from his accomplishments. The second half of Frederick's reign, however, was remarkably different from the first. To be sure, Prussia participated in the partition of Poland, and it briefly went to war with Austria over the crown in Bavaria, but the last years of Frederick's rule were peaceful. The king increased the treasury, implemented social reforms, and administered the details of running the kingdom. In the process, he let the readiness of the army deteriorate. Each year the army grew older and more entrenched in the ways of the past, while the king focused on other issues. Regardless of historical opinions, there can be no doubt that the great military king presided over the demise of his army.[190]

Prussia faced several problems that contributed to the decline of the army. Frederick felt that his primary job was to reestablish the Prussian economy. When England abandoned Prussia is the last stages of the war, it demonstrated that Prussia needed to be economically self-sufficient. Ever frugal, Frederick managed to emerge from the war without a major debt, and he spent the next twenty-three years improving Prussia's fiscal situation. When he died, he actually left the state with more money than when he came to office. The accumulation of wealth, however, did not come without a price. Frederick's frugality hurt the army, especially the cavalry. Every coin in the coffer came at the expense of the army.

A second problem centered on the class structure of the officer corps. The number of Junker casualties in the Third Silesian War produced a plethora of middle-class officers. Middle-class commissions were the solution to a demographic problem; the Junkers had few young men capable of military service in the final years of the war. Unable to fill officer slots with

---

[190]Reinhold August Dorwart, *The Prussian Welfare State: Before 1740* (Cambridge, Mass.: Harvard University Press), 77–90. See also Gerhard Ritter, *Frederick the Great* (trans. by Peter Paret) (Berkeley, Calif.: University of California Press, 1970), 178–79.

the declining numbers of nobility, Frederick opted to commission lawyers, merchants, tailors, teachers, and other middle-class professionals. This had not been uncommon prior to Frederick William I, but after 1712 the officer corps evolved into a gathering ground for the sons of the aristocracy. The middle-class presence was minimized and generally limited to the artillery. Frederick the Great had expanded the role of the middle class only by necessity, and when peace came, Frederick began to discharge the middle-class officers. Many of the talented officers who had answered the call to arms found themselves unceremoniously returned to their civilian occupations. Lucky middle-class officers were retained in second-rate duty assignments like fortress or artillery commands, but most were simply fired. In case after case, cashiered officers would appeal to the king, only to be sarcastically rebuffed. Even when officers pointed to their sacrifices or wounds, the king remained unmoved. Frederick completed the work his father started. The officer corps became the private domain of the nobles.[191]

Whatever motivated Frederick's actions, he placed the future of the Hohenzollerns in the hands of the nobility, and he had no qualms about it. If a noble officer failed in his duty, the king reasoned, he had no other livelihood. His failure was not only a personal disgrace, it ruined his reputation and destroyed his social status. Junkers were motivated to succeed, the king said, because suicide was their only recourse after failure. It was not so with the middle class. If a middle-class officer failed the test of honor, his education and training allowed him to pursue a new career. The king claimed that he could not rely on the middle class because of this career mobility—a difficult argument to understand in twentieth-century America! Although the middle-class officers had marched across the bloody fields of Frederick's wars, the king had little to do with them after 1763. With a few exceptions, the only officers Frederick wanted in the post-war army were men with *von* in front of their last names. Nobility linked the officer corps to the royal house.[192]

For all the proper pedigree of the Junkers, Frederick lost valuable combat experience by dismissing his middle-class officers. Young Junker offi-

[191]Michael Howard, *War in European History* (New York: Oxford University Press, 1987), 69.

[192]William O. Shanahan, *Prussian Military Reforms: 1786–1813* (New York: AMS Press, 1966; orig. 1945), 29–30.

cers learned from their noble lords, and most of the lessons were outdated. When the old generation left the service, the new generation of commanders accepted obsolete theories of war as holy gospel, resulting in several mistaken notions. For example, by 1790 many regimental and general officers believed that artillery was a necessary evil, not a decisive weapon. All battles, they believed, would be decided by Prussian infantry volleys. Since it was cheaper for regiments to practice individually than collectively, the highest ranking generals in Prussia had little experience in commanding more than a couple thousand troops. Schooled in such a system, the officers became bureaucratically rigid. The best measure of an officer became his ability to follow the rules.

The rapid decline of the cavalry was symbolic of the deterioration of the entire army. While Seydlitz lived, the cavalry remained razor sharp. As the commander of all Prussian cavalry, he brought not only a fighting spirit to each unit but also practical experience and unyielding training demands. Seydlitz knew that horses could not fight unless they were physically conditioned. His troopers, therefore, were required to ride continually. They mounted long-range patrols, picket duties, and simulated mass attacks each month. Yet, Seydlitz could not live forever, and at age fifty-two his hard-charging lifestyle got the better of him. Already debilitated by syphilis, Seydlitz suffered a stroke in August 1773. He died on November 8, signaling the end of an era. In an army characterized by growing command rigidity, the cavalry went into serious decline as money-conscious commanders curtailed Seydlitz's costly measures. The cavalry gradually slipped into mediocrity, and it would never again rise to the stature it had achieved under the devil-may-care leadership of Friedrich Wilhelm von Seydlitz.

The one branch that could have weathered the storm was the artillery. Since the use of cannons required technical expertise, artillery became the arm of the middle class. The son of a merchant with a university education could find employment with the guns, but Prussian artillery failed to live up to its potential. The reason was simple. Frederick believed that artillery, while indispensable, was a nuisance. It slowed the movement of troops, it was cumbersome, and he could not envision it as an offensive weapon. Frederick did not understand that he had nearly revolutionized the use of artillery at Zorndorf and Burkersdorf by creating batteries to support infantry attacks. Therefore, Frederick never fully developed the artillery, and he tolerated the presence of both the guns and the university-educated officers who fired them as damnable necessities. This attitude did not lead to tactical innovation.

Partitions and Potatoes

In 1772 Frederick joined Austria and Russia in the first partition of Poland. The patriotic Poles were in no position to resist the onslaught of three major European powers, and they were forced to acquiesce. While fervent revolution boiled beneath the surface, the Prussians attempted to ignore the problem by incorporating the Polish territory into their country and deeming the Polish serfs to be "Prussian." Frederick also used the occasion to raise five new regiments of Polish soldiers for his army. Although most of the officers could neither read nor speak German, they were placed in the Prussian order of battle. Frederick had used such tactics against Saxony with less-than-successful results, and the policy was illogical because Frederick's distrust of Prussian officers with Polish-sounding names was notorious. Apparently, however, the king could comprise principles to expand Prussia's borders.

Frederick would have been content to spend his remaining days in peace, but the situation in Germany would not allow that. In the early 1770s Frederick confided to Prince Heinrich that when the Bavarian elector died, Prussia would soon be forced to go to war. There was no Bavarian heir to the throne, and both Frederick and his brother feared that the Austrians would attempt to annex the mountain state at the first opportunity. Their fears were realized in 1777 when Austrian Archduke Joseph II, who was reigning with his mother, Maria Theresa, began making overtures toward Bavaria. Maria Theresa and Joseph believed that the aging Frederick was far too old and ill to go war over Bavaria, and they could usurp the throne virtually unnoticed. They were wrong. Frederick not only prepared for war, but he began making the diplomatic alliances necessary to ensure his success. In the Silesian wars, Frederick stood alone, but the passing years had changed political temperaments. Neither the European powers nor the German states wanted to see Austrian expansion in Germany, and Frederick took advantage of their attitude. As the spring of 1778 drew round, European and German opinions swung decidedly to Prussia.[193]

---

[193]Mitford, 274.

If the German states agreed with Frederick's attempt to limit Austrian expansion, the Prussian people did not. The blood-letting business of the Silesian wars had not faded from memory. As Frederick gathered his army for battle, he found that some of his subjects were reluctant, including Prince Heinrich and Ferdinand of Brunswick. Frederick would not be put off. Mobilizing over 150,000 soldiers, he added 20,000 Saxon allies to his ranks. In April Frederick moved into southern Silesia with more than 80,000 men. Heinrich led another contingent of similar size up the Elbe for Bohemia. It appeared as if the master had returned to the game.[194]

The War of the Bavarian Succession, however, was far from a masterful display by the Prussian army. Heinrich began the campaign by moving from Saxony into Bohemia in July. After initial localized Prussian successes, the Austrians fell back to defensive positions in Königgrätz. The extended supply lines revealed the weakness of the Prussian army; its logistical system was antiquated. As Heinrich tried desperately to supply his army, Frederick fared little better on the other side of the theater. He moved through the Silesian passes into Bohemia and found that he, like Heinrich, had to stop for lack of supplies. Although the road to Vienna lay open, Frederick had to retreat to Heinrich's position to feed his men. As a result, the Prussians and the Austrians spent the summer fighting small battles around Königgrätz.[195]

The Austrians gained an upper hand in around Königgrätz. The Prussians, as weak as ever in the "small war," suffered greatly, and their morale dropped as low as their dwindling supplies. Common soldiers were forced to cope with the fiasco by digging potatoes from local gardens, and they began to refer to the campaign as the "Potato War." Without the spirit of Seydlitz, the cavalry performed miserably. Horses, also suffering from short rations, became too weak either to harass the enemy or pull their loads. When winter came, the Austrians stepped up local actions. Unable either to force the Austrians to battle or to maintain his position, Frederick sought a solution in diplomacy. In November 1778, Catherine of Russia offered to mediate the dispute, and Frederick quickly accepted. When Frederick in-

---

[194]See summary by Christopher Duffy, *Frederick the Great: A Military Life* (London: Routledge, 1988), 267–78.

[195]Alan Palmer, *Frederick the Great* (London: Weidenfeld and Nicolson, 1974), 200 ff.

formed Vienna that he had enough votes to move the seat of the Holy Roman Empire from Vienna to Berlin, Joseph accepted Catherine's mediation. Prussia and Austria signed a peace treaty in May 1789.[196]

The War of the Bavarian Succession was indicative of the decline of the Prussian army. Although it was still impressive on paper, the army was not as powerful as it had been in the Silesian wars. The Prussians did not fight a single major engagement, but they lost 40,000 troops through small actions, sickness, disease, starvation, and desertion. Frederick used the army his father had built, but he did not leave the army in the condition he found it. Although not quite a paper tiger, the Prussian army could no longer run roughshod over the armies of Europe.

The Decline in Hohenzollern Leadership

From 1640 to 1918, the history of the Prussian state and of the army was essentially bound up in the destiny of the Hohenzollern family. The Great Elector had served the state well and founded the nucleus of both the future kingdom and the army. Frederick I was no military commander, but he allowed the Old Dessauer and others to maintain the army's standards. When Frederick William I came to the throne, his entire reign focused on the structure of the army, and Frederick II became "the Great" due to the inheritance from his father. Frederick's talents came in the tactical employment of troops, however, not building the army. During the last half of his reign the army underwent a constant decline, and this was complemented not only by Frederick the Great's apathy but by a sudden dissolution of Hohenzollern leadership.

Rumors of Frederick's sexual orientation flourished from his own time until the present day. He married the woman his father chose but maintained separate quarters. This was not uncommon among European royalty, but Frederick had no known mistresses and rarely entertained single female guests. His discouraged all his young male attendants from seeking female companionship, adding to the gossip around the court. The question of Frederick's sexual preference has no military importance, except for one point. For some reason, Frederick was unable to father a child, and this opened the throne to his nephew, Frederick William, a myopic military incompetent.

---

[196]Mitford, 276.

The new prince assumed the duties of king in 1786, after the Great Frederick died in the early hours of August 17.

The Prussian government worked under Frederick because he had both the physical stamina to endure endless hours of detailed work and because he had concentrated all power in the royal office. Ministers in Frederick's government were assistants. He made every crucial decision, and he needed no chief of staff for the army because he did all his own staff work. On the night before Hohenfriedberg, for example, it was Frederick, not his generals, who plotted the movement and drew up the orders for each battle unit in the field army. As long as Frederick remained on the throne—in good health—this system worked. When his nephew Frederick William II assumed power, things began to take a turn for the worse. Frederick William II was a sensual king. During his short reign, his most notable accomplishment was to maintain twenty mistresses in court at one time. He enjoyed being king and threatened anyone, including the Königsberg philosopher Immanuel Kant, who even hinted criticism of royal absolutism. In military affairs, he was hardly the man to bring the army back up to standards. While the king gave in to every desire, the army grew older and more narrow-minded as each day passed.

Gordon Craig pointed out that mature leadership in the Prussian army was not necessarily a bad thing. Older generals served the Prussians well—Blücher and Moltke, for example, were both at peak form in their seventies—and age alone did not inhibit the continued performance of many officers. The problem with the Prussian army in the early 1790s was not so much the issue of age as it was the attitude of age. Prussian leadership became inflexible. Generals who had cut their teeth on the Silesian wars were not capable of grasping new concepts.[197] The age problem was coupled by the inability of the rank-and-file to change. As the infantry locked itself into the tactics of the line and the cavalry continued to decline, the administration of the army did little to reverse the process. In fact, the status quo was encouraged. Innovation was not an attribute of the Prussian army, and Frederick William II was not the man to suggest that it should be.

---

[197]Gordon A. Craig, *The Politics of the Prussian Army: 1640–1945* (London: Oxford University Press, 1968), 26.

The Beginning of the French Revolution

While the Prussian army maintained its garrison mentality, the situation in France changed rapidly. Unlike Prussia, France became a hotbed of political activity for several reasons. First, the economic structure of mercantile nations—trading nations, such as France—had given way to capitalism. As a result a great deal of wealth transferred from the aristocracy to the middle class. Second, people began moving from the countryside to the cities, and the urban populace brought new ideas that challenged conventional social structures. Third, the Enlightenment had questioned the old political structure. The works of the French *philosophes*—social and political philosophers who caused a revolution in political thought—called for a new form of government, a government that existed for the benefit of the people. Finally, the French government was bankrupt. The King of France, Louis XVI, wanted to ignore political unrest, but he could not. He needed his subjects to produce revenue for the kingdom.[198]

Louis's options were extremely limited, and it seemed that everyone was against him. His own aristocracy sought to increase their manorial rights at the expense of the king, while the middle class decried royal power and the peasants spoke of revolt. Louis charged his finance minister with the task of raising money, but the minister was not a magician. The only method he could use to get money involved calling the national legislative assembly, the Estates General. The Estates had not been in session since the age of Louis XIV, and members wanted to air political grievances rather than solve royal financial problems. Louis did not want to assemble the Es-

---

[198]E. J. Hobsbaum, *The Age of Revolution: 1789–1848* (New York: Mentor, 1962), 74 ff. For a sampling of literature on the French Revolution see William Doyle, *The Oxford History of the French Revolution* (New York: Oxford University Press, 1989) and D. M. G. Sutherland, *France 1789–1815: Revolution and Counter Revolution* (New York: Oxford University Press, 1985). Earlier works that provide excellent introductions include: John Hall Steward, *A Documentary Survey of the French Revolution* (New York: MacMillan, 1951) (this contains a series of documents from the revolution, which have been translated into English) and Frank A. Kafker and James M. Laux (eds.), *The French Revolution: Conflicting Interpretations* (New York: Random House, 1968). (Although it is dated, this work contains an anthology of historiography. It is worth reviewing.). If you are not familiar with the French Revolution, an excellent beginning is David L. Dowd, *The French Revolution* (New York: Harper and Row, 1965). (The text is easy to read, and the book is loaded with diagrams and colorful illustrations.)

tates, but he had no other way to finance the realm and he was broke. In 1789 King Louis reluctantly agreed to call the Estates General into session.[199]

The Estates General were officially divided into three groups—nobles, clergy, commoners—and when they assembled in the spring of 1789, the mood was hardly conducive to royal needs. While the king wanted to raise revenue, the Estates General, especially the Third Estate (the commoners), wanted to raise grievances. Louis tried to prevent this by mobilizing the French army, but the Estates responded by forming a national guard. By the end of the summer, Paris was in open revolt. The king tried to dissolve the Estates General, only to find himself confronting the new national guard. Louis tried to regain power in the fall, but things had gone too far. Within a matter of weeks the Estates General established their own government, and, as horrified royalty throughout Europe looked on, Louis was taken prisoner by a revolutionary government.[200]

None of these events escaped the eyes of Frederick William II. If the common people could rise up in France and replace the monarchy, there was nothing to prevent them from doing so in other countries. Afraid of anything that suggested democracy, Frederick William dispatched the Prussian army to Holland to keep the revolutionary spirit from sweeping into Amsterdam. The Dutch had no national guard, and the show of force by the Prussian army stemmed the tide of revolutionary fervor. Prussia's generals returned to Berlin confident that the rigid, uncompromising army was capable of meeting any threat. Europe's royalty agreed. If a show of force could work in Holland, it would work in France, and the royal houses of many nations, including Prussia, formed a coalition to restore Louis XVI. Unlike the Dutch, however, the French revolutionaries had their own army.

The Rhineland Campaign and the First Coalition

By 1790 the French Revolution was in full swing. Nobles were being threatened or killed in the streets, and the royal family itself came under attack from the revolutionary government. Louis's imprisonment was the final straw. Spurred by England, the reigning monarchs of Europe pooled their

---

[199]Doyle, 66–85.
[200]Ibid., 86–111.

interests in a diplomatic alliance called the First Coalition. The purposes of the alliance were to restore order in France and to stop the spread of revolution. This was not mere diplomatic rhetoric; the coalition planned a three-pronged invasion of France, hoping to reach Paris and restore Louis. Britain would come from the Low Countries, Austria from Alsace, and Prussia from the Rhineland. In 1792 the conservative Prussian army mobilized for war.[201]

Prussia's leaders came to the campaign with mixed emotions. Despite his military incompetence, Frederick William II understood that there was little for Prussia to gain in a war with revolutionary France. The army could be destroyed far way from home, and, even if it won, victory would only restore the fortunes of the Bourbon family. By the same token, if the Prussian middle class revolted while the army was deep in France, Frederick William would have no defense for his own throne. A French war promised too little and risked too much. Prussia's senior generals not only reflected Frederick William's fear, they believed Louis XIV was their main enemy. They had no desire to have Prussian forces reduced for the sake of restoring a monarch they might have to fight. Duke Karl of Brunswick, the commander selected to lead the campaign, cautioned against the war. Only the younger officers, bored by years of garrison duty, were eager for the prospect of action. Having never experienced war, they did not know any better.

The Austrians and the English were not as quick to adopt a cautious attitude. Without coordinating their efforts, the British and their German allies moved into Belgium in the summer of 1792, while the Austrians prepared to cross into southwest France. Brunswick, very reluctantly, began moving against French holdings in the Rhine but refused to subordinate either his army or his efforts to the needs of the allies. When Brunswick encountered French resistance, he pulled the Prussian army back to the German side of the Rhine. The Austrians, irritated by Brunswick's refusal to serve under the Hapsburg flag, decided to call off their planned invasion and sit on the Prussian flank in Alsace-Lorraine. The Anglo-German forces in Belgium, isolated in Belgium, fought on alone.

---

[201]Curt Jany, *Geschichte der königlich Preussischen Armee bis zum Jahre 1807, Volume III* (Berlin: Verlag von Karl Siegismund, 1929), 259 ff. For a brief summary of the campaign in English see Peter Paret, *Clausewitz and the State* (London: Oxford University Press, 1976), 24–33.

The French national guard and the Prussian army were a statement of contrast.[202] The French army of 1792 was highly disorganized, poorly led, and hardly efficient. French forces were enthusiastic but inexperienced, and they could not take advantage of the squabbles in the allied camp. By contrast, the Prussian army appeared to be straight from the Seven Years' War. Its formations were linear, its prime strength was the infantry, and the infantry was defined by its rate of musket fire. The artillery and cavalry played supportive roles to the infantry volley. At first, the conservative Prussian officers appeared to be correct. When undisciplined French formations faced high volumes of fire erupting from Prussian lines, they scattered. French retreats, however, were only a surface issue. Realizing their inability to fight from the line, the French began developing new tactics, not from a desire to be innovative but to compensate for their own military weaknesses. As a result, the French gradually began fighting from very strange formations—a development that went unnoticed by the Prussian officer corps.[203]

Despite their tactical weaknesses, the French achieved success in the summer of 1792. After driving the allies from French soil, the revolutionary army crossed the Rhine and engaged Brunswick. Although the basic doctrine of the French army remained rooted in the line, French generals began to deploy light infantry in irregular formations. The Prussians frequently turned long lines to counter the strange-looking units, only to watch them scatter. Whenever the Prussians advanced, light infantrymen reformed to pick them off. It was frustrating but not definitive. When the Prussians were able to bring the French to a full linear battle, the French offensive stopped in place. In December Brunswick gathered reserves and prepared for a counter strike. By January 1793 the French were retreating.[204]

Brunswick was not inclined to open a full winter offensive against the French. Content with local offensives, he kept the French off balance and forced them to the west. He managed to lay siege to occupied German cities along the way, and by the end of July the Prussians regained the Rhineland. The allies talked about conducting another offensive from the north, but

---

[202]See Peter Paret, *Yorck and the Era of Prussian Reform* (Princeton, N.J.: Princeton University Press, 1966), 63 ff.

[203]Basil H. Liddell Hart, *Strategy* (New York: Praeger, 1967), 113.

[204]Alan Forest, *Soldiers of the French Revolution* (Durham, N.C.: Duke University Press, 1990), 40–43.

Austria balked. Instead of supporting the scheme, the Austrians began a new offensive in Alsace, and Brunswick, with only 45,000 men under his command, marched south to assist them. He spent the next few months marching in a forty-mile front along the French border, and although he won victories over the French at Pirmasens and Kaiserslautern, he allowed the French to retire in good order. There was an important undertone in these battles, however. Unable to form battle lines in a timely manner, the French fought from columns—the marching formation of *infantry* units. The French were on the verge of creating a new style of warfare.

By the end of 1793 Frederick William tired of the Rhineland fighting, and his attention focused on the east where Austria and Russia spoke of a second partition of Poland. He sent Field Marshal Wichard von Mollendorf to replace Brunswick in January 1794, hoping the new commander could end the fighting. Mollendorf invaded France in the spring, intent on going to Paris, but the French were unaffected. Abandoning conventional military logic, hordes of French infantry began to attack Mollendorf from positions so weak that no respectable eighteenth-century commander would have considered them. Shocked by French tenacity, Mollendorf withdrew to the Rhine to avoid risking Prussian casualties. By autumn, the Austrians were defeated in Alsace, and the English offensive in the north was completely out of steam. Frederick William II opted to end the war. Turning his attention completely to Poland, he signed a peace treaty with the French early in 1795. Although the Rhineland campaign did not produce a strategic success, the king declared victory. It was not.[205]

The Political Problems of the Prussian Army

The last half of the eighteenth century was a period of transition for the Prussians as much as it was for the French. Commanders took on a garrison mentality, and routine operations dominated military affairs. Two events would have alerted more able sovereigns, but not Frederick William II. After observing the War of the Bavarian Succession, Frederick William took no action to improve the quality of the army when he became king. Simply accepting losses in the "small war," he applied the same logic to the debacle

---

[205]Palmer, 212.

of the Rhineland campaign. The Prussian army was outmoded, and the king did not seem to care.

With the aid of hindsight, it is possible to see several weaknesses in the Prussian system of war. Prussia suffered from a shortage of officers. Since Frederick William maintained the ban on employing middle-class officers, the army turned to foreign recruiting. This was not altogether harmful—in fact, it brought some outstanding men to the Prussian officer corps—but it was indicative of the paucity of local talent among the Junkers. In addition, the officer corps was not an educated group. With a few notable exceptions, Prussia's officers were only schooled in drill and discipline. The Junkers had no appreciation of the political potential of war, and they scoffed at university education.

Another weakness of the army was the command structure. When the sensual Frederick William II died in 1797, he was replaced by his moralist son, Frederick William III. The new king inherited five separate military command structures: the Military Department of the General Directory, the Governors of the Garrisons, the General Inspectors, the General War Council, and the Adjutant General. Rather than cooperating with one another, these bureaucracies fought with each other for power. Like his father, Frederick William III did nothing to address this situation. He patiently listened to the director of each bureau and tried to keep peace among his generals.[206]

Gordon Craig points out that the General War Council had been created in 1787 to unite all military command in a single structure. The process failed. By the time Frederick William III assumed power, the General War Council was just another bureaucratic entity in the struggle for power. When the king convened the army for maneuvers, he was supposed to command all functions from the Military Department. Instead, he found himself perpetually arguing with members of the General War Council and the Adjutant General. In addition, local political figures maintained garrisons for economic benefits. They were not about to relinquish power to a central authority. The General Inspectors completed the picture of confusion. They were responsible for recruiting and training, yet they were completely independent of any centralized form of command.[207]

---

[206]Shanahan, 25 ff.
[207]Craig, 29.

Tactical Problems

The political confusion in the Prussian army was complicated by a host of tactical weaknesses. The Rhineland campaign had failed to demonstrate the basic problem of the army, the tactical inflexibility of the infantry. The bulk of the men were concentrated in musketeer regiments. Firing smoothbore muskets, the musketeers were responsible for delivering the deadly Prussian volleys. At 100 yards, 30 percent of their shots hit the target, but they made up for inaccuracy by rapid reloading techniques. In practice they could deliver six volleys a minute, and even though they could only fire about half this number in combat, they remained the fastest guns in Europe. Their role had not changed greatly since the innovations of the Old Dessauer.[208]

The cavalry had become exceptionally lax. On paper, cavalry forces were strong. Still divided into three separate units, cuirassiers, dragoons, and hussars, the cavalry added a new type of light cavalry composed primarily of Poles. Called Uhlans, these units carried lances in the tradition of Frederick the Great's Bosniaks. Adding light cavalry, however, could not make up for the deficiencies in training. In the days of Seydlitz, troopers lived on their horses. By contrast, in 1800 many Prussian cavalrymen were spending less than two weeks a year in the saddle. Riding the horses under combat conditions cost both money and horse flesh. Commanders ordered cavalrymen to keep their horses rested and "fresh." In a bureaucratic army, the conservation of resources makes fiscal sense.

If conditions were bad in the cavalry, they were absolutely horrible in the artillery. Officers saw the need for cannons, but they felt that the guns had only a limited role. Whether in siege operations or field battles, most Prussian officers saw artillery as a supportive branch. As a result, gunners were improperly or inadequately trained. Although the army had about 5,000 guns, cannons were parceled out to various units, and there were no provisions for coordinating artillery fire in battle. Most artillery officers spent their time abandoned in fortress duty. Things were different west of the Rhine. Napoleon Bonaparte, who steadily gained promotions in the army of revolutionary France, envisioned another role for artillery. While

---

[208]Shanahan, 19.

the Prussians intended to shoot muskets, Napoleon decided to volley with cannon.

Chapter 9

The Revolution in Warfare

The real meaning of the tactical victories and the strategic defeat in the Rhineland campaign was lost on the Prussian army. In the 1790s the methods for waging war underwent change as profound as the political changes of the French Revolution. The limited battlefields of Frederick the Great expanded into the arena of total war. Instead of relying on small select armies of professional soldiers, governments began to turn their citizens into soldiers. Although the Industrial Revolution was in its infancy, machine production became the basis for arming large armies of citizen soldiers. It is necessary to examine briefly this shift in military activities before continuing with the history of the Prussian army. The Prussians would be dragged kicking and screaming into the new age, but they would go. They would eventually become the leading authorities of industrial warfare.

The Influence of Early Industry

The full impact of industrialization did not culminate in the early nineteenth century, and its influence on warfare was not completely developed until the twentieth century. Still, changes in factory production transformed the armies of the French Revolution. Even though the factories of the late 1700s were modest by later standards, they created the conditions for mass production. Weapons, uniforms, and ammunition could be produced by the fledgling factories at an unprecedented rate. Production became the key to arming the citizen armies. The work of American inventor Eli Whitney symbolized the importance of the factory. While developing methods to produce standardized equipment, Whitney hit upon the idea of interchangeable parts. Whitney believed that if a mechanical part could be made to precise standards, it should fit on any item designed to accept the part. For example, a bolt produced to fit a specific type of screw would fit any screw matching the type no matter where it was used. Whitney's concept worked, and he searched for a piece of precision equipment to popularize the procedure. The mechanical instrument he chose was the military musket.[209]

Albeit a precursor to modern mass production, the concept of producing large numbers of standardized weapons completed the seventeenth-century reforms of Gustavus Adolphus. The Swedish king standardized the size and function of artillery in the Thirty Years' War, but he lacked the industrial base to implement complete uniformity. The military reformers of the late eighteenth century were given a tool Gustavus could not possess. The newborn factory system allowed them to order equipment made within interchangeable tolerances. Flintlocks, barrels, and wheels created to specific standards could be produced at a rate heretofore unknown. Cannon were cast from exact molds, and even the size of buttons and button holes could be standardized. When weapons malfunctioned, the weapon was no longer rendered useless. The dysfunctional part could simply be replaced. As a result, the importance of uniformity and conformity was emphasized in the citizen armies. Colorful individuality was subjugated to the collective will of military force in most instances. By the end of the Industrial Revolution, the best type of soldier would be the coldly rational bureaucrat.[210]

While weapons experts were considering the implications of increased production, chemists began to develop tougher types of metals. Again, the full impact would be felt in the next century, but chemical innovations improved the quality of weapons. The most important advance came in the area of steel. New techniques for refining iron ore produced stronger versions of steel, and this resulted in more effective, reliable cannon. The British, who were the early leaders in industrial developments, standardized their artillery pieces, shells, and firing mechanisms. In the Prussian Ruhr Valley a small forge began producing military armaments in 1782. A few years later the firm was acquired by the Krupp family, and it began produc-

---

[209]Lynn Montross, *War through the Ages* (New York: Harper and Brother Publishers, 1944), 444.

[210]J. F. C. Fuller, *The Conduct of War, 1789–1961* (Minerva Press, 1968), 42–59 and 77 ff.

ing solid shot for Prussia's artillery. Krupp steel would change the face of the Prussian fighting machine.[211]

The initial impact of industry can be overstated, because implementation of technological innovations on the battlefield was a slow process. Military forces tend to be extremely conservative, and when a new weapon is introduced—such as the tank or the aircraft carrier in the twentieth century—one or two generations of commanders usually complete their careers before the potential value of the weapon is realized. The same can be said of foundry-produced weapons of the late 1700s. To be sure, there were few commanders who understood the full implications of mass production. By the same token the importance of early industrial production cannot be overlooked. Without increased industrial production, the new large forces could not have been armed. The full effect of technology would not be felt until the mid-twentieth century, but early technology transformed the character of military force.

The Political Change in Military Force

If industrial production was in an embryonic stage during the wars of the French Revolution, the new political aspects of military force burst on the scene as a full-sized, terrifying adolescent. Technology changed war, but the ideological transformation responsible for creating the citizen army was an overnight development started on the streets of revolutionary Paris. The Great Elector and Frederick the Great had taken the field with small professional armies, and they had conducted operations on limited fronts. Frederick the Great once called his battles "cabinet wars." There was no need for the citizens to know, Frederick reasoned, when the country was at war. Only the king's cabinet needed to be concerned with battles and their outcomes. The French Revolution changed that. The French army increasingly became an army of the people. If the foundries could equip larger armies, the politics of France gave the citizens a reason to join the army.[212]

---

[211]William Manchester, *The Arms of Krupp* (Boston: Little, Brown, and Co., 1968). An older but highly readable history is Bernhard Menne, *Blood and Steel: The Rise of the House of Krupp* (New York: Lee Furman, Inc., 1938).

[212]Gordon A. Craig, *The Politics of the Prussian Army: 1640–1945* (London: Oxford University Press, 1968), 27.

As the revolutionaries of 1789 threw away the old regime in France, they swept away the structure and order of the French army as well. When England organized the First Coalition against France in 1792, the French were in a difficult position. Many royalist officers were in full sympathy with the invading European powers, and they wished to return the king to power. Not only did they fight against the revolutionary government, but they deprived the new French army of leadership. Faced with this problem, the government was forced to conscript citizens and to develop leaders from common soldiers. Any soldier of France had an opportunity to join the officer corps, if he could lead men into battle and win. French commanders were rewarded for developing innovative solutions to tactical problems.[213]

Many military historians have pointed to the genius of the French commanders for "inventing" new methods of fighting. French officers are praised for their foresight, their willingness to be innovative, and their ability to adapt to a radically new form of warfare. The reality of the situation may be somewhat less glamorous. No matter what the motivation, the conscripted battalions of the national guard could not fight like the conventional armies of Europe because they were not trained and disciplined in the same manner. It took too much time and training to teaching men to volley in musket lines, so the French began fighting from a marching formation, the column. While military analysts hail this tactic, they often fail to point out that it is immensely easier to teach troops to move, assemble, and fight from a column than from a line. Historians have praised the French for changing their tactics, but there may have been little else the French army could do. Faced with the inability to fight from a line, the French threw out the book on linear warfare. They needed some new ideas.[214]

The people who were conscripted into the national guard also differed from French soldiers in the past. First and foremost, they were citizens with full democratic rights. Even though most of the soldiers were conscripts, they "owned" both the army and the state. The professional pre-revolutionary soldiers were fighting for their king, and their function was to

---

[213]Alan Forrest, *Soldiers of the French Revolution* (Durham, N.C.: Duke University Press, 1990), 59–61, and D. M. G. Sutherland, *France 1789–1815: Revolution and Counter Revolution* (New York: Oxford University Press, 1986), 59–68.

[214]James Sheehan, *German History: 1770–1866* (New York: Oxford University Press, 1989), 228.

risk their lives for the royal house of France. Such service differed little from one European army to another, as professional soldiers transferred their services at will. Soldiers were not expected to serve any other purpose than the will of their sovereign, and foreigners could do the job as well as native sons. The revolution radically changed this method of soldiering. The revolutionary army was an army of political equals. All soldiers, from private to general, were subject to the laws of their state, a state that in theory existed to protect their rights. In 1792 many French citizens simply flocked to the national colors. Unlike any other army in Europe, French military service became an expression of nationalism. Even after Napoleon assumed the royal title of emperor, revolutionary French soldiers fought for their country.[215]

The social nature of soldiering also changed. The job of the French soldier was to serve the nation during a time of crisis, and soldiers expected to return home after a crisis had passed. Since their primary orientation was toward civilian life, revolutionary soldiers were not apt to obey the orders of superiors blindly. They needed to be convinced that their officers were actually qualified to lead and correct in their tactical assessments. In other words, French officers were expected to convince their troops to fight. To accomplish this, officers needed to gain the support of their men and see to their needs. This evolved into a sense of self-sufficiency for individual French units. Unlike the armies of the past, the new army comprised individual units capable of self-leadership. This radically improved the strategic capabilities of the French army.[216]

The Changing Face of Battle

The French army was more or less continuously engaged in battle from 1792 to 1800. Strategic doctrine emerged from this period as well as the genius of the Corsican general, Napoleon Bonaparte. When the Duke of Brunswick started his ill-fated venture toward Paris in 1792, the French were unable to meet him with traditional force. Their armies fought bravely,

---

[215]Forrest, 59–69. (Forrest demonstrates the reality behind the romanticism. Despite its revolutionary reforms, the French army was subjected to the same discipline and morale issues as the Prussian army.)

[216]Ibid., 81–82.

but under no circumstances could the French hope to exchange volleys with the Prussians. Therefore, the French had to do something new. Over the course of many battles they came to three conclusions: (1) artillery fire could cover areas better than rapid musket volleys; (2) columns moved much more readily than lines; and (3) semi-autonomous corps were more mobile than large armies tied to one supply line.[217]

French military leadership gradually implemented tactical doctrines based on their experiences, developing an army of unparalleled mobility. This new-found flexibility also reduced the multitude of orders necessary to fight from a line, and individuals units could be sent in the direction of the enemy with one general order: engage. If they were to do this, each unit needed to be self-sufficient, meaning that the unit had to contain enough infantry, cavalry, artillery, and supplies to fight on its own until other units arrived. The need for self-contained units gave birth to a new tactical formation, the corps, and each corps contained branches from all arms of service. The famous phrase "march to the sound of the guns" became the tactical guide for the self-sufficient, semi-autonomous corps of the French army, and mobility became the key to a growing number of French successes.[218]

The corps was also the organizational backbone of the new army, and its presence made it distinct from the armies of the past. Late in his reign, Frederick the Great had commanded relatively massive forces in the War of the Bavarian Succession. The king had suffered quite a few casualties due to the cumbersome nature of moving vast amounts of troops along one or two axes. The French concept of maneuver changed this situation. Since they relied on many small units to encounter and engage the enemy, the French could move along any number of routes. They had two basic requirements. First, the army needed a road system that would support multiple movement, and, second, each corps had to be close enough to its partners to give

---

[217]See Hans Delbrück, *History of the Art of War, Volume IV, The Modern Era* (trans. by Walter J. Renfroe, Jr.) (Westport, Conn.: Greenwood Press, 1985), 395–414. (This work had a strong influence on German militarism.)

[218]Theodore Ropp, *War in the Modern World* (Durham, N.C.: Duke University Press, 1959), 81.

support. The corps were actually small armies. They separated for movement and joined together for battle.[219]

As the French continued to struggle against the linear structures of their enemies, they eventually had to deal with the problem of the infantry. Despite their superior organization and the emphasis on artillery, cannon alone could not destroy an enemy army. The time still came when infantry had to meet infantry. The French infantry was not superior to that of other countries, and their discipline was not as good. It took quite some time for the revolutionary soldiers to master the techniques of infantry fighting, and while they were learning, they developed new methods. French infantrymen began to leave their large formations to fire at enemy lines from a safer distance. Many of these troops began using rifles instead of muskets, resulting in more accurate fire. The outcome was that the French began to emphasize a new type of soldier, the light infantryman. Light infantrymen, fighting from irregular formations, could do much damage to an enemy line.[220]

French infantry units also modified the method for attacking enemy lines. In Frederick's era, the purpose of battle was to bring infantry to bear against infantry. If the enemy line could be properly decimated, cavalry could be used to force the enemy from the field. In some circumstances, such as Seydlitz at Rossbach, cavalry could actually be used to sweep a broken enemy away. The irregular formations of the national guard had been unable to fight in this manner. Rather than engaging the enemy in the standard line-to-line struggle, the French eventually found that they could move their enemies back by throwing troops at a single weak point in the enemy line. In other words, it was not necessary to engage the whole line; it was only necessary to break the line at one spot and to exploit the break by continued attacks. Two new methods were devised for such exploitation.

The first method involved utilizing the infantry column as an offensive formation. In the early stages of the revolutionary wars, untrained French battalions of 500 to 800 men would line up ten to fifty men abreast and walk toward the enemy with fixed bayonets. This formation—a ready solution to the problem of training—could withstand repeated volleys, and, better yet, it was virtually impervious to cavalry. As with the ancient Spar-

---

[219]Jac Weller, *Weapons and Tactics: Hastings to Berlin* (New York: St. Martin's Press, 1966), 70.

[220]Ropp, 82.

tan phalanx, the sight of hundreds of men moving toward a single point on an enemy line could also strike terror in the hearts of the defenders. French commanders began to perfect the tactic until the column attack became the standard trade mark of the revolutionary army.[221]

The second method of exploitation involved the cavalry. While each corps had enough cavalry troopers to scout and defend against enemy cavalry, the French soon found that cavalry could be used for strategic exploitation. Cavalry units could not only exploit a local situation, but they could pour through the battlefield to eliminate the logistical support of a field army. No general understood this better than Napoleon, and he refined the concept to an art. If the purpose of massed infantry and artillery fire was to create a gap in the enemy line, the purpose of the cavalry was to swarm through the line and attack everything that might be used to support the enemy army. By combining all the new styles of fighting into a single system, Napoleon perfected the military effectiveness of the revolutionary army.[222]

The Napoleonic System of War

Unlike the generals of the professional European armies, Napoleon's generals came from all classes. He led an army of citizens that had but one objective in mind: destroy the enemy's ability to make war. Battles were not individual exercises, their purpose was to destroy a nation's military system, and Napoleon's wars represented the concept of total war. The democratic élan of the French army reflected of popular will. Even when Napoleon abandoned all pretext of democracy by crowning himself emperor, enthusiasm did not leave the country or the army. Napoleon's army believed in its country, its leaders, and itself.[223]

Napoleon's wars utilized popular armies, and conscription gave him forces of unprecedented size. By utilizing the corps structure, Napoleon found an effective method for moving large numbers of troops. This had two significant results. First, corps commanders were free to operate independently, but they remained subject to Napoleon's overall direction. Napoleon was responsible for moving the corps to designated areas, and corps

---

[221]Weller, 72.
[222]Fuller, 50.
[223]Delbrück, vol. IV, 421–39.

commanders were obligated to employ their individual units in battle. Napoleon's real genius came in his managerial ability to coordinate the movement of troops over vast areas and bring them together at a crucial point. Commanders were free for local movement, but they were always subject to Napoleon's strategic plan. Local successes were immaterial unless they supported the central objective.[224]

The second major change involved the rapid physical expansion of the battle front. Frederick the Great's battles were rarely fought on a front that extended more than four miles. He maneuvered along narrow lines, partially to defend against enemy forces and partially to stop deserters. Napoleon's advances were a different affair. The French army marched along a variety of mutually supporting axes, and the front could extend to a width of fifty miles. Maps of Napoleonic advances look as though they are a net thrown over a very sizable patch of ground. As the troops advanced, they fought their way to the main battle over the course of many miles. Under Napoleon's system of war, the limited battle front became a relic of the past.

Napoleon believed that artillery was the decisive arm of combat. Accordingly, he modified the structure of French guns. He increased their mobility by lightening the caissons, and many artillery batteries could move nearly as fast as the cavalry. Although his field artillery usually remained stationary during a battle, he could move it into position rapidly, concentrating fire after enemy infantry had deployed. French artillerymen practiced firing and movement as much as the Prussians drilled in the volley. Napoleon also concentrated his artillery in grand batteries, and it soon developed the ability to devastate an enemy line.

Napoleonic cavalry was colorful, dashing, and daring, and it was geared toward strategic exploitation. Napoleon understood that the cavalry was the least important of all the combat arms, and that under most circumstances it could not break the enemy infantry. But, cavalry—especially heavy cavalry—could be used to exploit a break in an enemy line. Napoleon was not satisfied simply to drive the enemy from a field; his major purpose was to destroy the enemy's army. He used heavy cavalry not only to attack broken formations but to seize their lines of retreat and the towns and fortresses where they could run for safety. Napoleon's cavalrymen became experts at

---

[224]Michael Howard, *War in European History* (New York: Oxford University Press, 1987), 76 ff.

isolating and destroying enemy units. Had Frederick the Great faced such tactics at Kunersdorf or Hochkirch, he would not have lived to fight another day.[225]

Napoleonic infantry columns complemented the whole affair. Columns of revolutionary soldiers came to view themselves as invincible, and—with the exception of the Iberian peninsula—they seemed to be so until 1812. Conscripted soldiers needed little training, and they could be easily fitted into the infantry column. Basically, a soldier simply needed to follow the person in front of him. Very few tactics were effective against Napoleon's infantry. British and German soldiers under Arthur Wellesley, the Duke of Wellington, managed to adopt defensive tactics that could stop the French infantry columns, but the French infantry usually ruled the day against other generals.[226] France had truly become a nation-in-arms, and Napoleon turned from defensive struggles against European monarchies to offensive conquests of foreign governments. If Prussia were to resist Napoleonic France, the army would need to revise radically its methods for making war. Unfortunately for the Prussian army, only a handful of officers understood this.

---

[225]John R. Elting, *Swords Around a Throne: Napoleon's Grand Armee* (New York: The Free Press, 1988), 227–47.

[226]Weller, 71.

Chapter 10

Attempted Reform

Some officers in the Prussian army were keenly aware of France's revolutionary approach to warfare. Representing the intelligentsia of the Prussian army, these men banded together and vowed to drag the army into a new era. They had one major goal—they wanted to transform the Prussian army into a modern organization. They sought universal military service through short-term conscription, changes in strategic and tactical doctrine, and—most importantly—the creation of a general staff that would do nothing but design, move, and supply the army. Although few of the reformers were political liberals, most of them realized that democracy was the key to uniting Germany. Therefore, military reform went hand-in-hand with political reform. A united Germany could field an army large enough to fend off the French. Despite the popular appeal of the reformers, most military officers, including the older high-ranking commanders, were skeptical about the proposed reforms. A growing number of conservative officers formed organizations to counter the reformers, and the conservatives had the king's ear.

Scharnhorst Enters Prussian Service

Gerhard Johann David Scharnhorst (later "von" Scharnhorst) was born in Hanover in 1755, the year before Frederick the Great opened the Third Silesian War. His father, a sergeant in the Hanoverian army, owned a modest farm in the English province. The meager income from the farm could not support all of the children, so young Scharnhorst left his family in 1773 to seek a military appointment from King George of England, the Elector of Hanover. Scharnhorst entered the Hanoverian army at age eighteen, joining the cadet corps of Count Friedrich Wilhelm zu Schaumburg, an articulate and highly accomplished soldier. Scharnhorst immediately fell under the count's spell, and Schaumburg's social and military theories continued to influence Scharnhorst for the rest of his life.[227]

Count Schaumburg was an enlightened soldier in an enlightened age. Rejecting the conventional military wisdom of the day, the count sought to build soldiers by molding their characters. An officer's character, the count explained to his cadets, was far more important than noble birth. Character, not nobility, entitled a person to lead units in combat. No one, officer or noncommissioned soldier, possessed an inherent right to command troops; leadership grew from respect. Only the most qualified cadets were suitable for command rank, and qualification came with diligent study, hard work, and the willingness to inspire others. Command was the reward for ability, not nobility.[228] As the young commoner Scharnhorst absorbed the count's wisdom, no social philosophy could have pleased him more. Scharnhorst hoped to rise through the ranks on his ability and capacity for hard work. He eventually won a commission in the Hanoverian army and joined the instructional staff at Schaumburg's military academy.[229]

Scharnhorst quickly developed a reputation for thoroughness and competency. He never became an exciting instructor—in fact, many of his students believed that he was rather boring—but he mastered the material to perfection. Unlike many officers in the German states, Scharnhorst was fully informed about the winds of change blowing through the streets of Paris. He was no revolutionary, but he believed that the Germans must keep pace with the French. Modern wars, he taught, could not be fought under the organizational structure of the Seven Years' War. In the wake of Frederick the Great's "Potato War," young Scharnhorst taught his students that the nature of war had bypassed the illustrious leader of Prussia. War had grown too complicated for one person. In the past a single commander could mastermind a campaign or move an army as a monolithic entity, and Frederick the Great had excelled at both. By the late eighteenth century, however, Fre-

---

[227]See German-language Max Lehmann, *Scharnhorst* (2 vols.) (Leipzig: Verlag von S. Hirzel, 1887). For summary see Peter Paret, *Clausewitz and the State* (New York: Oxford University Press, 1976), 56 ff.

[228]C. T. Atkinson, *History of Germany: 1715–1815* (New York: Barnes and Noble, Inc., 1969; orig. 1908), 372.

[229]Charles Edward White, *The Enlightened Soldier: Scharnhorst and the Militärische Gesellschaft in Berlin, 1801–1805* (New York: Praeger, 1989), 3–5.

derick's leadership style was out of date. The size of individual units, the need for coordinated movement, and the sheer numbers of men demanded more organization than a single commander could provide. Scharnhorst believed that future forces would be commanded by leadership teams, and he preached this message from the pulpit of Schaumburg's military academy. Although he was never dynamic, his emphasis on meticulous planning earned the respect of his students.[230]

Scharnhorst, like Schaumburg, stressed the importance of character in leadership. Officers who exhibited thorough understanding of culture, society, and human behavior could inspire others on the battlefield. This was more than simple military romanticism; it reflected the spirit of the Enlightenment and the impact of social theories on military forces. Reaching beyond the classroom, Scharnhorst fostered debates that encouraged junior officers to espouse their ideas even when they flew in the face of conventional wisdom. Such actions did not endear him to either conservative elements in the nobility or the Hanoverian officer corps, but Scharnhorst was not deterred. The German states needed to unite for defense against France, and they desperately needed a new type of army. By applying the social principles of the Enlightenment, Scharnhorst believed Germany could produce its own nationalistic army. Such an army would require an officer corps capable of leading troops through merit and inspiration, officers who possessed "character."[231]

The Enlightenment fostered Scharnhorst's approach to war, and rational arguments against offensive warfare struck a sympathetic chord in Scharnhorst's mind. The French philosopher Rousseau wrote that war and military forces were evil necessities within social structures, and Scharnhorst took Rousseau's thesis to heart. If humans were to behave morally, then military force could be employed only for defensive purposes. It was possible to launch an offensive war, Scharnhorst believed, but only in defense of the homeland. In other words, even the offensive should be morally guided by

---

[230]Ibid., 19.

[231]For an analysis of character, or *Bildung*, in military affairs see James J. Sheehan, *German History: 1770–1866* (New York: Oxford University Press, 1989), 204–5. *Bildung* was the key to liberal reforms. See Frederick C. Beiser, *Enlightenment, Revolution, and Romanticism: The Genesis of Modern German Political Thought, 1790–1800* (Cambridge, Mass.: Harvard University Press, 1992), 23–24.

defensive principles. Wars to add territory or aggrandize the realm were ethically unacceptable. Scharnhorst was no democrat, but democracy played a role in his thinking. Democracy provided the incentive for citizens to join the army to defend their homeland. He believed that citizens with individual rights would fight to protect those rights. Democracy made military sense.[232]

Count Schaumburg encouraged such innovative thinking, and Scharnhorst soon became the most notable instructor in the count's military school. Scharnhorst was given a free hand in curriculum development, and Schaumburg's academy soon developed a training program structured around three areas: military history, science, and the liberal arts. Scharnhorst used military history and science as practical subjects. The arts and humanities were added to build "character" among the future officers. Classes were supplemented by practical sessions in military functions, and every aspect of school focused on the development of leadership skills. While nobles poked fun at the academic methods, Schaumburg nodded with approval, and Scharnhorst's reputation grew. Stated simply, Scharnhorst told the students that they could learn the methods for commanding troops through the study of military history and by applying history to practical field studies. He demonstrated that the profession of arms was not merely a craft; it was a comprehensive discipline that could be mastered through systematic study. By 1782 Scharnhorst's reputation won him a promotion and an assignment to the new Hanoverian artillery school.

Despite his growing reputation, Scharnhorst remained a commoner in an army dominated by nobles. He published several articles and a military journal, and officers all over the Germanys read his theories and opinions. Even though his written work enhanced his standing, conventional Junker officers—whether in Hanover, Saxony, Bavaria, or Prussia—were not ready to accept a commoner among the high ranks of command, no matter how far his reputation soared. In a peacetime army, Scharnhorst was relegated to the middle-class branch of service, the artillery, and, like Napoleon, Scharnhorst began to develop tactical theories based on the deployment of

---

[232]Gordon A. Craig, *The Politics of the Prussian Army: 1640–1945* (London: Oxford University Press, 1968), 39–40. (Scharnhorst was not a democrat, but he saw democracy as a military tool. Clausewitz was neutral. Gneisenau was a true democrat, completely in sympathy with Stein, Humbolt, and Boyen.)

cannon. Yet, Scharnhorst was at destiny's doorstep. When Great Britain formed the First Coalition against revolutionary France in 1792, Scharnhorst received a field command in the Anglo-Hanoverian army, and he gained combat experience in Belgium. His performance drew the admiration of one of Prussia's leading soldiers, Duke Karl of Brunswick. After the war the duke asked Frederick William III to lure Scharnhorst into Prussian service.[233]

Scharnhorst seemed a most unlikely candidate for the Prussian army. In addition to his class status, Scharnhorst reacted to war with utter disgust. After witnessing combat in Belgium, he found that he passionately hated military conflict. He was especially appalled by the random killing of wounded men and captives, and he vehemently complained to his superiors about the murder of hapless soldiers and their civilian supporters. Scharnhorst understood battle on an intellectual level, but when practically confronted with wanton desolation, he absolutely hated it. Only his fear of the French allowed him to accept war as an evil necessity. Ironically, this made the Hanoverian officer all the more attractive to the Prussians. Brunswick and others were not concerned about Scharnhorst's views on the morality of combat. They were attracted to his command ability and his characterization of France. Under Brunswick's urging, King Frederick William III offered Scharnhorst a commission as a major in 1797. Scharnhorst declined, accepting a promotion to lieutenant colonel in Hanover. By 1800, however, he was talking to the Prussians again. The Hanoverians were not willing to let a commoner move beyond the bounds of artillery command, while Frederick William was willing to offer more. When Scharnhorst asked for a lieutenant colonel's appointment in the Prussian artillery, a substantial salary increase, and ennoblement—that is, placing the Junker *von* in front his last name—Frederick William was happy to agree. In May 1801 Lieutenant Colonel "von" Scharnhorst entered the Prussian army.

The Military Club of Berlin

Scharnhorst was not an immediate success in Berlin. Upper-class officers viewed him with suspicion, and the mere mention of basic educational requirements for the officer corps brought ridicule in Junker circles. Uni-

---

[233]Atkinson, 502.

versities, books, and journals were something for the middle class, not the nobility. Still, within weeks Scharnhorst was drawing some of Prussia's keenest military minds into his orbit. He accepted an invitation to join a military reform society in Berlin—called the Military Club—and he soon became its leader. Members of the club routinely discussed methods for modernizing the army, training, and leadership.[234] They fell under the spell of Scharnhorst's thinking, and he was quickly elected club president. The club grew rapidly under his leadership, and it became the forum for reform in the Prussian army. He called meetings every Wednesday night at 5 P.M., asking officers to present papers on the changing nature of war.[235]

The officers who fell under Scharnhorst's influence understood the significance of France's popular army, and a number of papers acknowledged the need to change Prussian tactics to meet new French techniques. One of the main topics focused on modernizing the linear musket formations of the infantry. The French were moving away from the line in favor of more mobile, semi-autonomous infantry groups, and members of the Military Club urged the Prussian army to follow suit. Papers that argued for augmenting musket lines with bands of free-floating rifle battalions won Scharnhorst's endorsement. Since the Prussian infantry fought in three ranks, Scharnhorst argued that the third rank should be abandoned and transformed into free-roaming light infantry units. These units could skirmish with the enemy while protecting the main line from enemy light infantry. Such an idea, however, seemed preposterous to most of the Junker high command. The famed discipline of the Prussian army would need to be modified to develop light infantry units, and such a suggestion sent shudders through many older officers. They began to perceive that the young Hanoverian was not ideologically suited for the Prussian service. Scharnhorst had little time for their objections.

Scharnhorst's concern went far beyond the bounds of the manner and style of infantry formations. At issue was the rigid caste structure of the Prussian army. The old-style Junkers were part of a system that solidified positions by family rank and social prestige. Prussian noblemen believed that they held their commissions by right of birth and that their authority directly reflected the power of the crown. Technical expertise was hardly

---

[234]Sheehan, 230 ff.

[235]White, passim, has the best English-language analysis of the Military Club.

the measure of a commander, according to conventional Junker wisdom, and the mere thought of middle-class education was abhorrent. Although the Junkers would accept promotion by merit among their own kind, they were only willing to allow officers with the proper pedigree into the highest ranks. *They were born to command.* Scharnhorst took the opposite view. Given his influence of Schaumburg's military academy, Scharnhorst argued that officers could be *taught to command.* Light infantry formations simply represented the tip of the iceberg for Scharnhorst. He wanted to build an entire army where officers and enlisted men would be called to make decisions on their own. Many Prussian officers could not tolerate this.[236]

Undaunted, Scharnhorst forged ahead. Unlike the officers reacting to his ideas, Scharnhorst realized that the French Revolution had changed not only the tactics of war but the very structure of military forces. Freedom in command was not merely a tactical anomaly, it was a requirement for democratic armies. Citizen soldiers demanded competent officers and the ability to rise through merit. They were not held together by discipline alone but through their spirit and their dedication to the state. While Napoleon was certainly not the master of democratic command, his army was unique, and French soldiers believed in the rhetoric of the republic. Further, Napoleon held his commanders accountable to the letter of his orders, but he allowed them to lead self-contained battle groups. Scharnhorst understood this and wanted to import the system into the Prussian army. This was no Enlightenment ideal. Modern armies moved vast bodies of troops over diverse geographical regions, coordinating their arrival at a single point of attack, or *Schwerpunkt.* Scharnhorst knew that this type of battle required commanders who were relatively free to act and soldiers who were motivated to obey. Unfortunately for Scharnhorst, such a system not only threatened the command structure in Prussia, it also challenged the very social fabric of the class system.

Frederick William III feared French aggression, but he valued his Junkers. Although some of them flocked to Scharnhorst's Military Club, the senior commanders were more skeptical. They revered Frederick the Great and believed that no power on earth—even French light infantry and massed artillery—could stop the Prussian infantry line. The defeat of the First

---

[236]Karl Demeter, *The German Officer-Corps in Society and State: 1650–1945* (trans. by Angus Malcolm) (London: Weidenfeld and Nicolson, 1965), 4–5.

Coalition had not fazed them. The Prussians commanders believed they were beaten because Prussia's interests were not at stake. Should France invade Prussia, the conservative commanders stated, the circumstances would be different. Frederick William agreed with this conservative thinking, while he sought to appease Scharnhorst and the Military Club. In an effort to resolve the conflict between the ideological camps, the king took the worst action possible. Instead of implementing and improving a single system, he let the reformers and the conservatives carry on a debate that tore the army apart.[237]

The First Reforms

From 1801 to 1805 Scharnhorst became closely associated with a group of reformers who not only proposed substantial changes in military structure but sought to change the essence of Prussian society. Serfdom was to be abolished, citizens were to be given the right to vote, and constitutional government was to be proposed. Baron Karl vom und zum Stein, a minister in King Frederick William's cabinet, led the call for political reform, and Scharnhorst became his military ally. The association of military reform with political change brought Scharnhorst into even deeper confrontation with conservative royalists who began to view him as a French revolutionary. Scharnhorst realized, however, that the Prussian army could not be modernized without changing the political structures that supported it. Scharnhorst was a pragmatist, not a democrat. Political reform suited the needs of the military.[238]

Aside from his views on the utilization of light infantry, Scharnhorst sought several practical reforms to enhance the strength of the Prussian army. He took command of the War School, the *Kriegschule*, in Berlin and demanded that future general officers develop competencies far beyond those required for linear warfare. He encouraged officers to seek university educations and attempted to attract educated men to the profession of arms. Since he believed that war was a matter of organization, he advocated the creation of a general staff, a group of officers who were to organize the

---

[237]Ibid., 7.

[238]Franklin L. Ford, *Europe: 1780–1830* (London: Longmans, Green, and Co. Ltd., 1970), 217 ff.

army for the commanding general. Scharnhorst also believed that the army should be restructured into independent corps, like the French army, and that each corps commander should have his own general staff. Finally, Scharnhorst advocated universal military conscription for short-term service. Conscription could come, Scharnhorst believed, only with political reform because universal service would create a nationalistic army that owed its loyalty to the state, not the king.

The conservatives did not oppose every issue Scharnhorst raised. For example, proposals for light infantry formations were not altogether unacceptable. The occupation of Poland revealed serious weaknesses in the army's ability to deal with partisan warfare. In many instances Poles, inspired by tactics American guerrillas used against the British, proved to be a match for the Prussian garrisons. As the Polish partitions gave Prussia more territory, Polish partisans stepped up their resistance. The Prussian army found that the standard linear formations did little good against the hit and run tactics of the Polish peasantry, and even the most conservative commanders thinned the musket lines to arm soldiers with rifles. By 1787 the army had formalized the practice by creating the first fusilier (rifle) companies, and this practice increased during the War of the First Coalition. By 1798 most Prussian infantry companies had groups of sharp shooters who were trained to fight independently of the rank and file. Conservative Prussian officers did not have a problem with the concept of skirmishers or many of Scharnhorst's other suggestions. Their problem came with the political ideology behind military reform.

Papers presented in the Military Club went far beyond the nominal creation of fusiliers, but controversy about the use of rifle companies symbolized the army's political division. Scharnhorst knew that if light infantry units were to be successful, they needed the ability to operate independently from centralized command. That is, they needed the freedom to hit and run. Both political and military conservatives did not wish to give fusiliers such freedom to act because it implied that units would also have freedom not to act. Conservatives were perfectly willing to station light troops on their front and flanks, provided that the troops fell under their direct command. Individual riflemen, however, needed the autonomy to select their ground, their targets, and their areas of combat. They also needed the ability to retreat when threatened by cavalry or heavy formations. The conservatives asked, what would keep soldiers from running away? The response came in the form of political freedom. Soldiers would remain at their posts, the reformers argued, not only as a result of discipline but because they were loyal to the state, the state that guaranteed their political freedom. The argument polarized the reformers and the conservatives. Reformers embraced democracy not from a Jeffersonian ideal but because it increased military effectiveness. Conservatives rejected increased military effectiveness in the name of political stability. It was most ironic.[239]

Scharnhorst soon had his fill of the conservatives. As Stein fought political battles for governmental reform, Scharnhorst returned to his earliest profession, military education. Rather than to argue ad infinitum with the conservatives, Scharnhorst thought it far better to enhance the education of the officer corps, and he opened a number of new fronts. While supervising the weekly meetings of the Military Club, Scharnhorst continued to publish. Military journals flourished in Germany at this time, and Scharnhorst published two of the most popular. They became a base for advocating military reform and education among the officers corps. Ever the pragmatist, Scharnhorst also wrote a book entitled *The Officer's Handbook*. Designed to fit into a saddle bag, the *Handbook* contained practical information on the deployment of infantry and artillery units. Many Prussian officers carried Scharnhorst into the field, even when they disagreed with his politics. Scharnhorst also sent younger officers to school.

The *Kriegschule*, the War School, was one of Scharnhorst's greatest areas of influence. He presented the king with a proposal for creating a staff school inside the war college, and the king accepted. Scharnhorst himself taught in the *Kriegschule*, drawing some of the brightest minds in the army into his circle. Two of his prize students were Karl von Clausewitz and Karl von Grolman. Scharnhorst also discovered August von Gneisenau, a colonel who frequently espoused not only the reform agenda, but the political policies needed to support it. A former mercenary who had fought in Hessian service during the American War of Independence, Gneisenau returned to Germany as a confirmed liberal. Scharnhorst placed his friend at the head of the *Kriegschule*. As Clausewitz and Grolman rose to the top of their class, Gneisenau and Scharnhorst grew ever closer. Major Hermann von Boyen

---

[239]White, 28–50.

also joined the entourage, and the staff school became a breeding ground for reform-minded officers.[240]

Frederick William III was delighted with Scharnhorst's work at the *Kriegschule*. Scharnhorst presented the king with the top student essays after the first year of study—including a promising work by the young Clausewitz—and the king shared them with his command staff. On the surface, the command staff liked the results. On a deeper level, however, they also knew that the *Kriegschule* was merely a front for Scharnhorst's educational ideology. When Scharnhorst formally proposed a general staff system for the Prussian army, a staff system composed of graduates from the *Kriegschule*, issues came to a head. Conservatives saw this as an attempt to undermine Junker authority by placing decision-making power in the hands of university-styled officers. Field Marshal Wichard von Mollendorf and General Friedrich von Zastrow went to the king to express their concerns.[241]

The staff system was at the heart of Scharnhorst's concept of modern war. If organization was the key to victory, Scharnhorst believed that the immense number of soldiers required for modern war went beyond the leadership capabilities of one man. It must be admitted that he underestimated the abilities of Napoleon, but Bonaparte was the exception, not the rule. For most military operations, Scharnhorst believed that a competent staff was an absolute necessity. By the end of the eighteenth century, regiments were augmented by two new formations, divisions and corps. Divisions were made up of combined brigades (or groupings of regiments) and placed together in corps. The corps, as previously mentioned, were independent mini-armies designed to fight on their own until supported by other corps. Scharnhorst believed that neither an army nor a corps could be effectively organized without the aid of officers assigned to staff functions. Conservative commanders disagreed. They felt competent to lead without the assistance of Scharnhorst's "damned clerks."[242]

In essence, Scharnhorst was dividing managerial duties into line and staff functions. Although his system became a model for all Western armies—and most industrial concerns—in the twentieth century, he was

---

[240]Sheehan, 232.

[241]White, 87 ff.

[242]William O. Shanahan, *Prussian Military Reforms: 1786–1813* (New York: AMS Press, 1966; orig. 1945), 71–73.

ahead of his time. If the job of the line commander was to run the operations, staff officers were to assist him. Staff officers worked in obscurity, making sure that soldiers were assigned to companies, that regiments were formed, and that the army was fed and quartered. When a commander wanted to move, staff officers produced maps and plans, and they organized the army for battle. Although the king listened to Mollendorf and Zastrow, he also liked Scharnhorst's idea. By 1803 staff officers appeared in the Prussian ranks, and by 1805 each corps had its own chief of staff.

Implementation was not without problems. To appease conservatives, the king refused to form divisions, and each corps was composed of an unwieldy number of traditional regiments. Realizing the problem, the king changed his mind in 1805, allowing Scharnhorst to organize regiments into divisions, but organizational changes alone did not ensure success. Scharnhorst and the officers he had trained knew exactly what the staff was expected to accomplish, but most of the conservative generals had no idea what to do with them. Some generals simply ignored their staffs, while others gave staff officers menial tasks. Only a few commanders, like Gebhard von Blücher, put the officers to work. When Scharnhorst complained about the situation, the king backed his generals, allowing each commander to employ his staff as he saw fit. As a result, most staff officers were forced to idle their time away.

Moving the army was another matter. Scharnhorst realized that mobility was as important as organization, but the army could not move with maximum efficiency due to upper-class domination of the officer corps. Citizen soldiers could travel lightly and live off the land, but nobles were not expected to do so. Many Junkers were hardly willing to sacrifice their positions to live like common soldiers. It was not unusual, for example, for junior officers to have five to ten horses when they went into the field, and most nobles were loaded with personal items—materials that encumbered the army's mobility. This problem was exacerbated by the existence of massive artillery trains. By 1800 the Prussians had over 7,000 guns, and their philosophy was to take as many into the field as possible, parsing cannons to individual units rather than concentrating them as artillery batteries. Scharnhorst did his best to reduce the artillery trains and the officers' baggage. Even the conservatives saw merit in this argument, and they supported Scharnhorst's attempts to increase mobility.[243]

There was one point where the conservatives could be pushed no further. They would not transfer their loyalty from the crown to the state. Frederick the Great created a officer corps composed of Junkers, and the army's loyalty was pledged to the king. Scharnhorst and the reformers posed a different idea. The reformers wanted to create a nationalistic army composed of short-term soldiers serving the country in the time of crisis. Both officers and enlisted personnel were to swear allegiance to the state, not the king. And there was more. In order to match the numerical strength of France, the reformers wanted to unite all the German states in a single government. They spoke of a Germany under a single, parliamentary constitutional monarch. No conservative could tolerate this, even with the threat of a French invasion.[244]

No one championed the idea of a united Germany more than Stein. A descendant of the Teutonic royal knights, Stein entered the Prussian civil service under Frederick the Great, and he soon proved to be an able administrator. Hailing from Nassau, a German state near the lower Rhine, Stein worked as a mining administrator in Rhenish provinces known for their relative political freedom, and the experience shaped Stein's political views. Stein rose through the Prussian ranks, believing that democratic cooperation between government and citizens produced a strong state. When he entered the highest levels of service, Stein fought to give all Prussian citizens basic rights. As the threat from revolutionary France grew, Stein felt that Germany must unite to defend itself. He believed that democracy was the tool to foster unification, and the military reformers believed that universal conscription would lead to democracy.[245]

To modern American college students, conscription may seem an odd way to create a democracy, but it made sense in early nineteenth-century

---

[243]Ibid., 82–84.

[244]Golo Mann, *The History of Germany Since 1789* (trans. by Marian Jackson) (New York: Frederick A. Praeger, Publishers, 1968), 34 ff.

[245]J. R. Seeley, *Life and Times of Stein, or Germany and Prussia in the Napoleonic Age, Volume I: 1757–1807* (London: Cambridge University Press, 1878), 40–73. See comments by Friedrich Meinecke, *The Age of German Liberation: 1795–1815* (trans. by Peter Paret and Helmuth Fischer) (Berkeley, Calif.: University of California Press, 1977), 35 ff. See interesting comments by Meinecke, 39 ff.

Prussia. Conscription could be used to move peasants from the status of "royal subjects" to "full citizens," if their loyalty were pledged to the state. Conservatives saw the need to increase the strength of the army, and they were willing to endorse a variety of schemes to fill the ranks. In 1802 the Duke of Brunswick even asked the king to create a national militia, but the radical nature of this proposal did not go unnoticed. The French militia, the national guard, had been the instrument of revolution in Paris a decade earlier. Yet, the Prussians needed troops. Major Karl vom dem Knesebeck, a prized student of Scharnhorst, produced a plan for universal conscription in 1803. At first, both the reformers and the conservatives endorsed the concept, but they soon began to part ways. Knesebeck saw conscription in terms of the last years of the Seven Years' War. Peasants and the middle class were to be incorporated in the existing Prussian army. Scharnhorst saw things differently.[246]

Scharnhorst wanted a national guard modeled after the French. With Stein's support, he countered Knesebeck's proposal by asking the king to create the *Landwehr*, an organization of part-time soldiers that would be mobilized for national emergencies. This was a far cry from simply expanding the army. The *Landwehr* would be composed of soldiers who did not surrender their rights of citizenship while under arms. While Knesebeck and the conservatives sought to expand the accepted system, Scharnhorst envisioned an army where people would willingly flock to the colors. Frederick William decided to expand the army in the safest manner possible. He accepted Knesebeck's proposal, creating a heated rivalry between Scharnhorst and his former pupil. Conscription with liberal exemptions became the law of the land, and the conservative Junkers continued to dominate military affairs.

The reformers would not be silenced. While Stein continued to lobby for national political change, Scharnhorst used his position at the Military Club to continue the debate. The conservatives won the first round, but the reformers remained in the bully pulpit. Although their talk brought heated reactions from conservatives in the army, they managed to keep military reform on the national agenda. Frederick William tried to please both sides, refusing to silence the reformers while allowing the conservatives to control

---

[246]Guy Stanton Ford, *Stein and the Era of Reform in Prussia: 1807–1815* (Gloucester, Mass.: Peter Smith, 1965), 72–79 and 104–18.

most day-to-day affairs. The French had little interest in the Prussian debate, and Napoleon began a series of actions that would ultimately bring all
the issues to a head. In 1805 he launched a campaign against Russia and
Austria.[247] The Military Club closed its doors as Scharnhorst and his students reported for active duty in the field.

---

[247]See the terse evaluation by A. J. P. Taylor, *The Course of German History* (New
York: Capricorn Books, 1962; orig. 1946), 39.

Chapter 11

Disaster

Despite the best efforts of the reformers, the French did more to modernize the Prussian army than internal attempts to liberalize the nation and its military forces. In 1805 Scharnhorst's worst fear came true: Napoleon moved east. The reformers begged the king to join an alliance against Napoleon, but the conservatives were of another mind. Led by men like Duke Karl of Brunswick and Prince Friedrich von Hohenlohe, the conservatives had no love for the French, but they feared Napoleon's power. Far from the bellicose position of the reformers, the conservatives urged caution. Frederick William III let the two sides debate for more than a year while the French swept the Austrian and Russian armies from the field. When the king finally issued a call to arms in 1806, Prussia's potential allies were gone. Napoleon fell upon Frederick William's army with more power than the Prussian commanders could imagine. Ironically, this would eventually provide Stein and Scharnhorst with an opportunity to implement the reform agenda.

Prelude to Disaster

Frederick William III was a decent, moral man, and the desire to act in Christian charity stood at the base of his policy toward Napoleon. The king hoped to coexist with France, and he longed to avoid war. Aside from moral considerations, battle brought another danger. Prussia could lose. Therefore, the king wanted to do everything possible to avoid war. Other political leaders did not share Frederick William's sentiments. William Pitt, Prime Minister of Great Britain, attempted to lure the Prussians into an Anglo-German alliance, while Francis III (later Francis I) of Austria and Alexander I of Russia urged their brother monarch to mobilize. The king's own military reformers were champing at the bit in their desire to strike France, and the resulting confusion brought Frederick William's most prevalent personality characteristic into full bloom. He was one of the most indecisive leaders in Prussian history. Whether dealing with matters of state, court, or war, Frederick William found it difficult to act with vigor and confidence. Wanting to be loved by all, the king usually agreed with the last person he encountered. As all Europe looked to Prussia for action, the king hemmed and hawed, leaving his country in the most disadvantageous position imaginable. It seemed as though the Hohenzollerns were content to rest on the laurels of Frederick the Great.[248]

The world did not wait for Frederick William. Fearing French expansionism and a threat to its markets, Britain's Pitt sought to contain Napoleon economically. When Russia, Sweden, and Austria offered to provide military support for Pitt's policies, Napoleon decided to set his own course of action during the early autumn of 1805. Britain's navy closed French shipping, but the French army could keep Britain off the continent. Napoleon also demanded that Austria and Russia close their lands to British trade. When they refused, Napoleon mobilized and headed toward Austria. The Austrians countered by sending an army, and the tsar dispatched a large force to support it. Great Britain also responded. On land Pitt mobilized the Anglo-German forces in west Germany and offered British assistance to an alliance dubbed "The Third Coalition." When British Admiral Haratio Nelson destroyed a combined Spanish and French fleet at Trafalgar in October, it appeared as though Napoleon's star might be falling, and Pitt wanted to follow the naval action with a decisive victory on the continent. Taking a page from Marlborough, Pitt send emissaries to Berlin formally to ask Frederick William for Prussian help.[249]

The king politely listened to the British emissaries, and they seemed to make sense. With the Russians and the Austrians massing to the south, Prussia's army could make the key difference. Napoleon might be able to beat the Austrians and Russians, but his forces surely could not withstand

---

[248]Dennis E. Showalter, "Hubertusberg to Auerstadt: The Prussian Army in Decline?" *German History* (3) October 1994: 308–33. See 322 ff. for critical comments on the thesis reflected in this chapter and a defense of Frederick William III. For an example of traditional historiography see Halo Holborn, *A History of Modern Germany* (New York: Alfred A. Knopf, 1964), 375–82.

[249]Paul W. Schroeder, *The Transformation of European Politics: 1763–1848* (New York: Oxford University Press, 1994), 263–64.

the added weight of the Prussians. In addition, Prussian forces could be used to augment the strength of the Anglo-German legions in the west, just as they had done in the Seven Years' War. Frederick William agreed that it would be in Prussia's best interest to join the Third Coalition—at least he seemed to agree until his conservative generals talked to him. The conservatives reminded the king that Prussia had very little to gain from a war and very much to lose. The king immediately shrank from Pitt's offer of alliance.[250]

The reformers, who sided with Pitt, saw the king next. If Frederick William failed to understand the situation, Stein was convinced that he could rectify the king's views. The Third Coalition was the best option to check growing French aggression, and the reformers believed that it was only a matter of time before Napoleon struck Prussia. With Prussian forces hopelessly outclassed and outnumbered, Stein argued that the alliance was a dream come true. With Russian and Austrian forces united, Napoleon would be forced to march deep into Central Europe to meet them. If the Prussian army were to join these forces, even Napoleon would be hard pressed to match the numbers. Stein agreed wholeheartedly with Pitt, urging the king to join the alliance. The reformers took up the call and demanded mobilization of the armed forces. The king felt that he had to take some type of action, so to make Stein and the reformers happy, the king mobilized the army.[251]

Prince Friedrich von Hohenlohe was a nephew of Frederick the Great and a popular officer in the Prussian army. When word of the mobilization came, Hohenlohe called on two friends, Field Marshal Wichard von Mollendorf and Duke Karl of Brunswick. They were appalled at the king's action, and they urged Hohenlohe to keep Prussia out of the impending war. Assured of their support, Hohenlohe used his royal title to gain the king's ear. The coming war was not a Prussian war, Hohenlohe told his sovereign. If Prussian blood were to be spent, it should be spent defending Prussian soil, and wasting it on an excursion to defend foreign interests was a risky business. The conservatives closed ranks around Hohenlohe because the

[250]Franklin L. Ford, *Europe: 1780–1830* (London: Longmans, Green, and Co. Ltd., 1970), 205 ff.

[251]Guy Stanton Ford, *Stein and the Era of Reform in Prussia: 1807–1815* (Gloucester, Mass.: Peter Smith, 1965), 93-100.

Junkers no longer feared the spread of the French Revolution as much as they feared their own popular army. They felt that Napoleon had grown to be more like Louis XIV than a radical revolutionary threatening to overturn the established social order. Conservative Prussian military leaders could tolerate this, and they made their peace with the French emperor. Impressed with the argument, Frederick William agreed with Hohenlohe. He canceled the plans for mobilization.[252]

Stein was infuriated with the king's shift in sentiment, and Scharnhorst openly vented his frustration. Storming into the king's presence, Stein demanded some type of action. The king changed his mind again. In an effort to appease the reformers, Frederick William agreed partially to mobilize the army for defensive purposes. When Hohenlohe vehemently objected, the king sought to appease the conservatives by promising to avoid the Third Coalition. The result was political chaos. By trying to please everybody, the king pleased nobody. Worse yet, the policy was completely ineffective. Napoleon, not being hampered with an indecisive mind, marched toward the Austro-Russian army, crossing Prussian territory in the process.[253]

In the military the situation was no better than in the political realm. The lethargic, outmoded Prussian army had discussed reform, but it had not changed. Not only was their timid sovereign unable to cope with diplomacy, but his failure to reform the army left the soldiers in a deplorable condition. Although Scharnhorst was able to place newly organized divisions (groupings of regiments) and corps (groupings of divisions) in the field, the new-fashioned configurations led to mass confusion. Commanders, who had never led more than two or three regiments in the field, hopelessly tried to move thousands upon thousands of troops. The size of the Prussian army, 140,000 men, overwhelmed their abilities. Large contingents of soldiers meandered through the countryside in search of food, shelter, and supplies. The senior commanders did not know that Scharnhorst's staff officers could help them organize logistical support, and divisions of hungry troops marched to unknown destinations only to turn and march to new unsupplied locations. Scharnhorst was beside himself with frustration. Now more than

---

[252]See analysis by Philip G. Dwyer, "The Politics of Prussian Neutrality: 1795–1805," *German History* (12) October 1994: 351–73. See specific comments, 370 ff.

[253]G. Ford, 107–10.

ever he saw the need to develop the general staff to assist corps command-ers.[254]

Scharnhorst was aware that the large Prussian army required organiza-tion. If organization for war was a science, commanders needed to under-stand scientific principles and have the authority to effect change. Scharn-horst believed that increased educational standards would eventually solve the leadership problem, but there was no time for long-term strategies while Napoleon openly marched through Prussian domains. The immediate solu-tion involved utilizing a general staff. Although senior officers did not have the organizational skills to move, quarter, and supply their troops, Scharn-horst had dozens of talented officers who could help them do so. Even some of Scharnhorst's rivals began to see the wisdom of his proposals, but the Prussian army had a problem. It was permeated with command bureaus possessing overlapping authority. Several leaders were in charge of major functions, but no single command staff or general held overall command. Only the king had the power to eliminate the duplicative bureaus, and he would not do so.[255]

The summer of 1805 revealed a dangerous deficiency in the command structure of the Prussian army. In theory, the king was to take the field and lead as Frederick the Great had done. In practice, Frederick William took command of nothing and allowed his subordinates to engage in limitless debate. Frederick the Great had taken personal command of every detail of the army, and commanders worked directly for him. Frederick William was in no position to do this. The king had too many political responsibilities, too many soldiers, and not enough military training. In addition, the king had too many advisors. Rather than consolidating the organizations compet-ing for control of the army, the king tried to listen to each special interest. The summer turned into a season of endless arguments as if no one were in control, and when the king finally made a decision, it was always a com-promise. Even when Frederick William appeared to act decisively, he could easily change his mind the next day. In the Prussian army, no one seemed to

---

[254]Charles Edward White, *The Enlightened Soldier: Scharnhorst and the Militäri-sche Gesellschaft in Berlin, 1801–1805* (New York: Praeger, 1989), 122–28, and F. N. Maude, *The Jena Campaign 1806* (New York: Macmillan, 1909), 74–75.

[255]Gordon A. Craig, *The Politics of the Prussian Army: 1640–1945* (London: Ox-ford University Press, 1968), 29–30.

know what was going to happen next. Things were very different in the French camp.

On August 31, 1805 Napoleon catapulted the French army eastward. In a series of lightning moves, he brought forces to bear against the Austrians by moving through Germany, including Prussia's western provinces. Frederick William dispatched strong notes to protest the violation of Prussian territory, but he sent no troops. By October 17, the leading elements of Napoleon's forces destroyed an Austrian army at Ulm along the Danube. Taking no time to savor the victory, Napoleon immediately struck the Russians, forcing them to retreat. By early November the French occupied Vienna. On December 2 Napoleon met the combined Austro-Russian army at Austerlitz and slaughtered them. The allies were devastated, and on December 26 the Austrians surrendered.[256]

Every commander in the Prussian army was suddenly jolted into reality. With Austria seeking peace and Russia in retreat, Prussia stood alone. Saxony and a few other German states expressed an interest in an alliance, but the Prussian army was now isolated. Stein sadly commented that Napoleon had changed the political fate of Europe, and he believed that Prussia's only hope for maintaining political autonomy was to unite immediately with Russia. The military reality presented a dark future for such a proposal. While the army—both conservatives and reformers—would welcome a Russian alliance, the Russian army was running away, and each day took it farther and farther from the Prussians. After Austerlitz the tsar might lend his moral support to Frederick William, but it would take months for a rejuvenated Russian army to reach Prussia. With a sense of fear, Frederick William took the only step he could take. He embraced the little Saxony as Prussia's new ally.

Napoleon, fully aware that he was now the master of Western and Central Europe, dictated the terms of peace to the Austrians as he consolidated French holdings along the Rhine. Frederick William was shocked as Napoleon annexed huge segments of Prussia's Rhenish provinces. Stein looked longingly at his beloved mining districts in Westphalia, watching them pass into the hands of France, and Scharnhorst continued to advocate war. Napoleon, however, offered Prussia generous terms. In return for Prussian territory, Napoleon gave Frederick William large sections of British Hanover,

---

[256]For impact on Berlin peace party see Schroeder, 282–85.

and the king accepted them, incurring the wrath of Great Britain. Frederick William thought that he could appease Britain by explaining the rationale for the actions, but William Pitt would not listen. In early 1806 Britain declared war on Prussia, leaving Prussia completely at the mercy of the French. Tragically, this was the moment the indecisive Frederick William chose to act.

Disaster

In 1805 Frederick William had ignored the advice of the reformers, but after Napoleon's actions in the spring of 1806 the king suddenly began to lend them an ear. Napoleon's occupation of the Rhineland was no mere trade of territory. The French moved in and began building troops, causing Frederick William to see the threat presented to Prussia. The reformers banded together to form a war party in Berlin. Scharnhorst now convinced Frederick William that France would not let Prussia exist in peace. The occupation of the Rhenish area was illegal, and Napoleon could no longer be ignored. To the delight of the reformers, the king ordered the French to leave Prussian territory, but neither Frederick William nor the war party realized the full gravity of the situation. Napoleon had anticipated Prussian belligerency. Instead of sending his army back to France after the victory at Austerlitz, he had quartered a good part of it in Bavaria throughout the winter of 1806. When Frederick William demanded that the French leave the Rhineland, he did not know that Prussia was outflanked in southern Germany.[257]

In the summer of 1806, Prussian units drilled with the hint that another mobilization would take place. Each time Napoleon ignored Frederick William's demands, the influence of the war party increased. By August the king would wait no longer. He called the entire army to arms, gave command to the Duke of Brunswick, and sent it to Saxony. The duke divided the army into two operational groups, hoping to increase mobility. He took command of the largest group and gave the other to Prince Hohenlohe. Each group was further divided into corps, and each corps had a chief of staff. Reorganization, however, was hardly effective. Brunswick and Hohenlohe

---

[257]See David G. Chandler, *The Campaigns of Napoleon* (New York: Macmillan, 1966), 443–51.

did not move in tandem, and the supply situation was as bad as it had been in the previous summer. The army blundered about as it moved slowly westward, and in late September Brunswick received shocking news. The French were coming from the south.[258]

Brunswick did not have the authority to consolidate all command power at his headquarters, and as a result he had to listen to an endless drone of opinions from a variety of officers. Nothing could have been more confusing. The king—who promised that he would take personal command of the army—did little to help, and by mid-September the army was bogged down in an enormous debate. Scharnhorst described the situation in a letter to his wife. The army was not operating as a team, he said, and each segment was its own little force. In war, Scharnhorst said, sometimes it is better to take action than to argue about the best course of action, but the command post was nothing more than a debating society.

Utilization of the general staff was another matter. The king allowed Scharnhorst to form a general staff, but most of the field commanders had no idea how it was to be used. Two of the corps commanders, Gebhard von Blücher and Ernst von Rüchel, were exceptions, and they saw the benefit of having a staff of highly qualified officers to help them plan and maneuver. Blücher even took a liking to Scharnhorst, and Rüchel utilized his new staff, finding that it allowed him greater flexibility in planning and operating. Blücher and Rüchel, however, were atypical. For the most part, corps commanders had no idea how to employ their staffs. Wichard von Mollendorf summarized their attitudes by claiming that Scharnhorst's theories of war were too complicated. Unfortunately, Mollendorf and many other generals did not understand that war itself, not Scharnhorst's theories, had grown too complicated. Scharnhorst proposed the solution, not the problem.

By the end of September, Brunswick's commanders were divided into three different ideological schools. One group advocated a retreat to Brandenburg. Since Napoleon had outflanked the army, they reasoned, it would be best to retreat and choose other ground for fighting. The Saxons, who would see their homeland abandoned by this plan, opposed any talk of withdrawal. A second group advocated a bold offensive toward the Rhine.

---

[258]Summary based on Maude, 78–180, and Chandler, 456–502. See also Curt Jany, *Geschichte der königlich Preussischen Armee bis zum Jahre 1807, Volume III, 1763 bis 1807* (Berlin: Verlag von Karl Siegismund, 1929), 561–93.

The logic behind this position was twofold. It would liberate the Rhenish provinces while forcing Napoleon to move west. The third school suggested neither advance nor retreat. Since Napoleon was forming to the south of the Thuringian Forest—a hilly, wooded barrier between the French and Prussian armies—advocates of the third position urged Brunswick to sit and wait for the French to move. Brunswick seemed to favor the third course of action, but as the debate continued he flirted with the idea of withdrawal. Rather than adopting a defensive stance, the Prussians formed in a vulnerable position about thirty miles southwest of Leipzig. In the meantime, Frederick William promised to leave Berlin to command the army.

Napoleon suffered from no such indecision. While 140,000 ill-supplied Prussians marched back and forth in Saxony, Napoleon reinforced his army in southern Germany until he had nearly 200,000 men under arms. Unlike the Prussians, the corps of the French army were ready to strike. Napoleon organized his army into five self-sufficient corps, dividing them into three columns over a fifty-mile front. The weaker forces took the flanks, while Napoleon kept the main striking power in the center under his personal command. His plan was to march north through the Thuringian Forest and find the Prussians. Any corps that bumped into the Prussian army was to attack, and the other columns were to wheel in the direction of the fighting. The purpose was to destroy the Prussian army in the field. If Napoleon could do this, he planned to exploit the defeat by sending his cavalry far into Prussian territory. Such a plan was beyond the conceptual abilities of many Prussian officers.

On October 8 the French army began to move from Bavaria at a rate of fifteen miles per day. French scouts roamed over the vast front in search of their Prussian enemies, and each column stayed close enough to support its partners. While Napoleon's formation was flexible enough to strike in any direction, the Prussians were organized at the opposite extreme. Brunswick knew Napoleon was coming but little else. He did not know, for example, Napoleon's location, his direction of travel, or how the French were organized. News of French movements merely started another tremendous debate in the Prussian camp, and when Brunswick decided that retreat was the only option, the King of Prussia arrived to take personal command of the army.

Frederick William was not happy with what he found. The Prussian camp was a comedy of errors with few supplies and no central organization. Some units were away on forced marches while others were quartered in the open ground without firewood. The Saxons threatened to mutiny because they had not received bread for three days. Brunswick seemed completely out of control, and actions at his headquarters resembled a debate tournament. Frederick William took the only decisive action he knew how to take. Summoning the corps commanders and each army bureaucrat, he asked for opinions on the next course of action. Scharnhorst fumed, and Napoleon continued to advance. When the forward French echelons crossed the Saale River below the Prussian positions, Scharnhorst begged the king to take action. Frederick William ignored Scharnhorst and patiently listened to each commander.

The first signs of impending disaster appeared on October 10. The king held court with Brunswick's main army of 85,000 men loosely scattered around Leipzig. Hohenlohe sat several miles away south of the village of Jena with 48,000 more troops. Brunswick began toying with regaining the strategic initiative and thought about attacking Napoleon. If the Prussians moved through the Thuringian Forest, the general thought, it might throw the French temporarily off balance. Brunswick ordered Hohenlohe to begin the operation, but Hohenlohe was not overly excited about weakening his front. His first order of business was to deal with the hungry Saxons. Personally assuring them that food would arrive, the Hohenlohe begged the Saxon officers to keep their troops ready for action. As Hohenlohe calmed the Saxons, Prince Louis Ferdinand, another nephew of the Great Frederick, reported that the French had crossed the Saale River in strength. Neither Louis Ferdinand nor Hohenlohe knew that Napoleon was in the process of turning the Prussian flank.

Hohenlohe received Brunswick's order to attack as he was trying to assess Louis Ferdinand's report, and he reluctantly dispatched Rüchel's corps of 15,000 men toward the Thuringian Forest. Some analysts, such as Colonel F. N. Maude, believed such an attack would have been successful if it had been fully supported and pursued with vigor. As it turned out, Rüchel's maneuver became a futile march in the wrong direction because Brunswick had a change of heart. Within a day of ordering the attack, Brunswick ordered Hohenlohe to recall Rüchel after hearing of scattered fighting around Saalfeld. Rather than attack, Brunswick opted for full retreat. Exasperated, Hohenlohe sent messengers to find Rüchel, now a full day's march away, and he began the long process of reassembling his army at Jena. Late in the night on October 10, another message arrived at Hohenlohe's headquarters.

The Prussian units at Saalfeld were in flight, and Louis Ferdinand had been killed. Hohenlohe sent the message on to Brunswick.

News of Louis Ferdinand's death filled common Prussian soldiers with gloom. The troops, already exhausted by seemingly pointless marches and lack of food, began to fear the impending battle. Their mood took hold of Brunswick, who now felt that war was no longer an abstract idea. He decided that the entire business had become too risky and planned to rejoin Hohenlohe behind the Elbe River, hoping to use the water as a defensive barrier. Ordering Hohenlohe to fight a limited action to support the retreat, Brunswick told his subordinate to avoid a decisive engagement. Hohenlohe gathered his forces at Jena, and he was ready to move by October 12, except for Rüchel's corps that was positioned in the Thuringian Forest. Hohenlohe decided to rest his troops for a day and march to the Elbe on October 14. In the meantime, Brunswick moved the main body of the Prussian army into the village of Auerstadt about eleven miles to the west of Hohenlohe. As evening fell on October 13, a thick fog enveloped the two wings of the Prussian army. The Prussian officers, no doubt, thought the fog would cover their retreat.

Napoleon was in a whirlwind of action on the other side of the fog. A day earlier he had not been exactly sure of the Prussian position, but on October 12 he knew the Prussians were on his left. Although he had slightly miscalculated their position—Napoleon believed that Frederick William was with the main army at Erfurt—the emperor knew that he was closer to Berlin than the Prussians. He spent the night of October 12 issuing orders to his five corps commanders. They were to swing west and prepare to attack. By October 13, four of his corps were on the way, while the fifth marched to the north of the village of Jena. Napoleon, outside of Jena, recalculated his enemy's position when French patrols reported contact with Hohenlohe during the night. With fog and darkness covering his actions, the emperor called his artillery forward.

Even though Hohenlohe's troops were supposed to have been resting on October 13, they did not have the equipment or facilities to bring them into fighting shape. The night of October 13 had been unseasonably cold, and the Prussians had neither shelter nor overcoats. With the haphazard supply system, the troops had not been issued firewood, and they were not allowed to commandeer local housing since they were in friendly territory. The officers had been ordered to execute anyone found looting or stealing materials from the locals. As a result, the Prussians spent a cold exhausting night on a plateau above the French army. They woke up shivering, arising from the frost-covered ground. While sentries peered through the damp fog to the ground below, soldiers grumbled as they formed for another march. Suddenly, the first cries of alarm jolted the men to action. A French army of unknown strength was assembled below their position.

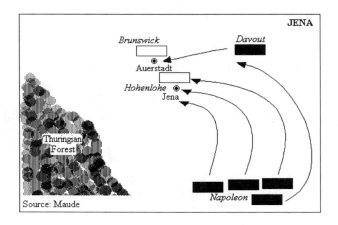

At first, Hohenlohe was not overly concerned. He knew the French were in the area, and he assumed the troops in front of him were the leading elements of a single corps. Remembering that his orders were to skirmish before retreating, Hohenlohe formed his army shortly before dawn. He had every right, he felt, to be confident. Even without Rüchel, he was able to muster 42,000 troops. He planned to attack, then continue the retreat, picking up Rüchel along the way. Assuming that he was facing no more than a corps, as soon as the fog was reasonably clear, Hohenlohe sent his infantry to drive the French away. As the Prussian troops attacked, the French guns began to lob artillery into their ranks. Hohenlohe moved back, not sure why a corps would have so many artillery pieces. Minutes later the prince received a frantic message from Rüchel. His corps was fully engaged and unable to join Hohenlohe at Jena. Apparently, the prince realized, he had encountered something stronger than a single French corps.

Things were going well for the French. Hohenlohe's early morning attack caught them off guard, but concentrated artillery fire sent the Prussians running. When Hohenlohe fell back, the French attacked Rüchel. As Hohenlohe formed a defensive line in front of the French position, Napoleon knew that the time was ripe to strike. He subjected the Prussians to a massive bombardment and sent his infantry forward. With the fog gone, Hohenlohe could now see his adversary: an entire French army attacking Jena. Although Brunswick had told him to avoid a decisive encounter, Napoleon forced Hohenlohe's hand. The Prussian prince was fighting for his life.

Louis Nicolas Davout had received orders to move his French corps in the direction of Jena during the night of October 13. A marshal of revolutionary France, Davout knew that Napoleon would be satisfied as long as his corps moved into the general area, so he moved to the north of the suspected Prussian position. His plan was to cut the Prussian line of retreat, but overshot the mark. Instead of marching to Jena, his 26,000-man corps ended up in Auerstadt around 9 A.M. in front of Brunswick's army. Realizing a tremendous opportunity, Davout made a daring bid. He attacked the Brunswick, sending requests for help.

Back in Jena, Hohenlohe's generals managed to form a strong line. When the French moved into musket range, their artillery could no longer support the attack, and the Prussians responded by firing volleys. For a short time it seemed as though the Prussian infantry line would prevail as disciplined fire poured into the French ranks, but Napoleon had no intention of trying to out-volley the Prussians. While the musket duel continued, Napoleon moved his artillery to higher ground. By mid-morning he returned the volleys of the Prussian line—volleys fired from heavy artillery. As the shells pounded the Prussian infantry, French soldiers arrived all through the morning, giving Napoleon 90,000 men by noon. Hohenlohe's line could not hold.

When morning had broken at Auerstadt, the Duke of Brunswick contemplated the sound of battle to his south. He believed that Hohenlohe would engage and run, but he was confused by the tremendous amount of fire. Despite his confusion, the duke decided to assemble his troops and continue the planned retreat. He sent repeated messages to Hohenlohe, but the messengers reported that they could not get through. Confused, the duke began to retreat eastward. Fearing that a large French force might be in the area, he ordered Blücher to set up a rear guard, but Blücher's corps, which had just finished a day-long forced march, was in no condition to move. Blücher sent word that he could not get into position quickly, just as Davout's French corps appeared where Blücher was supposed to go. Before Brunswick had time to react, his leading elements were fully engaged.

Scharnhorst had been banished from Brunswick's headquarters. The duke, irritated by the reforming officer's constant meddling, wanted to run the army without any of the "secretaries" from the general staff. Looking for some way to be of use, Scharnhorst moved forward and found the Prussian units engaged with Davout's corps. He rallied several units in the hope of launching an attack and asked Brunswick to send Blücher into the French position. Blücher's men marched forward but not under Brunswick's orders. The unfortunate duke was not in any condition to respond to Scharnhorst's request—or much of anything else. A musket ball struck Brunswick in the side of the head as he attempted to lead the retreat, and the ball passed through his sinus cavities, spewing his eyes onto the ground. Aides carried the blinded duke from the field, and he died shortly afterward. The main body of the army was without a leader.

Frederick William was watching the battle form Auerstadt when he received word of Brunswick's mortal wound. With Scharnhorst out of headquarters, there was no one left to take command, and the king wondered what he should do. For a moment, he thought about assuming control himself, but he believed that the position should go to a more experienced person. On the other side of the Auerstadt line French Marshal Davout was hardly as hesitant. Realizing that he was in a unique position, he sought to maintain the initiative. Although the Prussians outnumbered him, they were inactive, and they remained static in a long defensive position. Davout could not know of the king's indecisiveness, but he did see an advantage. If the Prussians were merely content to sit, he was going to attack them. Davout was not a man who let opportunity slip through his fingers.

The king might have been willing to wait, but individual Prussian commanders were not. Infuriated by Davout's attack, they began launching their own strikes. Yet with no overall commander, the attacks were uncoordinated. Davout beat the attackers back, and by 10:30 the Prussians were retreating. The Prussians—unused to war on this scale—were appalled by the large number of casualties. The king, who finally decided to assume control of the army, could take no more. Even though he had two corps that had not been committed to battle, he felt that it was time to leave. Unfortunately, instead of moving out of the battle, the king decided to move toward Hohenlohe. Frederick William had no idea that he was marching his army into the strength of the French position.

By noon, Hohenlohe was fighting to stay alive at Jena. Napoleon's artillery was now supported by wave after wave of French infantry, and thousands of Prussian casualties littered the ground. Each time the Prussians tried to stand, the position was blown to pieces by French artillery. Hohenlohe moved his men to a position—a final position—in the hope of organizing a retreat, but it was too little too late. The French were enveloping the Prussians. Rüchel, not knowing of his commander's predicament, fought to reach Hohenlohe's position. He arrived in the early afternoon, only to find that the Prussians were being surrounded. He ordered an immediate attack, but the corps was subjected to a merciless bombardment and Rüchel fell, mortally wounded. By mid-afternoon the Prussian survivors at Jena started surrendering. Those who could not tried to run from Napoleon's unceasing fire.

Temporarily free from Davout, Frederick William moved down the road to Jena where he hoped to link with Hohenlohe and organize a full retreat. His plans were dashed when he marched straight into Napoleon's advancing forces. Within minutes, the king's men were retreating—retreating back to the position held by Davout's tenacious corps. The Prussians tried to swarm over Davout's corps, but the fiery French commander held. By 4:30, advancing French forces struck Frederick William's rear, and Davout received reinforcements. The Prussian rout was now complete. With nowhere to run, Prussians started surrendering by the regiment. A few unorganized bands of men scampered back to the Elbe, but the bulk of the army was either killed, captured, or wounded. Napoleon was the undisputed master of the bloody fields around Jena.

The Aftermath of Disaster

On October 14, 1806, the Prussian army did not merely suffer a defeat; the twin battle of Jena and Auerstadt was a complete disaster. In a single day the Prussian army was obliterated. One by one the units were surrounded and destroyed as darkness fell. Scharnhorst, having joined Blücher's command on his own initiative, grabbed a musket and fell into the front rank of an infantry line. Although the action drew the respect of the common soldiers, it was only a symbolic gesture. Blücher was one of the few commanders able to retreat, but his command—Scharnhorst excluded—surrendered three weeks later on the Baltic coast. Most of the survivors followed suit. While a frantic Frederick William tried to rally officers and men in East Prussia, Napoleon sent Marshal Joachim Murat through Prussia in what can only be termed a cavalry blitz. Prussian fortresses began surrendering at the mere sight of French cavalry. Only one fortress—Kolberg under the command of August von Gneisenau—offered resistance. The action, incidentally, made Gneisenau a national hero.

Scharnhorst, who had been released in a prisoner exchange, and a few soldiers made their way to East Prussia to form a small corps, and Frederick William III sought safety under the protective wing of Tsar Alexander I. The Prussian government officially moved its seat to Königsberg, and the remnants of the army, under the indelible Scharnhorst, joined a small Russian army commanded by General Anton Wilhelm Lestocq. Scharnhorst and Lestocq hoped to be reinforced by a larger army from Moscow, but Napoleon circumvented their actions by continuing his offensive into Poland. The Poles, welcoming the French as liberators, flocked to Napoleon's tricolor, and the emperor set up headquarters in Warsaw while Russian reinforcements worked their way west. In November, Count Levin A. Bennigsen arrived from Moscow, and he absorbed Scharnhorst and Lestocq in his army. Bennigsen attacked Napoleon, believing the French would retreat and move to winter quarters. Napoleon had a surprise.

Desiring to bring an end to the Russian threat, Napoleon took his troops into winter campaign. He attacked the Russians on February 8, 1807, near the village of Eylau. Although battle conditions are always frightening, Eylau brought its own special version of hell. The fight took place in sub-zero temperatures in the midst of a blinding blizzard. Men slaughtered each other with swords, knives, and musket butts, because their firing mechanisms were frozen. Casualties were heavy, and wounded men froze to death where they fell. The results of the battle were indecisive for the leaders but not for the thousands of men who perished in the Arctic-like conditions. The Prussian corps played a small, significant role, thwarting the final French attack, but the battle was a draw. At the end of the day, both sides trudged away to winter quarters.[259]

That summer, Napoleon struck again, trapping Bennigsen at Friedland on June 14. This was the death knell for Prussia. Scharnhorst, completely isolated in East Prussia, could only maneuver when Joachim Murat made a triumphant entry into Königsberg with the French cavalry. With no Russian forces to come to their aid, the Prussians retreated eastward to the city of Tilsit, taking Frederick William and Queen Louise in tow.[260] On June 19 the French arrived and occupied the city. When Tsar Alexander agreed to accept French peace terms, there was little Frederick William could do. To reinforce the point, Napoleon would not even allow Frederick William to attend the peace conference, and he dictated the terms to the Prussians after the tsar signed the treaty. All the king could do was listen. His army—the only earthly power that could have stopped Napoleon—was strewn on the bloody ground between Jena and Tilsit. The Prussian army was weaker than it had been at anytime since 1640.

[259]Henry Lachouque, *Napoleon's Battles: A History of His Campaigns* (trans. by Roy Monkcom) (London: George Allen and Unwin, Ltd., 1966), 160–67.
[260]Ibid., 167–78.

Chapter 12

The Heyday of Reform

In the dark days after the Treaty of Tilsit a new light dawned on Prussia. The conservatives, fallen from grace, had not only failed the nation at Jena and Auerstadt, but defeat made national heroes out of the reformers. Gneisenau's refusal to surrender Kolberg and Scharnhorst's continued struggles were lionized by the lower and middle classes, while the conservatives were left with no heroes to champion their cause. During the summer of 1807 Frederick William gathered his beleaguered government in Berlin. Stein—who had been dismissed in January—was recalled to manage Prussia's financial affairs, and Scharnhorst received the most important post in the War Ministry. The question of whether to reform the army was a moot point. The main question was, how should the army be reformed? Although the king feared a popular army, he no longer ruled in west Germany, and Napoleon had torn Polish lands from Prussia. French troops were in Prussia, and they were going to stay until Frederick William paid a huge war indemnity. As a result, Scharnhorst and Stein came to the king with a daring plan. They asked permission to form a national army right under the very nose of Napoleon. It could be done, Stein argued, if ordinary folks were given the rights of citizenship. The king hesitated, but Stein urged him forward. As a result, a new Prussian army began to rise from the ashes of defeat.

The Terms of Tilsit

For Prussia, Tilsit was not a peace settlement: it was the culmination of humiliation. After the defeat at Friedland, Tsar Alexander believed that it was time to sue for peace, and without a Russian alliance Frederick William could no longer wage war. The Prussian king could only mumble his disapproval when the tsar announced plans to meet the French emperor. Napoleon, well aware of Prussia's powerless position, agreed to discuss peace with the tsar on a barge in the middle of the Nieman River near Tilsit. Frederick William and Queen Louise mustered their final reserve stocks of dignity to join the conference, only to be snubbed by the omnipotent Napoleon. Leaving the king and queen to await the outcome of the discussion on the banks of the Nieman, Napoleon summoned Frederick William after concluding the peace treaty with Tsar Alexander. All Frederick William could do was to listen.[261]

The terms of the treaty radically changed Prussian geography. With one sweep of the pen, Prussia lost all of her Rhenish provinces. Napoleon told Frederick William that the confused feudal configuration of autonomous German states hampered economic efficiency in Central Europe. Therefore, the emperor created the German Confederation of the Rhine, incorporating all German states west of Magdeburg into a single, French-dominated political entity. For all practical purposes the Holy Roman Empire was relegated to the dust bins of history. Prussia also lost territory in the east when Napoleon revoked the partitions of Poland, restoring Warsaw as a Grand Duchy. And the geographical reduction of Prussia was just a beginning.[262]

The military forces were next. Napoleon sought to neutralize Prussian military might in order to impose his will on the continent, and he insisted that the Prussian army be reduced to 42,000 men. Such a small aggregation of troops would hardly present a challenge to the French, and if the proper officers were installed, Napoleon even believed that he could conscript the army in times of crisis. He forced Frederick William into a military alliance and told him to select officers who would obey the terms of the Treaty of Tilsit. Wanting to rid Prussia of the most nationalistic officers, Napoleon sought to eliminate some of the most troublesome chiefs from Berlin. Blücher, for example, was exiled to Swedish Pomerania, nursing both an all-consuming hatred for Napoleon and a thirst for brandy. Despite Napoleon's purge, Frederick William was able to retain the services of Scharnhorst, and the reformer managed to hide his cohorts in out-of-the-way positions.

---

[261]See Paul W. Schroeder, *The Transformation of European Politics: 1763–1848* (New York: Oxford University Press, 1994), 320–22.

[262]Compare Schroeder's analysis to ethnocentric German views. See Heinrich von Treitschke, *History of Germany in the Nineteenth Century, Volume I* (trans. by Eden and Cedar Paul) (New York: McBride, Nast, and Co., 1915; orig. 1884), 333.

Economically, Napoleon imposed harsh conditions on Prussia, ordering Frederick William to pay a large indemnity to France and to bear the cost of occupation. Napoleon even reorganized the Prussian government so he could collect French tribute. The French occupied several fortresses in Brandenburg and Silesia, forcing Prussian locals to provide all supplies. The economic measures created fierce resentment among the common people, especially the middle class, and they grew angry with the abuses of French occupational troops. Napoleon's actions quite inadvertently began to fuel the fires of nationalism. When the king called Stein to Berlin in September of 1807, Stein officially revamped the tax system to pay off the French. Unofficially, however, the Teutonic knight implemented another agenda.[263]

The Beginning of Reform

Stein recognized that Prussia could never fight the French alone, but to gain its freedom Prussia needed to undergo significant political change. Most of all, Stein believed, it was time to start thinking beyond provincial borders. Prussia was merely a state among several other traditional German states, and nationalism—the only means of throwing off the French yoke—should be expressed as something greater than Prussian ethnocentrism. All Germans had to stop thinking of themselves as members of petty fiefdoms, Stein believed, and they had to start thinking that they were citizens of a greater group of German-speaking peoples. This could not happen overnight, and Stein realized that if such a miraculous political transformation were to occur, common Germans needed an incentive to change. He believed that the incentive to unite Germany was democracy.[264]

Germany would unite under a nationalistic banner, Stein felt, if all Germans were given full rights of citizenship. This was no small task in Prussia where eastern Junkers still housed serfs on large agrarian estates. Still, Stein believed in the power of citizenship, and the United States of America served as his example. The United States had developed a political system that allowed ordinary people to enjoy the rights of citizenship, and

---

[263]For summary see James J. Sheehan, *German History: 1770–1866* (New York: Oxford University Press, 1989), 238. Sheehan cites Georges Lefebvre, *Napoleon, Volume I* (New York: 1969), 273, for an English language summary of the treaty.

[264]Guy Stanton Ford, *Stein and the Era of Reform in Prussia: 1807–1815* (Gloucester, Mass.: Peter Smith, 1965), 137–45.

its leaders could call upon a nationalistic army in time of trouble. Citizen soldiers flocked to the colors in America. Inspired by this example, Stein argued that Germany should also be united as a democracy. Realizing that Germany could not eliminate royal power, he argued that a constitutional monarchy modeled after Great Britain would be preferable to the full democracy of the United States.

Stein's beliefs were not endorsed by the king. In fact, despite Napoleon's creation of the German Confederation, most Germans viewed themselves parochially, and they did not picture themselves as citizens of a larger country. In addition, democracy was a dangerous issue. The American democracy, an abstract concept in the minds of German royalty, threatened the social fabric of the nobility, and it seemed to be more akin to the ideology of revolutionary France than to the monarchies of Central Europe. Political and military conservatives referred to American democrats as Jacobins, the radical leftists of the French Revolution. Most Junkers believed that it would be folly either to unite Germany or to turn government over to the popular will, and Frederick William openly feared such a concept. Despite these constraints, Stein began working on a political reform agenda with three major goals: (1) fiscal stability, (2) increased nationalism through citizenship, and (3) the unification of Germany. All of these efforts, Stein believed, would allow Germany to liberate itself from the mantle of French servitude.

Stein's efforts met with modest success. Although a brilliant man, he tolerated neither opposition nor incompetence among his subordinates, and such qualities won him few allies—friends he needed to endorse reform proposals. The king frequently expressed contempt for both Stein's methods and his goals, and other ministers were openly hostile toward Stein in the king's presence. Yet, Stein's importance in fiscal matters made him too valuable to cashier—no matter how much the king desired to do so. On the other hand, Stein had a few important friends, many of them in military uniforms. His most important ally was Gerhard von Scharnhorst, and while Scharnhorst labored in the military realm, Stein worked in the political arena. In 1808 Stein achieved his greatest political triumph when Frederick

William freed the serfs. It almost appeared as if Prussia had taken the first step toward democracy.[265]

Scharnhorst also found himself in a new position at the beginning of the French occupation. The king might not have liked all of the reformer's politics, but he appreciated loyalty, and Scharnhorst's actions at Eylau had served only to strengthen Frederick William's esteem for the professor-soldier. As the king put his skeletal government together after the Treaty of Tilsit, he enthusiastically accepted Scharnhorst, promoting him to general. Even though several of the more conservative officers disagreed with Scharnhorst's ideology, they admired his courage and tenacity, and they knew that Scharnhorst was one of the few officers who fought until the very end. His name had become a household word throughout Prussia, and the king took advantage of Scharnhorst's popularity by appointing him to head the Military Reform Commission.[266]

At first, the purpose of the Military Reform Commission was to punish officers who failed at Jena. Unwilling to acknowledge Prussia's military problems, Frederick William wanted to purge the army's leadership. The king placed Scharnhorst in charge of the commission to do his bidding, but the new general, like the economist Stein, had his own agenda. The last thing Scharnhorst wanted to do was to unjustly punish the officer corps; he wanted to reform the army. The Prussian officers had not failed the country, the Prussian system had, and Scharnhorst realized that he was in a position to revamp the army. Popular with friend and foe alike, Scharnhorst gathered the reformers—including August von Gneisenau and Karl von Clausewitz—under the umbrella of his commission. They planned to expand and modernize the army, despite Napoleon's instructions to reduce its size. Scharnhorst's ideas went far beyond Frederick William's plans for the Military Reform Commission.[267]

---

[265]Ibid., 199 ff. See also Robert M. Berdahl, *The Politics of the Prussian Nobility: The Development of a Conservative Ideology, 1770–1848* (Princeton, N.J.: Princeton University Press, 1988), 115–23.

[266]Gordon A. Craig, *The Politics of the Prussian Army: 1640–1945* (London: Cambridge University Press, 1968), 38–39.

[267]Martin Kitchen, *A Military History of Germany: From the Eighteenth Century to the Present Day* (Bloomington, Ind.: Indiana University Press, 1975), 40–45.

To keep the Reform Commission from degenerating into a witch hunt, Scharnhorst overloaded the board with reformers, and Clausewitz proved to be an exceptional find. Not only was Clausewitz determined to revamp the army, he was fully committed to liberation. Unlike most Prussian officers, Clausewitz spent the dark days after Jena as a prisoner in France. This happened because his unit, under Frederick William's brother Prince August, had been isolated from the main body of Blücher's corps in the retreat from Auerstadt. Clausewitz's men fought a savage rear guard action, only to be forced to surrender with the prince. Napoleon personally interred the two men in France. While other officers were released, Clausewitz remained in France under the threat of execution. He returned after Tilsit with a deep-seated hatred for Napoleon. The German state, he concluded, was more important than anything—even the Hohenzollern throne—if it meant freedom for Prussia.[268]

The combination of Scharnhorst and Clausewitz produced a dynamic tandem capable of sustaining the reform movement. Clausewitz became both a sounding board for Scharnhorst's ideas and the man who could visualize methods for creating a nationalistic army. Always adept with pen and ink, Clausewitz began writing about the nature of a new army and the methods for organizing it. Frederick William did not find Clausewitz's philosophical mind altogether palatable, but the king would not abandon Scharnhorst. Since Scharnhorst took Clausewitz under his wing, Clausewitz enjoyed de facto protection from the crown.[269]

With Clausewitz's help, Scharnhorst introduced a number of changes under the guise of the Military Reform Commission. He increased educational standards for the entire officer corps to produce men who could run the general staff, and he rotated staff and command assignments, assuring

---

[268]Gerhard Ritter, *The Sword and the Scepter: The Problem of Militarism in Germany, Volume I, The Prussian Tradition: 1740–1890* (trans. by Heinz Norden) (Coral Gables, Fla.: University of Miami Press, 1969), 76 ff. See also Karl von Clausewitz, *On War* (trans. by Michael Howard and Peter Paret) (Princeton, N.J.: Princeton University Press, 1984), 479–84. For a brief introduction to Clausewitz, see Michael Howard, *Clausewitz* (New York: Oxford University Press, 1988). See also a sociological-historical analysis by Raymond Aron, *Clausewitz: Philosopher of War* (trans. by Christine Booker and Norman Stone) (Englewood Cliffs, N.J.: Prentice-Hall, Inc., 1985).

[269]Peter Paret, *Clausewitz and the State* (New York: Oxford University Press, 1976), 61–77.

that future commanders would know how to utilize their staffs properly. Since the standing army was limited to 42,000 men, Scharnhorst believed the backbone of Prussia's military strength would come from a militia—the *Landwehr*—modeled after the French National Guard. He secretly created a Prussian militia, utilizing the popularity of Gneisenau and another well-known officer, Hermann von Boyen, to sell the idea to the public. Most Prussians held the professional army in contempt, but they liked Gneisenau and Boyen. Scharnhorst also introduced measures for reducing the period of enlistment in the regular army and relaxing its rigid discipline. His greatest dream, however, was universal conscription. Since the French would never stand for such a law, he sought to circumvent their power by increasing the ranks of the militia.[270]

Despite his new-found power as head of the Military Reform Commission, Scharnhorst worked in a country controlled by a foreign government. He simply could not build, equip, and train a standing army with French troops roaming the countryside. The citizen's militia, the *Landwehr*, became the answer to his problem. Scharnhorst enticed recruits secretly to join local reserve units, and while giving the appearance that there were never more than 42,000 men under arms, he began rotating the reservists into the regular army. After a brief tour of duty, personnel were released to local *Landwehr* units. The militia men gathered each month to review military training and to go on short exercises, but they dispersed before the French could actually account for them. As each French excess brought more middle-class recruits to the *Landwehr*, Scharnhorst tried to create a large, well-trained reserve that could be mobilized when the time for liberation came.[271]

To the conservatives, the *Landwehr* represented everything that was wrong with reform. The units lacked the discipline and precision of regular troops, and they were politically unreliable. The most disturbing aspect of the *Landwehr* was the selection of officers. Since the *Landwehr* was a popular army, Scharnhorst looked for popular officers to lead it. He even allowed some units to elect their junior officers, drawing the unmitigated wrath of the king's conservative commanders. All military authority, the appalled conservatives argued, came from the king, not popular elections. Nothing represented the gap between the reformers and the conservatives

---

[270]Kitchen, 55.
[271]Craig, 46.

more than this issue. On the one hand, conservatives fully identified with
the power of the king, and they swore allegiance to the Hohenzollerns. On
the other, the reformers believed that the loyalty of the army should be di-
rected toward the state. The most liberal reformers, led by Gneisenau, believed that officers should take an oath of allegiance to a constitution, not a
king. The conservatives could tolerate no such nonsense, and the ideological
positions of the reformers and the conservatives came to a head over the
Landwehr. Quite unintentionally, the militia became the symbol of reform.[272]

Despite Scharnhorst's efforts, the reformers were not quite as successful
as a first glance might suggest. To begin with, the Military Reform Commission was expected to hand down indictments, not to change the social
characteristics of the Prussian officers corps. The king wanted officers to be
punished, and Scharnhorst had to do it. Eventually, over 200 officers were
charged with some form of negligence of duty.[273] In addition, the conservatives were not quite willing to roll over and play dead. While they tolerated
and even respected Scharnhorst, they actively lobbied against social reform.
Scharnhorst had to remain sensitive to the conservative views to avoid internal discord, and Napoleon did not turn a blind eye to the Prussian court.
In 1808, the emperor intercepted a letter in which Stein implored Prussians
to revolt against their French masters. Napoleon demanded that Frederick
William fire his finance minister, and the French confiscated all of Stein's
property. If Scharnhorst's plans ever became public, he would certainly suffer a similar fate.

While powerful, Scharnhorst's position on the Military Reform Commission did not give him control of the army, and Frederick William would
not go so far as to appoint another reformer to overall command. Yet, the
king had a problem. There were few traditionalists capable of leading the
new army, and since Blücher—the most likely compromise candidate—was
in exile, the king had to find a suitable leader. The man who eventually
came to the forefront was Lieutenant General Hans Yorck von Wartenburg,
a rather unusual officer. A conservative Junker by birth and a moderate by
political persuasion, Yorck steered a cautious path between zealots like
Scharnhorst and Gneisenau and the most conservative members of the offi-

---

[272]Ritter, 102–4.
[273]Kitchen gives results and analysis, 41 ff.

cer corps. He was sympathetic to many of Scharnhorst's ideas, yet his conservative temperament alienated him from the reform party. With Blücher
gone, Frederick William gave Yorck command of the 42,000-man army.[274]

All of Scharnhorst's reforms were completed under less than optimal
conditions. Even when reforms were implemented, some officers executed
them cautiously while others ignored them altogether. The *Landwehr* grew
but not with the speed the reformers intended, and it reflected the problems
that hamper most reserve armies. The *Landwehr* could not perform anywhere near the level of first-rank troops. Still, Scharnhorst's clandestine reforms continued. Drawing the attention of Napoleon in 1810, Scharnhorst
assumed a lower profile, but the army secretly trained. Yet, the reforms did
not completely change the Prussian army. Scharnhorst's best efforts notwithstanding, the Prussians could not become strong enough to challenge
French rule by themselves. To achieve liberation, they needed some event
that would shake the very foundations of Paris. That event came in 1812
when Napoleon assembled the Grand Army for an invasion of Russia.

1812: Overture to Liberation

By 1811 Tsar Alexander sat between two strong economic poles. At
one extreme, Napoleon demanded that Russia participate in the
"Continental System," a trading network designed to keep Britain out of
European markets and to provide an outlet for French goods. On the other
extreme, Great Britain used its navy to stop all French trade and smuggle
goods to and from the continent. The tsar was caught in the middle, suffering from Napoleon's closed system while needing British trade. Since the
terms of Tilsit officially bound Alexander to the Continental System, he secretly approached the British in 1810 with a plan. Within a year, Russia was
shipping and receiving goods through Great Britain. There was no way to
keep all the activities secret, but the tsar was prepared to take the risk. Since
France had no navy, Alexander believed that Napoleon could do little about

---

[274]See Peter Paret, *Yorck and the Era of Prussian Reform* (Princeton, N.J.: Princeton
University Press, 1966).

it. He did not believe that Napoleon would launch a ground war to stop Russia's commerce with Britain.[275]

Napoleon surprised the tsar. He spent the early months of 1812 assembling an army of unprecedented size, launching it toward Moscow on June 24. The army had nearly one-half million men, including 200,000 French troops and 250,000 foreign auxiliaries. Marching across Prussia, Napoleon crossed the Nieman River with the intention of surrounding and destroying the Russian army. As the Grand Army disembarked, Scharnhorst told his sovereign that Prussia's chance for freedom had arrived.[276]

While the Grand Army was the largest military force ever assembled to that point in European history, Scharnhorst and the reformers pointed out that it was marching in the wrong direction. It was headed away from Prussia, and this presented an opportunity to revolt. In addition, the loyalty of the French army was questionable. The majority of troops in the Grand Army, Scharnhorst told his king, were not French. Each flank was protected by foreign troops, the south being covered by an Austrian corps under Philipp von Schwarzenberg, while French Marshal Jacques MacDonald's northern command comprised foreign units that had no intention of fighting for the French. Finally, all Napoleon's supply lines would pass through Prussia. Scharnhorst gave the king plans for mobilizing the *Landwehr* while Stein, quite illegally, tried to open diplomatic doors in Vienna and St. Petersburg.

As a war party formed around Scharnhorst and Stein, the ever-cautious Frederick William weighed a demand from Napoleon. Yorck was to place the Prussian army, Napoleon ordered, at the disposal of Jacques MacDonald on the northern flank. To the reformers' horror, the conservatives, including Yorck, thought this to be the most prudent course of action. Despite Jacobin threats, the conservatives argued, Napoleon acted more like Louis XIV than a radical revolutionary. The peace terms for Prussia were harsh, but Prussian social order had not been overturned and the Hohenzollerns still ruled from Berlin. If the king revolted against Napoleon and lost, Napoleon might eliminate the Hohenzollern dynasty. Yorck also pointed to the dubious

---

[275]David G. Chandler, *The Campaigns of Napoleon* (New York: MacMillan, 1966), 739–49.

[276]Kitchen, 52. (Scharnhorst tried to resign from the army to join Blücher. The king refused.)

quality of the *Landwehr*, stating that it was no match for regular French troops. When Scharnhorst countered by stating that Austrian and Russian alliances would strengthen the Prussian army, Yorck pointed out that the alliances had not been formulated and that they only existed in the dreams of the reformers. Frederick William sided with the conservatives, and when the Grand Army moved out on June 24, Yorck reported to MacDonald for duty. The Prussians were to protect Napoleon's left flank.

The reformers responded to the king's decision in varying manners. Enraged when Yorck marched away, Scharnhorst resigned from the army, only to have Frederick William refuse the resignation. After strong words of protest, Scharnhorst sat in silent disgust, infuriated with a king would not risk his throne for the benefit of his kingdom and subjects. Stein, not officially in the Prussian service, countered the king's inactivity by creating a subversive diplomatic network. He eventually joined a group of Prussian administrators who assembled in St. Petersburg to advise the tsar. Gneisenau and Boyen, who were as upset as Scharnhorst, also practiced secret diplomacy, feeling out the possibility of military alliances with Russia and Austria. Nothing, however, matched the radical actions of Clausewitz. In a fiery letter to Frederick William III, Clausewitz resigned from the Prussian army, offering his services to the tsar. Alexander welcomed the disgruntled officer, giving him a position on the Russian general staff. Frederick William believed Clausewitz to be a traitor, and he never forgave him.

Napoleon's strategy in Russia was almost immediately frustrated. The Russians had two large armies along the border. Napoleon hoped to destroy them before they could retreat, but he was unable to bring them into a decisive engagement. By August 19 the Russians tried to make a stand at Smolensk—where Napoleon would have undoubtedly destroyed them—but their staff work was so slow that the army never assembled. Losing the opportunity to strike, Napoleon could only chase the Russians as they fled deeper into the east. On August 29 the tsar sent General Mikhail Katuzov to assume command of the army. The one-eyed Katuzov met Napoleon sixty miles west of Moscow on the field of Borodino on September 7. Even Napoleon was stunned by the magnitude of the battle, as more than 70,000 men were lost on a single day. Yet, when the victorious French advanced,

Katuzov's survivors ran away. By September 14, Napoleon was in Moscow, but the Russian army was still intact, east of the French position.[277]

Yorck von Wartenburg was a reluctant ally. He may have agreed to join the French, but he held no trace of loyalty to either Jacques MacDonald or Napoleon. Despite many orders to the contrary, Yorck continually refused to employ the Prussians against the Russians. There were a few scattered engagements as Yorck marched back and forth, but MacDonald suspected that all of Yorck's actions were mere illusions. Yorck always marched to the wrong location, usually several days behind the time schedule. When he finally took action he labeled it "active reconnaissance," and MacDonald noted that the operations always took place in areas previously secured by his own troops. In essence, Yorck was not extremely effective, and while Napoleon sat alone in Moscow, the Prussians were intact to his north. Had Yorck believed that the *Landwehr* could be successful, he might have agreed to side with Scharnhorst and the war party.[278]

By late October Katuzov was able to launch a counteroffensive. Hundreds of miles from friendly territory, Napoleon's long lines of communication were threatened by Katuzov's actions. The Russians had been defeated at Borodino, not destroyed, and they longed for revenge. Katusov's cavalry probed deeply along the French lines of communication, finding that local residents were happy to help them harass the French. Napoleon's troops had not been overly kind to Russian peasants during the invasion, and peasants flocked to the tsar's army at every opportunity. In frustration Napoleon attacked Katuzov at Maloyaroslavets on October 24. The Russian line held, and the French were driven back. With supplies in arrears and temperatures dropping, Napoleon decided to retreat from Moscow. It started snowing in early November.

---

[277]Summary based on Chandler, 861–954, and Henry Lachouque, *Napoleon's Battles: A History of His Campaigns* (trans. by Roy Monkcom) (London: George Allen and Unwin, Ltd., 1966), 269–323. See also Curt Jany, *Die Königlich Preussische Armee und das Deutsche Reichsheer 1807 bis 1914* (Berlin: Verlag von Karl Siegismund, 1933), 53–68.

[278]See comments by Kitchen, 49. See also Karl-Heinz Riemann, "Graf Yorck von Wartenburg—preussischer General zwischen feudalem Konservatismus und Patriotismus," *Militärgeschichte* (6) 1987: 585–91. (Riemann gives a leftist perspective. This is one of the most hotly debated actions in German military historiography.)

Yorck took these actions into account. While ostensibly under Mac-Donald's command, Yorck began acting more and more like an independent army commander. The Prussians retreated westward, acting as if they were following the French line of retreat, but they gradually began moving north. MacDonald complained, but he had no way of stopping Yorck. By the end of December, Yorck had retreated to Riga, several dozen miles away from the French army. Wanting to assure the tsar that encounters between Prussian and Russian forces were misunderstandings, Yorck sent emissaries to the exiled Prussians in St. Petersburg. Stein returned contact from St. Petersburg, asking Yorck to leave the war. Yorck hesitated, but did not leave Riga when his army was "surrounded" by the Russians. On Christmas Day 1812, a Russian general arrived at Yorck's headquarters in Riga to discuss terms for a Prussian surrender. Yorck recognized the officer in the green uniform. It was Karl von Clausewitz.

It was a most interesting dilemma for Yorck because Frederick William wanted Clausewitz arrested and tried for treason. Clausewitz could have had his own ax to grind—one of his closest friends had been killed by one of Yorck's hussars in the early fall—but patriotism proved to be stronger than personal feelings. Although both Yorck and Clausewitz felt the entire affair was a political mistake, they were enemies in different uniforms. Yorck proposed a solution. Since he was surrounded, the rules of field command allowed him to negotiate favorable terms of surrender. Wanting no part of a continuing war with Russia, Yorck was ready to do so, but he did not have the authority to do the one thing Clausewitz desired. Clausewitz wanted Yorck to sign an alliance with Russia. That was a policy matter for the Prussian crown, and Prussia was, after all, a French ally.

It could have been a tense moment, but Clausewitz brushed formalities aside. He was no more Russian than Yorck, he explained, and his Russian venture had been merely the most expedient way to fight Napoleon. Clausewitz wanted to continue the war in a Prussian uniform, and so did every other Prussian in the tsar's service. The real enemy of Prussia, Clausewitz argued, was France. The tsar was willing to fight France and assist with the liberation of Prussia. Clausewitz assured Yorck that the tsar had no territorial designs on Prussia; he simply wanted Napoleon out of Central Europe. All Yorck had to do, Clausewitz said, was to accept the tsar's offer of alliance and Prussia could be free. Although Yorck had no authority to agree, Clausewitz eventually began to wear him down. On New

Year's Day 1813 Yorck caved in and pledged his troops to the Russian army. When Frederick William heard the news, he issued a warrant for Yorck's arrest, not understanding that his generals had started a war for liberation.[279]

---

[279]Kitchen, 52.

Chapter 13

Liberation

From January 1813 to June 1815 the Prussians fought a series of on-again-off-again wars with the French. In German historiography these conflicts are known as the Wars of Liberation, and they were dominated by two themes. First, although the Prussians were fighting to oust the French from their homeland, their military role was limited. They always served as the junior partner in an alliance with other major powers. In their final battle, they joined Arthur Wellesley, the Duke of Wellington, at Waterloo, playing a part similar to the assignment given to the Old Dessauer at Blenheim. The Prussian army was an important actor but never cast in the leading role, even in its own war of liberation. A second theme developed from the changing nature of battle. As the Prussians rallied to the colors, the physical nature of war grew. Large, nationalistic armies battled one another for days on end. Casualties increased, reflecting the size of armies and the physical magnitude of battlefields. Fighting expanded into a series of days or weeks. Major battles, which would have consumed the army of Frederick the Great, became steps taken along the trail of a campaign. For the Prussians it was the first step into modern warfare.

Coalition Warfare

Yorck von Wartenburg was surprised at Frederick William's decision to charge him with treason. Clausewitz, with Stein's enthusiastic support, had convinced him that he was leading the first stage of liberation, and he had been assured that his decision to join the Russians would unite the ideologically divided Prussian officer corps. Lamenting over the king's decision, he wondered whether he had taken the proper action, but Stein did not give Yorck too much time to consider his position. After Yorck joined Clausewitz, Stein moved his entire entourage of exiled Prussians to Riga, and he told Yorck to forget the matter. The Prussians were fighting for liberation, and if the king could not realize that, *he* would be discarded after the French were removed from Prussian soil. Clausewitz joined the Prussian army as an official representative from Tsar Alexander, and Yorck moved to the west, not joining the French retreat but attacking his former allies as a Russian confederate. Jacques MacDonald could only notify Napoleon of the Prussian defection.[280]

Frederick William, feeling betrayed by Yorck's actions and the seeming willingness of his troops to follow the errant general, smoldered in rage. Every general's first goal was to preserve the Hohenzollern throne, not the nebulous "state" of Stein and Clausewitz. And Napoleon, though not much of a friend, had, at least, allowed Frederick William to rule from Berlin. Now his own subjects were taking matters into their own hands, and, fearing for his future, the king hesitated to support them. His sympathies were with the rebels, but if he joined them and they failed, he would lose his crown. Of course, Stein had no hesitation whatsoever. Burning with a terrible fever, the ailing former minister took up residence in Königsberg. Moving ahead of the army, he gathered Prussian patriots to the flag, organizing a de facto government. Stein told the East Prussians that loyalty to the state superseded loyalty to the Hohenzollerns. The king, he said, was merely a servant of the state. When Yorck arrived in February, he proclaimed East Prussia liberated in the name of the king. Frederick William wondered how Yorck could give the Hohenzollerns something they already ruled.[281]

Despite his fever, Stein was a flurry of action. He managed to convince the East Prussian Junkers to hold an assembly of the estates and to declare war. Stein told the Junkers that Napoleon was fleeing to Paris and that the Grand Army—at least, the few thousand souls who remained in it—was crawling westward through the snow. Yorck, Stein argued, was commander of the Prussian army and in action against the French. The only suitable conclusion, Stein passionately cried, was to take action. Although some of the lords wanted to wait for word from the king, the majority voted to mo-

---

[280]Paul W. Schroeder, *The Transformation of European Politics: 1763–1848* (New York: Oxford University Press, 1994), 450–52.

[281]Gordon A. Craig, *The Politics of the Prussian Army: 1640–1945* (London: Oxford University Press, 1968), 59. (Craig discusses the historiographical analysis of Yorck's action in footnote number 1.)

bilize local *Landwehr* units and the home guard, or *Landstrum*. Conceived by Scharnhorst and Clausewitz during the early days of reform, the *Landstrum* was nothing more than an organized mob of armed citizens. Ironically, the most feudalistic province in the kingdom endorsed the formation of a popular army, placing it under Yorck's authority.[282]

At court the king's entourage was aghast by the actions of the East Prussians. Forming a popular army and declaring war against the wishes of a king were Jacobin. The estates had the right to levy taxes but no right to make war. The legal argument, albeit correct, was lost on the citizens of Berlin, especially the middle class. Rumors of Yorck's actions spread through the streets, and the people expected that he would arrive when the snow melted. To the king's horror, the common people began flocking to local *Landwehr* units without his permission. The process was exacerbated when Frederick William learned that Scharnhorst and Clausewitz were raising a *Landwehr* army in Silesia. Clearly, the king had to take some type of action. With French troops still in Berlin, he joined Scharnhorst in February.

The process of liberation seemed to be underway. Karl von Hardenberg, the king's chief minister, endorsed the call for a national army, and word of the king's presence in Silesia solidified the nation. Clausewitz met with the king to beg his forgiveness. The king pardoned Yorck but still considered Clausewitz a traitor. Despite appeals from Scharnhorst, Gneisenau, and Hardenberg, the king would not forget the stormy manner in which Clausewitz left the Prussian army. He dropped the charges of treason, allowing Clausewitz to join Scharnhorst as a Russian attaché, but he refused to re-appoint the general in the Prussian army. Aside from this one disappointment, the reformers were elated. Within weeks Prussian troops under the command of the Russian General Ludwig von Wittgenstein entered Berlin. On March 16, 1813, Frederick William followed the lead of his subjects and declared war on Napoleon.[283]

---

[282]Martin Kitchen, *A Military History of Germany: From the Eighteenth Century to the Present Day* (Bloomington, Ind.: Indiana University Press, 1975), 54–56.

[283]Hajo Holborn, *A History of Modern Germany: 1648–1840* (New York: Alfred Knopf, 1978), 424–27.

The First Round for Liberation

With all of the German states in an uproar, Baron vom und zum Stein had the presence of mind to assess the situation fully. Prussia could not be liberated without an alliance, and Stein spent his energies convincing his friends at court to seek foreign liaisons. The Russians were the most obvious choice since several Prussians—including the exiled dignitaries, Yorck, and the East Prussians—had already joined the tsar. Blücher, who had taken command of Scharnhorst's Silesian army, enthusiastically endorsed Stein's advice, placing his army under Russian command. Tsar Alexander repeated an earlier promise, stating that he had no territorial aspirations in Germany and that he was only committed to the overthrow of the Confederation of the Rhine. Following Stein's advice, Hardenberg convinced Frederick William to sign a formal alliance with Alexander.[284]

The Prussians were not alone in the race to Russia. Caught in a crisis of leadership, Sweden searched for a successor to the royal throne, and Swedish nobles cast their hopes on a former Napoleonic marshal, Jean Baptiste Bernadotte. An accomplished French military commander, Bernadotte could become the crown prince due to marriage, and he bolted from the Napoleonic ranks when issues began to go sour in Russia. Accepting the future crown of Sweden, Bernadotte joined the Russo-Prussian alliance to cement friendly relations with St. Petersburg. Great Britain also joined the alliance, even though the country had been officially at war on Prussia over the issue of Hanover since 1807. As soon as word of the anti-French affiliation came to London, the British king mobilized all his German legions and promised monetary support for Prussia.

As Hardenberg and other Prussian officials listened to Stein's counsel on the alliance, they realized that Austria had a crucial role. The Prussians, Swedes, and Russians would be formidable, particularly if the Prussians were backed by British sterling, but the real power in Central Europe was Austria. Officially, Francis I was emperor of Austria, but foreign policy was guided by Prince Klemens von Metternich, a diplomat with full appreciation for the gravity of the impending war. The French might be retreating, Metternich thought, but they had handed the Austrians two devastating defeats in the last ten years. Metternich did not relish the possibility of another

---

[284]Craig, 59–60.

defeat, and he did not want to lose influence in German affairs. Austria played the central role in German politics, and Metternich was not anxious to remove French influence only to see it replaced by Russia or Prussia. Francis's sympathy was with the allied cause, yet Austria needed to do more than win the war; it needed to win the peace. Therefore, Metternich sat on the fence. Austria would join the allied coalition only if Metternich believed the allies could win.[285]

To his credit, Frederick William no longer chose the fence. Gone was the reluctant revolutionary of December, and the once-timid sovereign threw his full weight into liberation, knowing full well that there was no turning back. If Napoleon were triumphant, it would spell the end of the Hohenzollern dynasty. He signed orders to mobilize Prussian reserves, forced complaining Silesians to the colors by conscription, and even made a patriotic appeal to the citizens of Prussia. In contrast to the indecisive days of Jena, the king appointed generals who wanted to fight and gave them staffs to help them do it. He ordered reformers and conservatives to end their squabbles and gave subordinates enough latitude to whip their units into fighting trim. It was a new army for a new day, and Scharnhorst was delighted.

The list of the king's warriors was impressive. Frederick William placed Gebhard von Blücher in command of the Prussian army and made Scharnhorst his chief of staff. Yorck received the primary corps in Blücher's army, and Gneisenau was given a staff position. Generals Friedrich Heinrich von Kleist, Friedrich Bogislav von Tauenzein, and Friedrich Wilhelm von Bülow each received a corps and a chief of staff. Scharnhorst's former students—Boyen, Grolman, Rauch, Reiche, and Rothenburg—received staff positions, although when Colonel von Rauch reported for his assignment, Yorck greeted him by saying that he had no need of a chief of staff. Rauch stayed despite the ungracious welcome and helped organize the corps. Only Clausewitz, who officially remained in the Russian army at Frederick William's insistence, was rebuffed. The embittered Clausewitz continued to work for Prussia in a foreign uniform.[286]

---

[285]Holborn, 428–30.

[286]Craig, 62. See Peter Paret, *Clausewitz and the State* (New York: Oxford University Press, 1976), 233–38.

When the snow thawed the Prussians were ready to do battle. Their strategy was simple and dictated by the regrouping French forces. The Grand Army had limped back from Russia, leaving Napoleon no choice save to establish a defensive line through northern and central Germany. Using a string of Prussian fortresses, Napoleon hoped to fight a strategic defensive along the Elbe and Oder rivers. The fortresses would serve as strong points, and Napoleon hoped to meet any breach in the line with local offensives. The Prussians and their allies hoped to break the French line with either of two major armies, a Russian army under Wittgenstein or a Swedish-Anglo-German army under Bernadotte, now Crown Prince Charles John. Blücher and most of the Prussians served under Wittgenstein in Silesia, while the remaining Prussians assembled under Charles John in Pomerania and Brandenburg. At first, the allies were hopeful because Napoleon was rumored to be facing a domestic revolution, but they were sadly disappointed when Napoleon arrived later in the spring with a new army of fresh troops.

Napoleon came back to Germany in April with more than 200,000 men. Ill-trained and lacking equipment, the army was a combination of French, Italian, Polish, and German soldiers. It looked more like the armies of the revolution than the Grand Army. The French troops were not in the best shape, receiving only four weeks of training before reporting for duty and additional orientation from their individual regiments. The soldiers from other nations received less preparation, but it was the best the emperor could do. He needed men and he needed them quickly. Napoleon moved toward Leipzig in early May with the green army poised for battle. The coming war promised to be a battle of citizen soldiers.[287]

If the troops were not up to the standard of the old days, it was hard to tell so from their appearance. The pace of their march made it look as though the Grand Army had returned to Germany for a repetition of Jena. Napoleon moved toward Leipzig in three long columns, each commanded by a veteran officer of high ability. If the troops would stand in battle, Napoleon's genius could handle the rest. As he marched forward, one of his most trusted subordinates, Marshal Michel Ney, guarded the right flank with the

---

[287]See David G. Chandler, *The Campaigns of Napoleon* (New York: MacMillan, 1966), 865 ff., and Henry Lachouque, *Napoleon's Battles: A History of His Campaigns* (trans. by Roy Monkcom) (London: George Allen and Unwin, Ltd., 1966), 324 ff.

northern-most column, and Ney felt that the army was back to its old form. Green troops trained as they marched, and Ney knew that the allies had their share of first-time soldiers. He believed that when Napoleon appeared on the battlefield, the allies would whither. Unfortunately for the French, Ney's confidence was not borne out by solid information. He had been unable to conduct a thorough reconnaissance against the allied lines due to an outbreak of disease among his horses, and the cavalry could not train as his corps moved. Unable to see his enemies, Ney pressed ahead in ignorance, with Napoleon's approval.

General Count Ludwig von Wittgenstein was forming his own opinion about Napoleon's new command. Utilizing a combined formation of Russians and Prussians, the Russian general had been keeping close tabs on Napoleon's movements. He was pleased with the response of his own troops and the attitude of the Prussian army. Not knowing of Ney's diseased horses, he also noticed the lack of aggressiveness on the part of the French cavalry. Ney's flank was completely exposed, and he was doing little to guard it. Wittgenstein reckoned that it could mean only one thing; the French were not ready to give battle. Deciding to take matters into his own hands, Wittgenstein conferred with Scharnhorst to assure Prussian cooperation and decided to attack the French at Gross-Gorschen. The assault began on May 2, catching the French entirely unaware.

Since the Prussians were fighting under Russian command, there was little for the senior officers to do other than to encourage the men. Blücher and his commanders agreed that the best place to inspire the soldiers was in the ranks, so they moved to the front of the Prussian army. Wittgenstein's initial attack was a complete surprise. He hit Ney while the French forces were arrayed in columns along the open roads. The opening bombardment was devastating, and the French were forced to retreat. When Ney was able to rally the troops along a defensive line, Wittgenstein ordered the Prussians forward. Blücher, Clausewitz, Gneisenau, and Scharnhorst went in with the first wave. Ney unlimbered his own artillery, but the Prussian assault was too strong. The Russians moved forward in support, and Ney's command was decimated. Blücher was knocked to the ground and left with a severe contusion, Grolman was slashed with a bayonet, and Clausewitz used his sword to fend off an attacking French infantryman. The Prussians were inspired by such actions, and between the spirited Prussian assault and Wittgenstein's surprise, there was little the French could do. Ney's command disappeared in the smoke. By noon, the northern flank of the French army was in complete disarray.

Napoleon watched the opening events from the central French column a few miles away on the historic battlefield of Lutzen. Seeing Ney in crisis, he assembled his untried troops and marched toward the sound of the guns. As Ney's men reeled under Wittgenstein's assault, Napoleon established a second defensive line. The emperor brought the main artillery battery forward, and when Wittgenstein's assault hit the second French column, Napoleon's artillery stopped the allies dead in their tracks. This reversed the tide of the battle, and, by early afternoon, it appeared as if Napoleon would sweep Wittgenstein from the field. Yet, the French troops were raw and untested in battle. When Napoleon tried to advance, his own force came under disciplined fire. Rather than simply retreating, Wittgenstein formed his own defensive line and stopped the French counterattack. He then left the field in good order, leaving the French unable to pursue. Some military historians believe that Napoleon's brilliant efforts turned disaster into victory. Many others think the battle of Gross-Gorschen was simply a draw. Scharnhorst had another opinion. He considered the day to be an allied victory.

Scharnhorst was not merely waxing romantic about the battle of Gross-Gorschen; he was speculating about political ramifications. The battle may have been a draw, but Ney's wing had been defeated, and Napoleon's counterattack had been stopped. In the larger picture, Scharnhorst assumed that he now had something to offer Metternich. If the Austrians wanted to enter the war on the winning side, Scharnhorst held the battle of Gross-Gorschen as evidence to prove the allies could win. Sharing the day's dangers with the other generals, Scharnhorst resolved to go to Vienna as soon as the battle ended. Before he left the field, however, a French musketeer fired in the direction of the Prussians. The round caught Scharnhorst below the knee and sent him plummeting to the ground. Although it was painful, Scharnhorst ordered a doctor to wrap his leg, and he boarded a coach for Vienna. With Blücher's blessing and concern, the Prussian general rode through the night.

A devoted family man, Scharnhorst always confided his innermost feelings to his wife and children. They had vicariously experienced the ebb and flow of the reform movement through him, and they knew the passions that burned in his soul. He had once written his daughter about the meaning of liberation. Who would not willingly give his life, Scharnhorst wrote, if it meant an end to French tyranny? Perhaps he recalled these words as he rode through the night, or perhaps he only gritted his teeth in pain. Regardless, by morning Scharnhorst's wound was infected, and in a few hours his body was racked with fever. Ignoring the need for medical attention, Scharnhorst rode on to seek Austrian assistance. It was his last ride. By the time he allowed a surgeon to examine the wound, his leg had turned a blackish green. Over the next month nothing stopped the fatal spread of the infection. On June 28, 1813, Gerhard von Scharnhorst died a soldier's death, a casualty of Gross-Gorschen. Like so many Prussian soldiers, he did not live to see the day of liberation.[288]

For Napoleon Bonaparte the deaths of Scharnhorst and a few thousand other Prussians were insignificant. His objective was to destroy Wittgenstein's army. He reassembled the French forces and pressed forward into Germany. On May 20 he reached Dresden to establish supply base. Knowing that the allies were in front of him, he sent Ney to the north with the bulk of the army, while moving the remaining forces to Wittgenstein's front. His plan was to encircle the allies. On May 21 Napoleon attacked the Russo-Prussian force near the village of Bautzen as Ney moved to close the escape route. Wittgenstein was fortunate. Ney crossed the allied flank, completely exposing the rear, but he failed to attack. Wittgenstein reacted by establishing defensive forces in front of Ney, and, when Ney finally attacked, the allies were in a position to throw him back. By afternoon, Wittgenstein was able to disentangle his forward elements from Napoleon's frontal assault. The next day he retired in good order, with the allied army completely intact. Napoleon began to realize that the French army was in strategic peril.

Scharnhorst's ride may have cost him his life, but his logic was accepted in Vienna. Emperor Francis was not a zealous partner, yet he was impressed with Wittgenstein's resistance. On June 1 he ordered Prince Philipp von Schwarzenberg to take a strong Austrian army into Bohemia, placing it just south of the French positions in Saxony. Schwarzenberg could either observe the French from this position or possibly even enter the war. An enthusiastic Frederick William contacted Francis and promised to reinforce the Austrians if the emperor opened hostilities. At the same time Sweden's Charles John moved to the south from Berlin with 120,000 men, including a Prussian corps commanded by Friedrich Wilhelm von Bülow

---

[288](Gneisenau and Clausewitz wrote Scharnhorst's obituary.)

and his chief of staff, Hermann von Boyen. The tsar also took decisive action. Even though Wittgenstein's retreats saved the allied center, Tsar Alexander wanted the Russo-Prussian force to take the offensive. Since his army was operating on Prussian soil, he agreed to place the Silesia army under Blücher's authority. Blücher salivated at the thought of bringing Napoleon to battle, and the tsar liked the Prussian general's attitude. Surrounded by three major armies, Napoleon sought a truce. He received it on June 4, 1813.

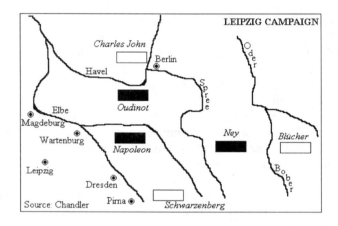

Neither the allies nor Napoleon believed the truce would hold. The French emperor needed time to prepare his inexperienced army and to gather more troops. The allied commanders wanted to assess the situation and redistribute their troops. The allies were poised to attack, but they could not fully coordinate their efforts. Frederick William, Alexander, and Francis wanted to develop a common strategy, and they also wanted to give Napoleon an opportunity to offer a permanent peace. In the meantime, they continued to surround the emperor with three large armies. The Army of Bohemia under Schwarzenberg had 230,000 men, while Blücher's Silesian Army numbered 195,000. The crown prince had another 110,000 men in Berlin. Not oblivious to the hazard, Napoleon quickly cemented alliances with Saxony and Bavaria, fielding a total of 300,000 men. He had no intention of offering peace, and by August allied patience was wearing thin. Francis took the first step on August 12, declaring war on France.

The Leipzig Campaign

Surrounded by three armies, Napoleon found himself in a situation similar to Prussia in the Seven Years' War. Fighting on "interior lines," his primary goal was to keep the allies from uniting, and he formed three units to counter the allied position. The Army of Berlin, under French Marshal Nicolas-Charles d'Oudinot, was to deal with the Swedish crown prince's Army of the North. Napoleon gave Ney the Army of the Bober, named for the river separating the French from Blücher's army, ordering him to protect Dresden from both Silesia and southern Saxony. Napoleon established a third army under his personal command to respond to the greatest threat, and knowing the aggressive tendencies of Marshal Blücher, he marched toward the Silesian Army to join Ney, hoping to drive Blücher away from the other two allied positions.[289]

After Napoleon joined Ney, Oudinot also took the initiative. In an effort to keep the crown prince from reinforcing Blücher, Oudinot moved north toward Berlin with the intention of either disarming the city or destroying it. On August 23 his troops ran into a determined Prussian corps in the village of Grossbeeren south of Berlin. Although reluctant to engage his former master, Charles John ordered the village defended, and he sent word to Blücher that the allies were under attack. Bülow's corps, entrenched inside Grossbeeren, fought off the French all through the day, and at night counterattacked in a driving rainstorm. Oudinot was forced to retreat. It was hardly a decisive victory, but for the first time since Jena, the Prussians had forced the French from the field. Prussian moral increased.[290]

---

[289]Summary based on Chandler, 865–941, and Lachouque, 324–80. See also Curt Jany, *Die Königlich Preussische Armee und das Deutsche Reichsheer 1807 bis 1914* (Berlin: Verlag von Karl Siegismund, 1933), 89–103. For an informative popular history, complete with pictures and maps, see Peter Hofschoer, *Leipzig 1813* (London: Osprey, 1993). (The series is edited by David Chandler. It is not only easy to read—like Chandler's other work—it is also highly informative for general audiences.)

[290]See Heinrich von Treitschke, *History of Germany in the Nineteenth Century, Volume I, The War of Emancipation* (trans. by Eden and Cedar Paul) (New York: McBride, Nast, and Co., 1915; orig. 1884), 569 ff.

Blücher achieved another small victory a few days later. After Napoleon joined the Army of the Bober, he moved forward toward the Katzbach River. Blücher, not willing to let Napoleon advance unchallenged, cautiously probed his enemy's front and was noticeably surprised by the French response. They seemed none too eager to fight. What Blücher did not know was that Napoleon had moved back toward Dresden to counter a threat from Schwarzenberg's Bohemian Army. Instead of facing a combined force, Blücher had only a single army to his front. He conferred with his new chief of staff, August von Gneisenau, and both men decided that the time was ripe for an attack. As rainstorms continued to soak the area, Blücher crossed the Katzbach River and struck the French. Jacques MacDonald, who had replaced Ney, was unable to resist and fell back. The real battle, however, had taken place about forty miles to the west of Blücher's victory.

When Napoleon moved east, Schwarzenberg moved north. His target was the French supply base at Dresden. Although Napoleon wanted to deal with Blücher, he had no choice but to turn around to protect his rear from Schwarzenberg. Leaving MacDonald with a small screening force along the Katzbach, the rain-sodden French army marched back to Dresden as fast as the muddy roads would allow. On August 26, just as Blücher's troops were crossing the Katzbach, Napoleon heard that Schwarzenberg's entire army was on the outskirts of Dresden. Schwarzenberg had over 200,000 troops compared to 70,000 French soldiers, but the allies had a political problem. All three of the allied sovereigns were attached to Schwarzenberg's camp, and he needed their permission to attack. While the morning was fretted away, the French prepared defensive positions along the city's walls and Napoleon marched to their relief. When Schwarzenberg finally received permission to attack, the French were prepared to receive him.

Schwarzenberg launched a massive attack in the late afternoon, and it appeared as though the delay was not going to hurt him. Outnumbered, the French troops were forced to retreat, and each allied assault brought a new French withdrawal. When night fell, Schwarzenberg continued the attacks, hoping to break through the city's defenses in darkness. The plan would have worked had Napoleon given his relief forces an opportunity to rest. He did not. Marching through the night, Napoleon and his reinforcements arrived in Dresden around midnight. The allies were forced to call off the attack. The next day found the allies in a delicate position. Not only had Napoleon reinforced Dresden, but he had slipped another corps, under General

Dominique Vandamme, around Schwarzenberg's rear. Casting all timidity to the wind, Frederick William argued for a continued attack. Schwarzenberg agreed with the Prussian king, but the other monarchs were not quite so sure. Napoleon settled the debate by counterattacking from Dresden in the afternoon, throwing the allies back toward Vandamme's corps. The Army of Bohemia was surrounded.

For the second time in two days, Schwarzenberg decided to use the darkness. Napoleon broke off the attack in the early evening, and Schwarzenberg moved his men back. Wet and exhausted after two days of fighting, the allies moved throughout the night, and by morning they arrived at the fortress of Pirna, the position held by Vandamme's corps. For some reason, Napoleon had not pursued Schwarzenberg, but the exhausted allied soldiers did not know this. Hopelessly outnumbered, Vandamme waited in vain for Napoleon to come to his relief, and when his scouts brought word that the emperor was staying in Dresden, Vandamme ran from the allies as fast as possible. Schwarzenberg and his commanders thought it strange that Vandamme should retreat, but they did not pursue. It could have been some type of Napoleonic trick.

Friedrich Heinrich von Kleist led the Prussian corps at the front of the allied retreat. He knew that the French were blocking the path, and, like the other allied commanders, he was convinced that Napoleon was poised and ready to strike the allied rear. When Kleist's men bumped into Vandamme's corps on August 29, the Prussians quickly unlimbered their artillery and fired everything they had. Kleist fully intended to sacrifice everything to keep Vandamme from stopping the army. The Russian and Austrian forces in Schwarzenberg's army heard the shooting and believed Vandamme to be attacking. Without waiting for orders, the Austrians and Russians formed defensive lines on Kleist's flanks, awaiting the main French attack. Only Vandamme was aware of the predicament. Far from being the aggressor, the allied "defensive" lines surrounded him, while Kleist's troops clawed forward like caged animals. By the end of the day, Schwarzenberg realized what had happened. Napoleon had not pursued; he had remained in Dresden. The next day Schwarzenberg's troops captured Vandamme, and the surviving members of his corps surrendered to the allies.

Napoleon left Vandamme to his fate because he believed that Schwarzenberg was running, and the French were in no position to advance deeply into enemy territory. Presuming that Vandamme would proceed to either Dresden or the Bober, Napoleon returned to the Katzbach to meet the advancing Blücher. The wily Prussian general had by now heard of the French victory at Dresden, and Blücher had no intention of letting his own small victory on the Katzbach go to his head. He had attacked when the French were not ready to fight, but when Napoleon returned spoiling for a fight, Blücher thought it best to slip quietly back into Silesia. The Prussian commander correctly guessed that Napoleon could not leave his defensive position to chase an allied army. For his part, Napoleon did everything he could to tempt Blücher from his lair, but the Prussian would not be moved. Blücher held his position for three weeks.

Blücher was waiting for a coordinated offensive. In September, Charles John slowly moved his army south from Berlin, and Napoleon was forced to send Ney to meet the new threat. Ney found the crown prince on September 6 around the village of Dennewitz but soon discovered that he had bitten off more than he could chew. The crown prince's troops refused to budge under a series of French assaults, and by evening Charles John counterattacked, threatening to overrun the French center. Once again, Bülow's corps lead the way. Oudinot came to the scene and assumed command of the center, trying to rally the forces in defense. It was too late. As Ney weakened other parts of the line to strengthen Oudinot's position, the crown prince sent a combined Swedish-Prussian force into the French right flank. It was more than the French could bear. The right flank gave way, followed by the collapse of the center, and Ney was forced to retreat. Charles John, the former Marshal Bernadotte, knew the French game.

Charles John's victory at Dennewitz had strategic consequences. Growing increasingly alarmed at the inability of the French to win a decisive victory, Napoleon's two major German allies, Saxony and Bavaria, became disgruntled. The Saxons, who had never been enthusiastic French allies, had only joined Napoleon because they thought the French were unbeatable, but after Dennewitz some officers talked about joining the Prussians. In addition, Napoleon's supplies from Dresden were starting to run low. With Saxon loyalty in doubt, Dresden appeared to be in jeopardy. A more severe political disaster brewed in Bavaria. A traditional French ally, the Bavarians voted to leave the Confederation of the Rhine after hearing of the crown prince's victory at Dennewitz. Napoleon tried to tighten his control of the German state while keeping his supply lines open, but this was no easy task. By early October Napoleon was having doubts about his ability to stay in Germany. Blücher decided it was time to attack.

Gebhard von Blücher correctly assessed the nature of the situation, but he had a major problem. Napoleon as well as the other two allied armies were on the west side of the Elbe River, while Blücher sat on the east. In short, he was on the wrong side of a formidable barrier, and Napoleon's main force separated Schwarzenberg and Charles John. Since the Elbe was difficult to cross, Napoleon did not need a large army to defend against Blücher, and the French even held all the bridges. Blücher decided to take matters into his own hands. Ignoring the French monopoly on river crossings, Blücher assembled a number of pontoon bridges at Wartenburg and crossed the Elbe on October 3. Wartenburg was defended by German Confederation and Italian troops with a small French reserve, but their resistance was stiff. Blücher was able to establish a small bridgehead, but the entrenched German and Italian troops kept his men pinned down on the river. The old cavalry general became irritated. Unwilling to see his troops massacred along the banks of the Elbe and unwilling to fail to cross the river, Blücher brought cavalry across the pontoons and threw them into prepared defensive positions. The shock of the attack caught the German-Italian force by surprise, and the French allies broke and ran. When evening came, Blücher's unconventional action allowed the entire Silesian Army to cross to the west side of the river.

Napoleon understood that he must destroy Blücher. If the Silesian Army were to join any of the other allied armies, Napoleon would be outnumbered. He moved the main French army in Blücher's direction, only to find the Prussians retreating north toward the crown prince. Reluctantly, Napoleon marched north to block Blücher's move, realizing that Schwarzenberg would probably use the opportunity to reenter Saxony. When Schwarzenberg moved, Napoleon pulled back from Blücher, unable to stop the allies from joining. It was the critical point of the campaign, and Napoleon faced two choices. He could either stand and fight at strategic disadvantage, or he could withdraw to France before he was forced to battle. The people of Paris would not tolerate another failure, so the die was cast. The Corsican chose to make his final showdown in Saxony.

Throughout the first days of October Schwarzenberg followed the French army as it lumbered north. Each day brought him closer and closer to Blücher. On October 13, Schwarzenberg's lead elements brought exciting news to headquarters. The Army of Bohemia was in contact with Blücher's Silesian Army. Blücher sent word to Schwarzenberg, proposing an offensive, and the three allied sovereigns, still attached to Schwarzenberg's headquarters, gave their approval. Schwarzenberg and Blücher decided to strike Napoleon in tandem, from the east and south, hoping that the crown prince would arrive from the north. According to plans, Blücher waited while Schwarzenberg moved into position. On the morning of the fourteenth the vanguard of Schwarzenberg's army, commanded by the veteran Russian officer Wittgenstein, found the French rear guard at Liebertwolkwitz. It was to be the start of a five-day battle that would lead to Napoleon's downfall.

The Cavalry Clash at Liebertwolkwitz

Napoleon was moving to the northeast in the direction of Leipzig. Hoping that he still might be able to divide Blücher and Schwarzenberg, he created a strong, independent group of four corps and ordered them to guard the rear of the army. Marshal Joachim Murat, the famous cavalry general, took command of the group and set up a defensive screen to guard Napoleon's march. The emperor intended for Murat to delay Schwarzenberg, then to join the main army in the vicinity of Leipzig. If all went well, this would give Napoleon a chance to prepare his troops for a defensive stand around the city. In theory, Napoleon could still separate the two allied armies. If Schwarzenberg were delayed, Blücher might chance an assault on his own. This logic gave Napoleon a slight ray of hope in an otherwise dismal campaign. His plan notwithstanding, the emperor could not afford to lose Murat's men, and he made the point clearly. He told his field marshal to maintain light contact with Schwarzenberg and to avoid a set piece battle. A cavalrymen like Murat seldom listened to such cautious instructions.

Riding a few miles in front of Schwarzenberg's main army, Wittgenstein had surmised that Murat was in the area of Liebertwolkwitz. When his leading troops discovered Murat's positions on October 14, he assumed that Murat was part of the retreating rear guard. With Napoleon in flight, Wittgenstein expected no resistance from Murat. The situation seemed strange, however, because Murat was formed in a battle line on the high ground overlooking the path of Wittgenstein's advance. The Russian general decided to let the matter pass and sent his troops marching ahead. He commanded two Russian infantry corps, a Russian cavalry corps and an attached

Prussian cavalry corps. Even if Murat wanted to fight—which Wittgenstein doubted—the allies had enough troops to deal with him. For Murat, the logical course of action would be to move slowly along the path of Napoleon's retreating army and to delay Schwarzenberg's vanguard. To Wittgenstein's surprise, Murat's next moves defied common military logic.

As the first Russian cavalry elements closed on Murat's position, the French did not retreat. Instead, they fired a salvo from an artillery battery gathered on a hill in front of Wittgenstein's units. Caught in the fury of the bombardment, the Russians wheeled and rode away from the fire. Murat then sent his cavalry forward. What followed turned out to be the largest single engagement of cavalry in the history of western warfare. Allied cavalry units in Wittgenstein's column arrived and saw the leading Russian units fleeing from the French at full gallop. Without waiting to form, the Russo-Prussian cavalry units counterattacked, driving the French back to their own lines. Murat could not stand for this. Rather than using his infantry musket fire to drive the allied cavalry away, he counter attacked with his own cavalry reserve. Thousands of cavalrymen fought in a melee in front of the French position. Not to be outdone, Prussian uhlans and cuirassiers arrived on the field and attacked the mob. The impact of their charge eventually cleared the French from the field.

It appeared as though the issue were over. With hundreds of dead and wounded strewn along the ground, it was time for the rear guard to retreat. As the Prussians formed to ride back to their lines, Murat took the opposite action. He unleashed a third French attack. This caught the Prussians completely by surprise, and they retreated. In a confused series of actions, reformed Russian cavalry units surged forward, joining in the fracas. The disconcerted affair lasted for more than an hour. Wittgenstein's last cavalry unit, the Prussian Neumark Dragoons, had been traveling at the rear and flanks of the vanguard. They arrived at the scene only hours after the initial confrontation, and they were still fresh. With both the allied and the French cavalry exhausted, the Neumarkers charged forward swinging their sabers. The sight of this was more than the French could bear. They swarmed back to their lines before the Prussians could cut them down. The engagement took over three hours and involved a total of seven full-blown cavalry charges. With the French cavalry spent, Wittgenstein's infantry forced Murat from the high ground. Schwarzenberg had an open road to Leipzig.

The Battle of Leipzig

By October 15, Napoleon's position was critical. With Murat's rear guard effectively neutralized, he decided that it was time to make a stand. Assembling his army around Leipzig, he took up a position just to the north of Wittgenstein's vanguard. Ever the optimist, Napoleon had no intention of fighting a static defense. On October 16, he unleashed the entire French army in the direction of Schwarzenberg. Napoleon felt that this would delay Schwarzenberg's advance and give the French more time to prepare Leipzig. The gambit failed as the Austrian general held his positions. By evening, Napoleon regarded further combat as a fruitless waste of effort, and he pulled the army back to Leipzig. Aware that nothing could stop the unification of Blücher and Schwarzenberg, the emperor received once last piece of bad news. Crown Prince Charles John, his former friend, was in sight of Leipzig. Napoleon was surrounded on three sides.[291]

Theologians who argue that God sheds tears over fields of battle could use the Leipzig campaign as an example. It seemed as though the entire two months of the affair had been accompanied by miserable downpours, and the morning of October 17 was no different. The day opened with a driving rainstorm. Soaked and cold, Blücher and Schwarzenberg moved forward for a final assault. To the north Charles John marched through the mud in their direction. Napoleon placed his army in a defensive ring, hoping to keep the allies from overrunning his positions. He then sent emissaries offering to negotiate a new peace treaty, but the allied sovereigns were no longer interested in peace on the Corsican's terms. They rebuffed his efforts to talk. Napoleon shortened his defensive positions and awaited the inevitable. It began the next morning at 9 o'clock.

Leipzig was truly a "battle of the nations." Schwarzenberg's Bohemian army struck from the south while the crown prince's army hit from the opposite direction. Blücher's Silesian army moved to block escape routes. The multinational allied army faced a French army composed of troops from all of Napoleon's allies. Almost every major nation in Europe was represented in the battle. The Prussians played the role of ally, with Prussian units attached to each major allied command. Napoleon's forces fought valiantly, but they fell back with each allied assault. By evening French forces held

---

[291]Ibid., 583 ff.

only the outskirts of Leipzig, and the defensive ring was collapsing. Since the city had not been fortified, darkness promised no relief. The allies halted their attack only to rest. The allies resumed operations at 10:30 A.M. on October 19. Two Russian corps struck from the north while a third hit from the south. General von Bülow's Prussians moved directly toward the center of Leipzig. Napoleon's forces fought for every street, but the allies were too strong. Once again the French line began giving way. As the retreating French army moved closer and closer to the center of the city, the allies took a page from Napoleon's book on strategy. They blanketed Leipzig with massive artillery fire, turning the interior portion of the city to rubble. The French tried to resist in house-to-house fighting, yet by afternoon large groups were moving into the streets. As the cannon fire massacred the retreating soldiers, Napoleon's Saxon allies decided that they had had enough. They pulled out of the battle and surrendered. There was little the French could do. Assessing the situation, Napoleon decided that discretion was the better part of valor. He followed the Saxon lead, escaping with a few French soldiers in a hasty gallop toward the Rhineland.

In the early afternoon Frederick William rode into the market square in the center of Leipzig with Tsar Alexander. Charles John was already there, and a smiling Prussian king offered his congratulations to the Swedish crown prince. Looking beyond the smoking rubble, the king also saw Blücher and Gneisenau. Leaving the tsar, Frederick William joined his two commanders. Blücher was openly jovial, but the wounded soldiers on both sides paid no attention to the highly ranked gathering. Most of them did not care who won or lost; they were concerned with their own pain. Many of them were aware that they would die from infections in the following days. Common soldiers who emerged unscathed also turned to their own affairs. They were happy to be alive. Looters from both armies moved through piles of bodies to rob the dead and dying, often murdering soldiers of any nationality who offered resistance. Prisoners were at the mercies of their captors. The lucky ones survived. Blücher continued to beam. He had seen it all before. This was victory.

Napoleon, however, had tremendous powers of resurgence. By late October he was on the Rhine, and by early winter he began to assemble a new army in France. These actions sent fear throughout the German states. Frederick William, relieved when Tsar Alexander decided that it would be necessary to invade France, promised Prussian assistance. In fact, most of the German states, including Saxony and Bavaria, joined the new alliance. The allied army crossed the Rhine River and marched toward Paris. By the spring the allies were deployed on the outskirts of the city. Meeting with a delegation of French emissaries, the tsar said that he had no quarrel with the French people and no designs on their territory. Napoleon was another matter. The people of Paris took the cue. Gathering under the banner of the opposition, they ordered Napoleon to abdicate. He did so on April 6, 1814, in favor of his son. Neither Frederick William nor Tsar Alexander would agree to this concession, and a few days later on April 11 Napoleon handed the throne back to the Bourbons. The allies exiled the emperor to the Mediterranean island of Elba thinking they had seen the last of him. They were wrong.

The Waterloo Campaign

In many ways, the battle of Waterloo mirrored the previous engagement at Leipzig on a smaller scale. Like Leipzig, Waterloo was not the result of a single day of battle; it involved a series of smaller battles culminating in a general engagement. Napoleon was once again in a desperate position, and his main objective was to keep his enemies from uniting. At the same time, Waterloo was different from Leipzig. The allied force that faced Napoleon was much smaller, and it contained only two major armies, an Anglo-Dutch-German force commanded by the Duke of Wellington and a Prussian army commanded by Blücher. The campaign was also much shorter. At Leipzig the emperor had been forced to the defensive, but offensive thoughts dominated his actions at Waterloo. Finally, Napoleon commanded a French army at Waterloo. It was his last bid for power.[292]

Napoleon escaped from Elba in February 1815, secretly boarding a ship for France. Upon his arrival, thousands of old comrades flocked to their emperor's banner as Napoleon marched toward Paris. On March 20 Louis XVIII fled, leaving Napoleon free to assume officially the reins of power. He immediately offered an olive branch to the crowned heads of Europe, but his startled neighbors—who wanted nothing to do with peaceful overtures—began to mobilize their forces to do battle with France. Frederick

---

[292]Chandler, 1007–90, and Lachouque, 413–59. Also see footnote 16.

William moved decisively. Although his attention was focused on the re-configuration of Germany and the policies of his army, he immediately sent Blücher toward the Rhine at the head of a Prussian army. The king knew that Great Britain would respond, but it would take time for Austria to act, and Russia would spend most of the summer getting to the theater. Ergo, Blücher would be one of Napoleon's first obstacles.

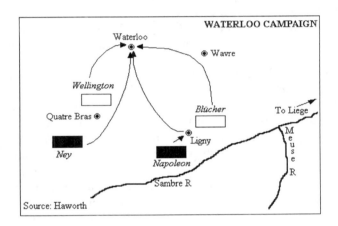

Another obstruction in the emperor's path was to be Arthur Wellesley, the Duke of Wellington. Wellington arrived in Belgium at the head of a coalition army much like that of Marlborough a century earlier. Despite the picture painted by popular histories in Britain and the United States, the majority of forces under Wellington's command were not British. Most of Wellington's redcoats were Hanoverians from the king's German Legion. Nevertheless, the duke's army was under British command, and it represented Great Britain's major effort against the resurgent Napoleon. More importantly, Wellington knew the overall plan. He was to invade France from the north, hoping to link with Blücher. The Austrians were to cross Alsace-Lorraine, while the Russians would follow suit. The duke's main task was to find Blücher and attach him to the British army.

Napoleon deliberated over his dilemma. He could not expect to match the allied numbers. While the French could put more than 200,000 men in the field, Wellington had over half that number in northern Belgium, and these forces were complemented by another 100,000 from Blücher coming from the Rhine. The Austrians were reported to have an army of 200,000, while the Russians would follow with another 150,000. With odds like this, Napoleon could only hope to keep his enemies from uniting and to defeat them one at a time. This meant that he needed to focus on the immediate threat, and the most pressing peril came from the army of Wellington. After studying the map, the emperor believed he had discovered the chink in the British armor. Logistically, the Prussians were linked to the Rhine, while the Anglo-Dutch army was attached to Ghent. If he could force each army to return to its base of supply, Napoleon could prevent Wellington and Blücher from uniting.

Blücher and Wellington understood the situation as clearly as Napoleon. Although Blücher vowed to fight the French, alone if need be, he recognized the value of the British alliance. More importantly, Gneisenau saw the need for a united effort, even though he had a low opinion of British military might. His experience with the British during the American Revolution caused him to distrust their competence. In addition, he regarded Wellington as a colonial officer whose service in India and Spain had not prepared him for the scope of battle he was about to face. Gneisenau laid the ground for cooperation with Wellington, accepting that the Prussians might well need to fend for themselves. Wellington, too, had his reservations about the Prussians. He wondered if they would actually abandon their supply bases and move into Belgium.[293]

The Prussian army was organized as it would have been had Scharnhorst been present. Blücher served as its commander and Gneisenau as his chief of staff. Blücher had four corps at his disposal, and each corps had its own general staff. Bülow commanded one of the corps while the others were led by General Johan von Thielemann, General Hans von Ziethen—a descendent of the heroic "Hussar King's" clan—and General Georg von Pirch, in charge of a reserve corps. Clausewitz, who had recently been reinstated in the army, was Thielemann's chief of staff. Blücher sent Ziethen into Belgium to keep an eye on the French border while he brought Thiele-

---

[293]See Gerhard Ritter, *The Sword and the Scepter: The Problem of Militarism in Germany, Volume I, The Prussian Tradition: 1740–1890* (trans. by Heinz Norden) (Coral Gables, Fla.: University of Miami Press, 1969), 90. (Clausewitz, on the other hand, held Wellington in esteem.)

mann and the main part of the army across the Rhine. Gneisenau was responsible for establishing contact with Wellington, and he arranged for British and Prussian officers to serve as liaisons on each other's staffs. Blücher and Wellington agreed on a line of demarcation between the two armies and pledged mutual support should Napoleon attack. A skeptical Gneisenau did not believe Wellington's promise, and there were some British officers who did not trust the Prussians. Such is the nature of coalition warfare.

Napoleon was also skeptical of the Anglo-Prussian alliance. He gambled on supply lines, guessing that neither Blücher nor Wellington would move too far from their respective bases. Leaving portions of the army behind to guard against possible Austrian actions in Alsace and Lorraine, Napoleon moved a force of 124,000 men to the north in early June. The campaign began in earnest on June 14 when the French crossed the Sambre River and ran headlong into Ziethen's corps. Ziethen sent word to Blücher, who in turn gave word to Wellington. Blücher then ordered Ziethen to fall back as he brought the main Prussian army forward, maintaining contact with the Rhenish supply bases. Wellington, cognizant of his own supply lines in northern Belgium, moved south in search of Blücher. The die was cast for Napoleon's last campaign.

On June 15, Blücher established his army in a defensive posture around the Belgian village of Ligny. Visiting Blücher's position in the late afternoon, Wellington fretted about Blücher's exposed infantry, pointing out that Napoleon's cannon would play havoc with the Prussian line. Blücher was unmoved, and he had every reason to be confident. His line stretched for nearly three miles. It included three of his corps, with the right flank anchored in the village of St. Amand and the left protected by a series of small streams. The fortified village of Ligny held the center of the Prussian front, and Blücher had taken pains to fortify every village and farm house along his line. Despite Wellington's misgivings, Blücher was in a strong position and he intended to hold it. With his own forces still scattered, Wellington promised to aid Blücher if Napoleon struck Ligny, and Blücher made a similar pledge if Napoleon moved north. Wellington then rode away toward his advance guard, seven miles to the west at Quatre Bras.

With Blücher and Wellington converging, Napoleon had no intention of letting the initiative pass to the allies. Dividing his own forces, he sent Ney to Quatre Bras in an effort to block Wellington, while taking personal com-

mand of the main French army. Aware that Blücher was in the vicinity of
Ligny, Napoleon planned to attack the Prussians, then turn to meet Welling-
ton. He struck Blücher in the morning on June 16 with the vanguard of the
French army. The French were actually outnumbered when they found
Blücher's positions, but Napoleon ordered all available troops to Ligny. By
early afternoon, the emperor had 68,000 men to match Blücher's 84,000.
Since Blücher was content to remained anchored to his defensive positions,
Napoleon decided that he could build numerical superiority at the point of
attack. Believing that Wellington would be frozen at Quatre Bras, Napoleon
decided to break Blücher's static defense and force the Prussians back to the
Rhine.[294]

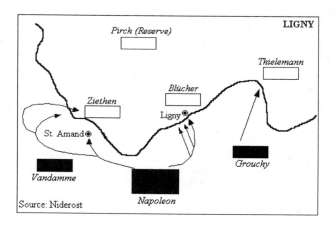

Blücher's line was anchored by two strong points, Ligny in the center
and St. Amand on the right. Napoleon planned to take these points by two
frontal assaults, while tying down the Prussian troops on the left with a series of smaller actions. This would keep Blücher locked into his positions,
and if Ligny and St. Amand fell, Blücher would be forced to retreat. Once
the strong points were taken, the emperor would call on Ney to swing

---

[294]Chandler summarizes Ligny on 1033 ff., and Lachouque does so on 424 ff. See
also Eric Niderost, "Desperate Victory Overshadowed," *Military History* (4) August
1987: 19–25.

around behind the Prussians. Such a threat would not only maul the beaten army, but it could well force Blücher out of the theater. This would leave the emperor free to focus all his attention on Wellington. It was a good plan, but Napoleon might have changed it had he known that Ney was heavily engaged with Wellington at Quatre Bras.

The French launched a full-scale attack on St. Amand in the early afternoon. Advancing through a wavy wheat field, the French were caught by surprise. Their artillery, which usually decimated opposing infantry, had not shaken the entrenched Prussians. The forward elements of Ziethen's corps welcomed the French with a salvo of canister followed by volumes of musket fire. The second wave of French soldiers surged forward as the depleted first ranks staggered into the outskirts of St. Amand. Despite heavy casualties, the French infantrymen were able to focus their own musket fire, and after a few minutes of savage fighting, the Prussians wavered. The French took advantage of the opportunity and breached the first line of trenches. The Prussians responded with bayonets, and Prussian commanders sent requests for reinforcements. Blücher was not able to respond, however, due to his own problems at Ligny, and St. Amand fell to the French after bitter hand-to-hand combat.

As Ziethen's men were being forced from St. Amand, Napoleon's second attack reached the Prussian positions in Ligny. The French advanced, despite the heavy casualties caused by Prussian fire, and the village was soon bathed in black powder smoke. Napoleon's local numerical superiority proved crucial, and Prussian units were driven from their defensive positions by waves of French infantrymen. Just when it appeared that Ligny would fall, Blücher received pleas for help from Ziethen. He thought of making a stand with his reserves at St. Amand, only to receive word that the village had fallen into French hands. The old field marshal understood that he could not afford to lose his second key position. He ordered the reserves forward to Ligny, realizing that this position was the key to success or failure.

Napoleon also realized that Ligny had come to be the turning point of the battle. Deciding that it must be taken at all costs, the emperor moved several artillery pieces within range of the village. The ensuing bombardment set the village on fire as the attacking infantry renewed efforts to oust the Prussians. The Prussians fought on through the smoke and confusion, while Blücher committed his last infantry reserves. Finally, around 5 P.M., the Prussians were breaking, and individual French units began moving through the Prussian line. Fearing encirclement, the Prussians retreated, and the French took Ligny. Local Prussian commanders, however, were far from ready to surrender the village. As soon as they were sure that their units were intact, the Prussians counterattacked. Hand-to-hand fighting broke out and lasted for the next hour. Although they were short of ammunition, the Prussians threw the French out of Ligny around 6:30. Broken French units retreated, and about an hour later buckets of rain began pouring from the heavens, rendering flintlocks useless and causing thousands of wounded to suffer in saturated misery. Blücher was not sure, but he believed that Ligny had held.

Napoleon could not afford to lose, and he brought more artillery forward. If flintlocks could not fire, the cannon could. While the infantry re-formed, Napoleon reopened a concentrated barrage on the Prussian troops in Ligny, and the cannon fire shook the resolve of the Prussians. As the artillery fire slackened, a contingent of 6,000 French infantrymen moved forward with bayonets lowered. Blücher had no more reserves, save the cavalry, and the Prussians in Ligny could not hold. By 8:00 P.M. Ligny was in French hands. Unwilling to accept defeat, Blücher called for the cavalry reserve. Drawing his sword, the old horse soldier placed himself at the front and ordered an attack. Initially, the French reeled under the pressure, but the French infantry quickly formed squares. With the rain no longer a factor, they took dry cartridges from their pouches and fired into the charging cavalry. Several Prussians went down, including Gebhard von Blücher. Spilling into the mud, the general's horse rolled over on him, pinning him to the ground.

As French cavalrymen rode over Blücher's bruised body, Napoleon called for Ney, but Ney could not answer. He was fighting at Quatre Bras where a single British infantry brigade had managed to hold off 20,000 French soldiers through the entire afternoon. As darkness fell, Ney intended to break the determined Britishers, and he called for the entire French reserve, a corps stationed between Quatre Bras and Ligny. Unknown to Ney, the reserve corps had already been moving under Napoleon's orders, and it was about to move to a position where it could cut off the Prussian army. When the commander received Ney's plea for help, he assumed Napoleon's order had been countermanded. He turned toward Quatre Bras, leaving the

Prussians free to escape. Napoleon was furious, but neither Ney nor the reserve responded to his order to surround the Prussians.

Although they thought their commander had been killed, the Prussians knew it was time to retreat. As saddened officers brought news of Blücher's supposed death to headquarters, Gneisenau assumed temporary command of the army. Having discerned that retreat was inevitable, Gneisenau pondered over the direction of march. If he were to move to the supply bases on the Rhine, the Prussians would be safe, but Wellington would be on his own. Unaware that Wellington had been engaged at Quatre Bras, Gneisenau assumed that the duke had failed to keep his word about reinforcing Ligny, and he was tempted to go to the Rhine. By the same token, Gneisenau knew that if the army retreated north to Wavre, the Prussians could possibly link up with Wellington in the coming days and face Napoleon together. Around midnight, a bruised and beaten Blücher staggered into Prussian headquarters. Far from dead, the old warrior immediately demanded to be told why the army was moving north to Wavre. Delighted to see his commander alive, Gneisenau told Blücher that he was searching for Wellington. Blücher nodded his approval.

Napoleon believed that he had given the Prussians a drubbing sufficient to remove them from the theater. Blücher lost 20 percent of his force as casualties and another 8,000 west Germans to desertion. To any trained observer it appeared as though the Prussians were whipped. Napoleon ordered Marshal Grouchy to take a reinforced corps and harass the Prussians all the way back to Liege, their supply base on the Rhine. If he could not destroy them, at least the emperor could keep the Prussians at bay while he attacked Wellington. Grouchy moved east on the road to Liege, firmly convinced that he should not break contact with the Prussians. His resolve to follow the original orders never wavered, and as the Prussians broke off contact to move north, Grouchy continued east in a most determined fashion. Napoleon turned his attention to Wellington who was said to be in a defensive position at a Belgian village named Waterloo.[295]

---

[295]Volumes have been written about the battle, and Wellington is now accepted as an adopted American hero, even though Andrew Jackson fought the British six months earlier in New Orleans. For an excellent introduction and one of the most profound essays on the meaning of battle see John Keegan, *The Face of Battle: A Study of Agincourt, Waterloo, and the Somme* (New York: Penguin, 1978), 117–206. For a most readable

Wellington was indeed in Waterloo. His small army was in a defensive line running along a ridge, anchored on the left by a fortified series of buildings called the Chateau de Hougoumont and in the center by a farmhouse known as La Haye Sainte. Napoleon attacked the position on June 18. Assembling a grand battery of artillery, over 200 guns, in the center of the French line, the emperor began a diversionary assault on Hougoumont around 9:00 A.M., but the British Guards refused to yield. The main French assault, under French General D'Erlon, fell on the middle of Wellington's line shortly before noon. D'Erlon encountered determined resistance from British regulars and the King's German Legions, and he was forced to retreat. As they had demonstrated at Quatre Bras, the defensive resolve of the British army proved impervious to frontal assaults.

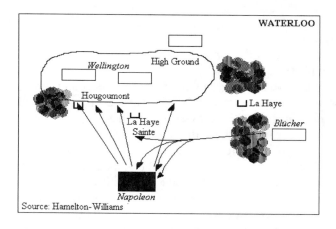

After D'Erlon retreated, Lord Uxbridge, the commander of the allied cavalry, launched a counterattack. One unit, the Scots Dragoons, became over enthusiastic and drove on to the main French line. When their horses were thoroughly exhausted, fresh French cavalry struck them, and the dra-

summary of primary sources see David Howarth, *Waterloo: Day of Battle* (New York: Antheneum, 1968). A newer in-depth assessment is David Hamilton-Williams, *Waterloo New Perspectives: The Great Battle Reappraised* (New York: John Wiley and Sons, Inc., 1993). (Wellington once said that he did not believe anyone would be able to determine the course of events or write a history of the battle.)

goons were massacred. Leaving the field with severe intestinal problems, Napoleon placed Ney in command of the French army, and Ney immediately launched his own cavalry toward the allied lines. Wellington's infantry formed squares and held off a series of cavalry assaults. When Napoleon returned to the field, most of his cavalry had been spent. At this point, about 4 P.M., both Wellington and Napoleon spotted large blue columns marching from the east. It was either Blücher or Grouchy.

After the retreat from Ligny, Blücher spent most of June 17 reassembling the army. He sent scouts to find Wellington, and by evening he arranged a defensive position near Wavre. The next morning cavalry scouts brought news that Wellington was fighting along a ridge line to the west at Waterloo. Blücher and Gneisenau conferred, agreeing that they must dispatch as many troops as possible in Wellington's direction. Leaving a corps in Wavre, Blücher moved the army toward Waterloo. Around 4:00 P.M. on June 18, his columns spotted the battle and marched in the direction of Wellington's right. Napoleon and Wellington spotted Blücher at the same time and wondered whether the blue uniforms were Prussian or French.

As the blue columns drew closer, Napoleon slowly began to realize that their color was Prussian blue. He had one last chance to carry the day. If Wellington's line could be broken before Blücher arrived, the Anglo-German-Dutch army could still be forced from the field. The Emperor of France called on his Imperial Guard and ordered it forward toward Wellington's line. Napoleon even picked up a standard and marched at the front of his men, but he abandoned them before they came into musket range. On the allied side of the line, several hundred British soldiers had been ordered to lie down in a rye field to avoid the thunderous French artillery. As the French guards marched into the allied positions, the British infantrymen rose up from the wheat and fired volleys into the massed French column. The surprised French staggered back. At the same time, Blücher's first cavalry units attacked the right flank of the French line, and the Prussians were not taking prisoners. Napoleon's attack crumbled as Wellington ordered a general advance.

The last French gasp came at Wavre. Assuming that the Prussians were preparing something, Grouchy attacked Wavre with the full weight of his reinforced corps. The Prussian rear guard was forced out of the village, and it retreated along the road toward Blücher. Only then did Grouchy realize that the main Prussian army had moved in Wellington's direction, and he broke contact as quickly as possible. Swinging toward Napoleon's lines at Waterloo, Grouchy's advance guard arrived near the collapsing French right late in the evening, and they found the French army in a state of panic. Soldiers fleeing from Napoleon's army informed Grouchy's men that the Prussians were killing everyone from the wounded to the stragglers.

The stories of the horrified French soldiers were not exaggerated. Determined to never face Napoleon again, Blücher's orders were clear. The French were to be exterminated. There would be no prisoners and no mercy. As Wellington's line advanced toward the center of the field, Blücher's Prussians came in from the right. As the French fell in front of them, the Prussians methodically bayoneted both the wounded and the dead. Frenchmen unlucky enough to be separated from their units were murdered by Prussian cavalry. By dusk, French soldiers in front of the Prussians dropped their muskets and ran. Grouchy's command, the last obstacle, joined the rout before Blücher could attack. As darkness fell, Blücher found Wellington, and the two officers exchanged greetings. Napoleon was finished.

Blücher was lionized by the British. The animosity between certain members of the officer corps vanished as the Prussians and British embraced each other as allies. The faculty at Oxford University even awarded Blücher an honorary doctorate. The field marshal—who was keenly aware of his intellectual limitations—found this very amusing. He had once confided to Gneisenau his immense respect for Scharnhorst's education, and Blücher claimed that he had wasted away his time on alcohol, women, and gambling. If he had it to do over again, the Prussian field marshal said, he would have studied and made something of himself. As Dr. Blücher received his degree from Oxford, he accepted it with good humor. In his graduation address, he told the faculty and students that he could not serve as a doctor, unless they agreed to make Gneisenau his pharmacist.

Chapter 14

The Triumph of Tradition

The period of time between 1815 and 1848 was eventful for Europe, Germany, and the Prussian army. Although it was a time of peace, the age brought new roles and expectations for all the armies of Europe. During the first two decades of a "concert in Europe," traditional European monarchs controlled the political destiny of the continent, and liberalism was in retreat. Yet, economic changes were afoot, and a growing middle class could not be perpetually ignored. Throughout Europe new members of the middle class demanded a voice in government, and the aristocracy—wary of shifting political and economic tides—began to fear a repetition of the French Revolution. They increasingly called on their armies for protection, and as a result many European armies began to act more like police departments than military forces. The Prussians army was no exception.

The *Boyenische*

The Congress of Vienna was a conservative affair designed to restore the power of the aristocracy. The Austrian Foreign Minister, Klemens von Metternich, sought political stability, and most of the diplomats who gathered in Vienna in 1815 to reshape Europe fell under his spell. Constitutions, they believed, were the outcome of revolutionary activities. Governments needed neither constitutions nor the consent of the people they governed to maintain power, and the key factor for the future was political balance—both at home and in foreign affairs. Rulers should be, as Prince Metternich so aptly explained, in concert. Like an orchestra, each country could play a portion of a symphony, and while each part was limited, each nation could be featured so long as the orchestra stayed intact. Such notions did not produce a climate conducive to liberalism or change.[296]

Yet, change was foremost in the minds of the new Prussian Minister of War, Hermann von Boyen, and the reform-minded generals who had contributed to the defeat of Napoleon. In the period immediately following Waterloo, it was hard to keep such popular heroes from public office. They reflected the common people, and their *Landwehr*—in the opinion of the middle class—had freed the country from French tyranny. The territorial settlements from the Congress of Vienna served only to increase the stature of these Prussian liberals. Prussian borders were restored, extending from the Rhine deep into Poland. In essence, Prussia occupied the north German plain and was second only to Austria in its influence among the German states. With the Holy Roman Empire gone forever, a German Confederation of thirty-nine autonomous states emerged in Central Europe. Many of the Prussian reformers hoped to see all interior German boundaries dissolve and the German-speaking people unite in a single constitutional democracy. Such thoughts worried Metternich.[297]

The former mavericks of the military took control of the Prussian army in 1815 as Frederick William III elevated Scharnhorst's former student, Hermann von Boyen, to the position of war minister. The king offered the position to Gneisenau, the most popular national hero of the liberation after Blücher, but Gneisenau wanted a field command. Clausewitz would have been the next logical choice, but Frederick William remained angry over Clausewitz's decision to join the Russian army. He passed over Clausewitz to choose Boyen, but he could not ignore Clausewitz forever. After a few months of wrangling, the king eventually, albeit reluctantly, gave Clausewitz charge of the *Kriegschule*. The philosopher-general began to train future commanders and to refine a manuscript called *Vom Kriege* (*On War*), a work that would ensure his fame beyond the next century. Clausewitz held the position of commandant until he assumed his last

---

[296]See Anthony Wood, *Europe: 1815–1945* (New York: David McKay Co., Inc., 1964), 6–48; Paul W. Schroeder, *The Transformation of European Politics: 1763–1848* (New York: Oxford University Press, 1994), 517–82; and A. J. P. Taylor, *From Napoleon to the Second International: Essays on Nineteenth-Century Europe* (New York: Penguin, 1994), 91–104.

[297]Schroeder, 594.

command with Gneisenau in Poland in 1831. Grolman, another Scharnhorst student, became chief of the general staff. All in all, the post-war Prussian army was cast in the mode of Scharnhorst and the reformers.[298]

Although they held many command positions, these men—the commanders who had saved the nation from France—did not enjoy Frederick William's unbridled favor. The king was quite content to play Metternich's tune from Vienna, and he was not alone. The majority of European monarchs agreed with Metternich. A return to despotism meant peace, and it ensured the stability of the royal households. The growing economic power of the middle class threatened the power of monarchical governments as much or more than the French Revolution. European royalty supported Metternich not only because they liked his message but because his policies rationalized their existence. The last thing the aristocracy desired was democratic reform because it smacked of social revolution. Yet, reform was at the essence of the new Prussian army. Boyen and his followers believed that national elections, popular assemblies, and monarchs limited by constitutions made military sense. Fearing that such an attitude would lessen Austria's influence in German affairs, Metternich played to Frederick William's bias against reform.[299]

The king could not simply dismiss the reformers, and he genuinely liked some of them. After all, they were national heroes. He had no quarrel with their military abilities, but he feared their talk of constitutions and pan-Germanism. The reformers themselves remained as enthusiastic as they had been under Scharnhorst. As they flocked to their new war minister, more conservative officers began to refer to them as the *Boyenische,* loosely translated as "Little Boyens" or, more figuratively, as "Boyen's Boys." The name caught on, and the reformers wore it as a badge of honor. The beginnings were rather shaky. Boyen fought a duel with cultural minister Wilhelm von Humbolt while in Vienna, but both men managed to avoid injury and return to Berlin as fast friends. They became the most high-ranking advocates of democratic reform. Humbolt brought a myriad of civilian intellectuals to the military reform group, while Boyen and many of his officers

---

[298]Gordon A. Craig, *The Politics of the Prussian Army: 1640–1945* (London: Oxford University Press, 1968), 68–79.

[299]Jacques Droz, *Europe Between Revolutions: 1815–1848* (trans. by Robert Baldick) (New York: Harper Torchbooks, 1967), 144 ff.

joined Humbolt's German unification movement. The king was suddenly saddled with two ministers who spoke of a united German democracy and military reform in one breath, and the *Boyenische* were born.[300]

European aristocrats, alarmed by Boyen's behavior, watched Berlin with wary eyes, and Frederick William was aware of the coolness with which his military appointments were viewed in other courts. The tsar expressed concern with the number of reformers assigned to prominent places in the army, and British representatives in Berlin believed that Boyen and his cronies were merely a step away from being revolutionaries. Metternich's concern—expressed in Vienna and later in Berlin—was that peace could be maintained through balancing power, and if the reformers were successful in creating a democracy, the balance of power in Central Europe would be threatened. In essence, Prussian reform would lead to war, and the reformers, Metternich warned, were espousing revolutionary goals. The tsar went so far as to tell some of his generals that Russia might be forced to attack Prussia to protect Frederick William from his army.[301]

The opposition was not limited to external sources. A growing number of Prussian generals resented the growing power of the reformers, and they did not use the term *Boyenische* with a great deal of affection. Traditional aristocratic officers had no desire to change the status quo, and the majority of the king's ministers balked at talk of a united German democracy. Like the king, the traditionalists were willing to tolerate the reformers because of their role in the Wars of Liberation, but they were willing to tolerate the people, not the ideology. Conservative military societies—which originated as a reaction to Scharnhorst's Military Club in the days prior to Jena— began to reemerge and expand their activities, and these associations soon came to represent the most effective opposition against Boyen. Since the king had little sympathy for reform, the conservatives found favor in Frederick William's court. But favor was not enough. The conservatives wanted to

---

[300]For an English-language biography of Humbolt see Paul R. Sweet, *Humbolt: A Biography* (Columbus, Ohio: Ohio State University Press, 1978). A synopsis of Humbolt's thought appears in Frederick C. Beiser, *Enlightenment, Revolution, and Romanticism: The Genesis of Modern German Political Thought, 1790–1800* (Cambridge, Mass.: Harvard University Press, 1992), 111–37.

[301]Gerhard Ritter, *The Sword and the Scepter: The Problem of Militarism in Germany, Volume I, The Prussian Tradition: 1740–1890* (Coral Gables, Fla.: University of Miami Press, 1969), 106. (Gneisenau reported this to Boyen.)

stop the reform movement before the social structure was reconstituted and their class power diminished. In order to stop the reformers, they had to find the Achilles' heel of the *Boyenische*.

The conservatives did not need to search long for the reformers' weakest point. The issue that proved to be the undoing of the military reform movement, and subsequently for Prussian and German democracy, was the *Landwehr*. Embodied in popular sentiment, the *Landwehr* soared to legendary proportions in the plebeian mind. In the public's eyes—at least, in the eyes of the middle-class supporters of the *Boyenische*—the regular army had failed in its gambit against Napoleon. It was responsible for the debacle at Jena, and it had allied with Napoleon in the invasion of Russia. The *Landwehr*, on the other hand, had rallied to the colors. When Blücher led the Silesian Army, his striking power came from the *Landwehr's* militia units. Although both its role in the Wars of Liberation and its battlefield effectiveness were grossly overstated, the *Landwehr* was the army of the people. The middle class viewed it as the army of a constitution.[302]

It was this point that so frightened the conservatives. Indeed, the *Landwehr was* the army of the middle class's aspiration for a constitution, and myths about the glory of the *Landwehr* only served to increase its stature in the popular eye. This terrified the conservatives. The army, they believed, was an extension of the power of the sovereign, and each Prussian officer swore loyalty to the king, not to a group of citizens and certainly not to their would-be constitution. Citizen soldiers could demand rights, and such rights would come at the expense of the king and his fellow Junkers. The conservatives were not about to hand over these rights to Boyen and the *Landwehr*. In an age when European monarchs were forced to rely on the loyalty of their armies, the *Landwehr* was completely unreliable. Nevertheless, the *Landwehr* turned out to be the savior for the conservatives. In the years that followed Waterloo, the liberals attached their fate to the *Landwehr*, and this allowed the conservatives to find their Achilles' heel.

Wilhelm von Humbolt was not the man to lead the fight for a constitution. He lacked the political savvy to grasp rationally the reins of power, but the liberals liked his style. By 1817 he became the prophet asking for

---

[302]Ibid., 101 ff. See also comments by Martin Kitchen, *A Military History of Germany: From the Eighteenth Century to the Present Day* (Bloomington, Ind.: Indiana University Press, 1975), 60–62.

change; the voice crying in the wilderness, and, surprisingly, nothing could have delighted the conservatives more. Like most academicians, Humbolt was not a practical man, and his visions of logic and endless philosophical discussions had little to do with the real world of power. He welcomed debate, exposed all his weaknesses to enemies, and championed the cause of the constitution. He completed a draft constitution and sent it through the halls of Prussian academe for debate. When his proposal made it to the court in 1819, the king was absolutely shocked. Humbolt had designed a plan that undermined the power of the monarchy and established a parliamentary democracy. Prussian faculty senates might have enthusiastically endorsed such a proposal, but faculty assemblies did not make policy. The people who held power were horrified, and the most horrified person of all was Frederick William. Humbolt was confused by the reaction.[303]

The conservatives used the opportunity to pounce on the political liberals and the military reformers. They knew that the king would never agree to Humbolt's constitutional designs, and they also knew that Boyen's military plans were inexorably locked into the proposal. Boyen believed that the regular army should be reduced while the ranks of the *Landwehr* should be increased. All officers, both regulars and reserves, were to swear allegiance to the constitution, not to the Hohenzollerns. To top it off, the proposed constitution arrived at court during an annual inspection of the *Landwehr*. The inspection was a disaster for the reformers, and even the *Landwehr's* most ardent supporters found the reserve force totally unfit for combat. When Frederick William recoiled from Humbolt's proposal, the conservatives urged the king to act. The king rejected Humbolt's constitutional and decided to incorporate the major portions of the *Landwehr* into the regular army. The remaining reserve units were to swear fidelity to the king. The reformers were stunned.[304]

Movement against the *Landwehr* was too much for Boyen. Incensed, the war minister sent the king an angry letter offering his resignation, only to find that the king was happy to receive it. Publicly, Frederick William feigned sorrow, praising Boyen for his service, but the king accepted

---

[303]See James J. Sheehan, *German History: 1770–1866* (New York: Oxford University Press, 1989), 363 ff. for an assessment of Humbolt and culture in the revolutionary era.

[304]Ritter, 106 ff.

Boyen's resignation without hesitation. Boyen's fall in 1819 sounded the death knell for the reform movement. The Prussian army would remain the property of the Hohenzollerns, not the people. The king followed Boyen's resignation with an edict known as the Carlsbad Decrees, effectively ending talk of limitations on royal power. Scharnhorst's policies had defeated Napoleon, but they could not defeat the Prussian military conservatives. After nearly two decades of struggle, the traditionalists had beaten the military reformers.[305]

Boyen did leave one lasting legacy for Prussia, a nationalistic army. Although he felt the army needed to be tied to a constitution, he also believed—as the other reformers believed—that universal conscription was the key to Prussia's military strength. All male citizens were subservient to the state; hence, they were all eligible for military service. Although the conservatives fought against it, Boyen pushed a military conscription bill through Frederick William's court in 1817, calling for the universal service of all able-bodied Prussian males between 18 and 40. After a period of initial duty, the soldier was required to participate in the active reserves, and after seven years in this assignment, he was placed in an inactive reserve unit that could be mobilized in a national emergency. University students could be exempted from the process by joining "one-year units." When their one-year period of training was completed, they could enlist in a reserve unit and continue academic endeavors. Boyen's law remained in effect until the Austro-Prussian war of 1866. It brought many soldiers to the Prussian ranks, but did very little to foster democracy.[306]

Military Advances 1819–1840

For two decades the political adaptability of the Prussian army stagnated, but its military effectiveness increased. Excellence in military leadership and precision became synonymous with "being Prussian," and the army made steady progress under its politically conservative leadership, in both increased technology and greater competency in the Prussian officer corps.

---

[305]See comments by Craig, 77–79. See also Franklin L. Ford, *Europe: 1780–1830* (London: Longmans, Green, and Co., Ltd., 1970), 303 ff. For an analysis of the constitutional question see Matthew Levinger, "Hardenberg, Wittgenstein, and the Constitutional Question in Prussia: 1815–22," *German History* (3) October 1990: 257–77.

[306]Kitchen, 58.

The conservatives found that the Scharnhorst-styled staff system was the most efficient method for organizing large groups of fighting personnel, and they emphasized the importance of the general staff. Prussian staff work was second to none, and the General Staff of the Prussian army drew the attention of other countries. Changes in organizational structure and technological innovations also improved military proficiency.[307]

One of the major keys to the efficiency and the effectiveness of the army was found in its organization. By 1820 most *Landwehr* units were absorbed in the regular army, and the units that remained outside it were placed in a restructured reserve system. Unlike the overlapping structure of the days before Jena, command authority centered in the war minister's office. The chief of staff—not a command position until the 1860s—and the commanding generals reported to the war minister. As a result, command authority flowed from the king to his war minister to the generals. Organizationally, this resulted in an extremely efficient bureaucracy, and commanders accepted centralized authority, ultimately placing themselves and their units at the disposal of the king.

The enlisted men, drawn primarily from rural areas where one-year student deferments were rare, also reflected this loyalty to the Hohenzollerns. Peasant lads were estranged from the people in the urban centers that hosted their garrisons, and they received their political opinions from their officers. The king's word was law, and the system resulted in an army that never thought to question the will of the king. During the last twenty years of his reign, Frederick William not only had the luxury of an efficient command structure, but he led an army that was unquestioningly loyal to the house of Hohenzollern.[308]

Technology also served to increase the efficiency of the army. Inklings of industrial change had seeped into production techniques during the Napoleonic period, but the Industrial Revolution boomed in the early and mid-nineteenth century. For the individual soldiers, the most important advances came in firearms. Flintlocks were replaced by more reliable percussion

---

[307]For a thorough analysis see Dennis E. Showalter, *Railroads and Rifles: Soldiers, Technology, and the Unification of Germany* (Hamden, Conn.: Archon, 1976), 11–31.

[308]Brian Berce and Jonathan R. White, "Roadblocks to Democracy: The Prussian Army as a Police Force in the 1848 Revolution," *Academy of Criminal Justice Sciences Annual Conference*, Chicago, 1994.

caps. Muskets fitted with these mechanisms were fired when the hammer struck a cap placed over a hollow nipple leading to the powder and shot at the rear of the barrel. As a result, the rate and reliability of musket fire increased. It was no longer necessary to prime the gun with powder from the cartridge, and the caps would fire in all types of weather, including rain. By 1835 most Prussian infantry units were armed with percussion cap weapons.[309]

Complementing the percussion cap was an innovation that allowed the soldiers to abandon the musket altogether. Rifles were nothing new to military forces, but they were exotic, slow-loading, and hard to fire. Even though they were extremely accurate because twists in the barrel spun the bullet as it was fired, after five or six shots with black powder, the barrel was often so dirty that a ball could not be forced through the twists. Riflemen even carried wooden mallets to hammer their ramrods down the barrel when shooting for long periods. A Frenchman solved the problem by inventing a conical bullet that expanded when fired. It was small so that it could be forced down the barrel without clinging to the sides, but after being fired it expanded to "take" the twists and leave the barrel with spiraling accuracy. A German gun maker named Dryse improved the concept by changing the size of the barrel, employing a tapered muzzle to put a better spin on the bullet as it left the gun. The result was a "needle-nose" gun that was accurate up to 400 yards. When cannon makers found that the same process could be used in the production of artillery pieces, rifled barrels were gradually introduced to infantry and artillery units.[310]

Improvements to weapons were important for individual soldiers, but the greatest technological advance came on the strategic level. In 1824 some Prussian officers argued that railroads had military potential, and by 1830 the army's General Staff created a permanent railroad department. The railroads offered two advantages that had never been present in the history of warfare, the opportunity for rapid troop movements and the means for supplying large armies. The Prussian General Staff embraced railroads with enthusiasm. For the first time in its life, the army had a system that would allow it to defend a multiplicity of fields on the rambling Prussian frontiers, and soon all mobilization, transportation, and supply plans incorporated

---

[309]J. F. C. Fuller, *The Conduct of War, 1789–1961* (Minerva Press, 1968), 86 ff.
[310]Showalter, 75–90.

railroads. The army encouraged the government to increase the rail network as it evolved into a force that moved and supplied itself by rail. By 1850 no army in the world could match Prussia's ability to employ railroads.[311]

Centralization of the command structure and technological improvements were enhanced by increased professional standards in the Prussian officer corps. The *Kriegschule* remained intact and even increased its entry standards. Promotion and entry to the General Staff became extremely rigorous, and each branch of service developed schools for the instruction of junior officers. All this was due, in no small measure, to Clausewitz's tenure as commander of the *Kriegschule*. Clausewitz accepted nothing but excellence, and his attitude was reflected in his graduates. Victory was the process of combining the will to win with proper organization, according to Clausewitz, and hundreds of graduates carried Clausewitz's concepts to command staffs, far-flung outposts, and local training schools. By 1830 the General Staff developed schools for teaching topography, strategy and tactics, and military history and eventually created a command center to centralize training. It was a major step beyond the *Kriegschule*, and it produced outstanding officers.[312]

Expanded professionalism had political ramifications. The reformers had defined professionalism in terms of *both* understanding technical proficiency *and* behavior in the name of the state. Indeed, this was exactly the reason Scharnhorst believed in liberal education; it gave officers the character needed to represent the common good. Such officers could see far beyond the military, but after the fall of Boyen, these concerns were virtually non-existent in the Prussian officer corps. The army defined professionalism in terms of proficiency and loyalty to the crown, causing many officers to become contemptuous of the public, especially the middle class. This resulted in a growing gap between the military leadership and the civilian population. As the officer corps deepened its commitment to the crown, it moved away from the people. Many political leaders, especially the urban liberals who stood in opposition to the king, started to view the army as a

---

[311]Ibid., 141 ff.

[312]Peter Paret, *Clausewitz and the State* (New York: Oxford University Press, 1976), 271.

tool of repression. The professionalism of the officer corps increased this perception.[313]

Foreign affairs had as much to do with the developments in the Prussian army as anything else. Unlike the violent years of the French Revolution, the Prussian army was at peace after 1815. Apart from its role as a domestic police force, the army was used for frontier expansion into Poland and garrison duty. The policies of the king and the General Staff were essentially peace time occupations. Combat experience came almost exclusively at the expense of Poland, and in 1830 a Polish uprising necessitated the mobilization of reserve units. The General Staff called the aged, and retired, Gneisenau back to service, and he brought Clausewitz from his teaching duties to serve as his chief of staff.

Those who believe in military glory would do well to hear of the end of the military reformers, Gneisenau and Clausewitz. Both men died a soldier's death, not on the battlefield, but of dehydration during a cholera epidemic. Until the twentieth century, sickness and disease claimed the lives of many more soldiers than combat, and it was not unusual to lose five to eight times as many men to sickness as to bullets. Gneisenau contracted the disease and died during the summer of 1831 in Poland, while Clausewitz returned to Prussia and became ill in November. Like his dear friend Gneisenau, and thousands of unnamed Prussian soldiers, he died in feverish agony while uncontrollably vomiting and defecating. Many Polish peasants faded into history with the same illness, but no one remembers their names. This style of death tends to go unnoticed in military history.

Liberals and Constitutions

The eminent nineteenth-century German historian Heinrich von Treitschke said that 1840 was a crucial year in Prussian history. The 1830s seemed to be nothing more than a continuation of the old ways, but many things changed in 1840. All the excitement was due to the accession of the new king, Frederick William IV. Prussian middle and working classes celebrated his coronation with glee, and traditional democrats in the Rhineland, an area incorporated into Prussia after the Congress of Vienna, joined the festivities. The Prussia of 1840 was not the country of 1819, and talk of

---

[313]Kitchen, 70.

German unification was growing. Change and progress seemed to be waiting to break forth into a new Germany, and many middle-class people believed Frederick William IV would begin the process. He was said to be sympathetic to democracy.[314]

The first signs of the new reign were promising. Frederick William granted all political prisoners amnesty, affording them a new opportunity to bury old animosities and to rejoin the realm. The new king promised political reforms to the merchants and citizens of Prussia's growing cities, telling them he would build the infrastructure necessary to run an industrial economy. After Humbolt's dismissal in 1819, Prussian schools and universities stagnated under legalistic bureaucratic management, but Frederick William IV pledged to restore academic quality to Prussian schools. And there was one promise that drew the attention of everyone working to establish a Prussian democracy. The king declared that he wanted to be closer to his people, and his subjects—breathing fresh hope—knew that nothing would bring them closer to their sovereign than a democratic constitution. In 1840, Berlin was full of promise.

Prussia had changed in the midst of the Industrial Revolution, and it was on an economic course quite independent of royal authority. If booming industrial growth had an impact on the military, it merely reflected what had happened throughout Prussia and all the other German states. In 1815 Prussia had been an agrarian society, but in 1840 it was home to industrial enterprise. The Rhine Valley, long known for its rich farmlands, held coal, the fuel for industry. Prussian entrepreneurs turned their eyes from Central Europe to the West and the industrial potential of the Rhine. Rhenish factories gave Prussia the capacity to become an industrial power, and coal from the Rhine's abundant stores promised to fuel it. Increasing commerce accompanied industrial growth, as well as a population shift from the countryside to the cities, and the agrarian pattern of royal absolutism no longer

---

[314]Heinrich von Treitschke, *History of Germany in the Nineteenth Century, Volume VI*, (trans. by Eden and Cedar Paul) (London: G. Allen and Unwin, Ltd., 1919; orig. 1886), 293 ff.

seemed applicable to the urban environment. Prussian society had changed.[315]

Frederick William IV could not comprehend the political and economic attitudes of the urban classes, and no sooner had he taken the throne than his colors unfurled to the disappointment of his subjects. The new king was neither a constitutionalist nor a democrat. The king was a paternalistic romantic, wanting to be close to his people through divine right. No political authority, Frederick William informed his new cabinet, would be allowed to inhibit the king's freedom. He was the ultimate political authority, the final judge and arbiter. Prince William, a professional soldier and the king's brother, agreed, and his position reflected the attitude of the army. The officer corps pledged its loyalty to the new king without hesitation. The population might be in the mood for democracy, but the king and his army were not. There would be no constitutional democracy in Prussia.[316]

Despite his self-delusions about divine right, there were many factors the king could not control, including economic market forces. Industrial Europe was subject to the rise and fall of prices, and poor economic conditions were to become the bane of Western governments. In 1834 Prussia led the north German states by creating a Customs Union. Like it or not, the new king had to work within the commercial system of the Customs Union or Prussia would go bankrupt. The king needed the merchants, and the merchants needed a stable economy. Yet, the industrialists feared France as much as they feared economic crisis. In 1830, Louise Phillipe was crowned King of France, and he made threatening noises about French resettlement of the German Rhineland. Such bellicose talk made all Germans nervous, including the liberal business supporters of constitutional democracy. They wanted a constitution, but they wanted protection from the French even more. If Frederick William allowed the merchants to generate wealth, and if he protected Prussia from France, his subjects could accept him. If he failed at either task, however, divine right would not guarantee his continued reign.[317]

---

[315]Robert M. Berdahl, *The Politics of the Prussian Nobility: The Development of a Conservative Ideology, 1770–1848* (Princeton, N.J.: Princeton University Press, 1988), 286 ff.

[316]Craig, 82–83.

[317]See Sheehan, 588 ff. for an analysis of the political culture.

The new king tried to overcome his subjects' initial disappointment. One of his first actions was to recall Hermann von Boyen—as popular and as committed to the *Landwehr* as ever—to the War Ministry. Much to the king's dismay, Boyen's public appearances spontaneously transformed themselves into liberal political rallies, and enthusiastic crowds greeted him with tumultuous cheers, often breaking into *Landwehr* marching songs. This response irritated the General Staff and the king's brother. To the army, envisioning a time when it would need to fight internal political enemies, the *Landwehr* was as unreliable as ever. At the same time, it represented hope to the civilians who viewed the army as an instrument of repression. The king recognized the danger. Keeping the old minister on his cabinet, Frederick William excluded Boyen from all important military decisions. Boyen sadly realized that he was retained solely for the benefit of his popularity and that he served as a figurehead.[318]

The Army as a Police Force

Prussia may have been an autocratic state, but it was a hotbed of political ideology. Urban wage-based workers and merchants were subject to the fluctuations of the economy, and the Prussian economy was influenced by government policy. When times were hard, the blame fell on the government and revolutionary rhetoric increased. Faced with economic problems at the beginning of his reign, Frederick William called a Prussian National Assembly to raise taxes, and he searched for methods to provide economic stability. It was a pragmatic necessity. In poor economic times, unemployed urban workers took their grievance to the streets, and many laborers committed themselves to a variety of political doctrines, including the newly emerged concepts of socialism and communism. When workers became too disenchanted, their frustration sometimes turned into violence. It had happened in 1830 and 1835, and Frederick William feared it would happen again. In times of public disturbances, city leaders were forced to turn to the police.

Modern German police agencies were in an embryonic stage in the mid-nineteenth century, and they developed much differently from forces in the Western democracies. For example, in London's Peterloo Square a

---

[318]Craig, 85–92.

group of unemployed workers created a disturbance in 1819, prompting a British cavalry unit to charge the workers and kill nearly two dozen of them. The British parliament was appalled and responded by creating a civilian police force that was expected to use the minimal amount of force necessary to maintain public order. In 1844 the United States followed suit by modeling the New York City Police Department along the lines of the London Metropolitan Police, but the Prussians had no such model. The king appointed a police president to major cities, and the president commanded a small number of officers, composed mainly of retired enlisted military personnel. By 1840 the large cities like Berlin had a standing police force augmented by a rural paramilitary police force and a number of uniformed investigators. Yet, this police power was merely an extension of the king's military power. When faced with disturbances beyond their control, the Prussian police routinely called for military support. By 1840 this was the normal procedure.[319]

When faced with urban disturbances, the Prussian army responded much as it would have to any other threat, and the rural lads who filled the bulk of the Prussian enlisted ranks demonstrated zeal when quelling urban disturbances. They were trained to use bayonet and butt against the enemies of the king, and it mattered little to them whether the enemies wore uniforms or hid themselves in the political adornments of liberalism. Unemployed workers were as good a target as Polish peasants. The General Staff and the officer corps did much to reinforce this type of behavior, and military officers, working in conjunction with local police officials, thought it was their duty to keep track of potential political enemies. Local garrisons were frequently transferred to keep the soldiers from becoming too friendly with the population, and the policy worked. When disenchanted subjects went into the streets, the army was there to confront them, and the loyalty of

---

[319]The best study is Alf Ludtke, *Police and State in Prussia: 1815–1850* (New York: Cambridge University Press, 1989). See also Herbert Reinke, "The State, the Military, and the Professionalization of the Prussian Police in Imperial Germany," in Clive Emsley and Barbara Weinberger (eds.), *Policing Western Europe: Politics, Professionalism, and Public Order, 1850–1940* (New York: Greenwood Press, 1991). For the etymological development of *police* and the Hapsburg influence see Roland Axtmann, " 'Police' and the Formation of the Modern State: Legal and Ideological Assumptions on State Capacity in the Austrian Lands of the Hapsburg Empire: 1500–1800," *German History* (1) February 1992: 39–61.

the rank-and-file soldiers was never in question. The army provided the muscle behind police authority.

Frederick William IV had occasion to take advantage of the loyalty of his troops. In 1845 the economy went into a slump, and by 1846 Prussia was in financial trouble. Unemployment rose, incomes dwindled, and violence increased in urban centers. Things had not improved in the summer of 1847 when unemployed and dissatisfied workers organized demonstrations in Berlin. The city's police president watched the workers grow into an angry mob, and, fearing the worst, he called on the army for help. The army moved quickly as hundreds of troops poured into the city to support the police president's 150 men. The army stayed in Berlin until August when it appeared as if the situation was well in hand. Workers might challenge the local gendarmes, but they were no match for the disciplined soldiers of the regular army—at least, that is what most observers believed. Neither the army nor the king was prepared for what happened next.[320]

Prussia's economic woes were part of a general European depression, and other countries also experienced worker dissatisfaction. Nowhere was the stress more acute than in Paris. The French middle class, disillusioned with the policies of King Louis Philippe, took to the streets of their capitol city in the summer of 1847, much like their counterparts in Prussia. Unlike the Berliners, however, the Parisians did not go home. Demonstrations grew and the workers elected makeshift assemblies. Referring to their extended demonstrations as "banquets," the French citizens denounced their government and "passed" a series of economic reforms by street votes. While the Prussian army cleared the streets of Berlin, the demonstrations in Paris grew larger, and they continued through the fall into winter. On February 22, 1848, the Parisians gathered for a massive demonstration, the largest assembly to date, and it grew far beyond the government's ability to control it. Students and workers sealed off streets with barricades demanding Louis Philippe's abdication. The demonstration had been transformed into a revolution.[321]

---

[320]For economic analysis see Berdahl, 300.

[321]See Priscilla Robertson, *Revolutions of 1848: A Social History* (Princeton, N.J.: Princeton University Press, 1952), 11 ff.; Raymond Postgate, *Story of a Year: 1848* (New York: Oxford University Press, 1956), 11–53; Melvin Kranzberg (ed.), *1848: A Turning Point* (Boston: D.C. Heath and Co., 1959) (dated, but excellent essays); and Schroeder,

Louis Phillipe's government toppled, and within days revolutionaries throughout Europe took to the streets as if a wave of revolution had spread from Paris. Rulers in Germany were caught in a bind. Normally, they could ask Prussia or Austria for help to quell political disturbances, but Berlin and Vienna were worried about their own futures. Smaller states were suddenly engulfed in massive demonstrations with no means of restoring order. Some local sovereigns fled while others made concessions to the revolutionaries. In the west and south, where democratic traditions were stronger, rulers agreed to call for constitutional assemblies, with a few sovereigns even welcoming the opportunity. Prussia was not immune. In the face of growing political agitation, Frederick William assembled his cabinet and military advisors in early March. Wanting to avoid a repeat of the new French Revolution in his own land, the king sought advice. Most of the cabinet favored preventive military action, and their view was supported by a request from the police president of Berlin. Alarmed by reports of rioting in other German cities, the president asked for military reinforcements when demonstrators flooded the streets of Berlin. The king agreed, and on March 13, 1848, regular army forces marched into the Prussian capitol.[322]

The king was no democrat, but he wanted to avoid bloodshed. Placing Berlin under the command of a military governor, the king ordered General von Pfuehl to maintain order in the city and clear the streets. Pfuehl, a moderate who wanted to do everything possible to avoid a confrontation between the army and the citizens of Berlin, quickly found that events were moving out of control. Regular soldiers marched into the city, only to be shocked by a rude reception from the Berliners. Most of the soldiers were greeted with jeers while some were pelted with stones. Tension spilled into general violence on one of Berlin's historic thoroughfares, *Unter den Lenten*, when a cavalry unit trying to move up the street encountered a hostile mob of civilians. As the troopers dodged bricks and insults, the officer in charge reached the end of his rope and ordered his unit to charge. The soldiers broke up groups of resisting civilians, intentionally wounding many

---

764 ff. For a summary of events in Germany see John L. Snell, *The Democratic Movement in Germany, 1789–1914* (Chapel Hill, N.C.: University of North Carolina Press, 1976), 76–100.

[322]James J. Sheehan, *German Liberalism in the Nineteenth Century* (Chicago: University of Chicago Press, 1978), 51–58.

of them with their swords. Word spread on the street, causing increased violence and armed confrontations. Pfuehl appealed to the king, and as Frederick William tried to restrain the army, a message arrived from Vienna. Klemens von Metternich had been thrown out of office and was fleeing to England. Frederick William realized that he was in the midst of a revolution.

The king was determined to retain power and bring an immediate end to the violence. On March 14 he informed his cabinet that he intended to call for a Prussian National Assembly to draft a constitution, and he offered to meet with the demonstrators. Unfortunately, the king made no provision for control of the military and the police. Pfuehl, noting the king's omission, confined the army to their barracks and planned to keep them there, unless there was a general withdrawal. The king's military advisors were not so quick to act. General Edwin von Manteuffel urged the king to make a show of force, arguing that negotiating with rioters would only fan the revolutionary fires. General Albrecht von Roon agreed, warning the king not abandon the army to join the people. While Frederick William listened to the debate, street fighting increased. Worse yet, there were reports of casualties, and the king learned that civilians had been killed. This was too much for Frederick William. Overriding Manteuffel and Roon, the king said that he would address his subjects on the eighteenth and tell them of his decision. Not only would he convene a constitutional convention, he would grant popular elections. Realizing that fighting might still intensify, he replaced Pfuehl with a more seasoned commander, General von Prittwitz, and Prittwitz quietly promised Manteuffel and Roon that he would immediately move the troops back to the streets, if the king's safety were jeopardized.[323]

March 18 turned out to be a disaster. The king scheduled his address for the Castle Square shortly after noon where hundreds of Berliners, mostly middle-class business people and workers, had flocked to hear the speech. Rumors of elections and constitutions were in the air, and the mood was festive. The local military commander felt differently, and as the crowd swelled, his anxiety about the event increased. Sensing that it might be dangerous to place the king in front of a crowd that had been rioting the day before, the commander summoned a cavalry unit. As the troopers rode to the opposite side of the Castle Square, the crowd booed and taunted the

---

[323]Summary based on Kitchen, 72–87.

cavalrymen. This caused the commander to request infantry support, and he sent word advising the king to delay the address. When the infantry arrived, things quickly went from bad to worse. The civilian crowd openly defied the military units, causing the commanders to begin clearing the square. The Berliners felt betrayed, and the king was nowhere to be found. When cavalry moved forward, the people resisted. Not wanting another incident, the cavalry pulled away, but the infantry formed a battle line and ordered the crowd to disperse. Somewhere two shots rang out, and the startled infantrymen responded by firing a volley directly into the crowd. With cries of treason on their lips, the enraged Berliners ran and armed themselves. Within minutes Berlin was a battleground.

By evening the army was losing control of the city, and the king's officers joined him to discuss the situation. The army was not trained to fight in the streets, the military advisors said. The soldiers stormed barricades with frontal assaults, only to have the citizens run down side streets and build new barricades. Casualties were high, the troops were exhausted, and ammunition was running low. One general suggested that the army retreat, regroup, and retake the city, but Prittwitz talked of withdrawing and bombarding the city with artillery. Looking at Prittwitz in disbelief, Frederick William recoiled at the suggestion and came to his own conclusion. He ordered the army to leave the city, while he alone would stay to meet the citizens. It must have taken a minute for the generals of the king's army to comprehend the decision, for certainly not one of the officers could stomach such an order. They begged the king to reconsider, appealing in the names of the soldiers who had given their lives in the fighting thus far. But the king would not be moved. Frederick William would cast his fate to the citizens of Berlin.

The command staff was completely disgruntled, and some high-ranking officers, like Roon, openly expressed their disgust, complaining that the king had abandoned the army. Regardless, Frederick William was the king, and his officers were sworn to obey him. Having no choice, the generals marched their troops out of the city, leaving the Berliners free to swarm into the king's palace the next day. They set up a revolutionary council and summoned their king. Forcing Frederick William to wear the French tricolored hat—the symbol of revolution—the council paraded the king around the city and compelled him to stop and salute at each site where the dead had been gathered. The king rose to the occasion. Saluting the fallen citizens, Frederick William made a public apology and promised that, as soon as order was restored, he would call a National Assembly for the purpose of drafting a constitution. The citizens were elated. It was one of the most confusing springs in Berlin since the French occupation. The king hastily called the Prussian National Assembly in May, instructing them to draft a constitution while sending delegates to an all-German assembly at Frankfurt-on-the-Main. A conservative delegate, Otto von Bismarck-Schönhausen, was elected to the Prussian National Assembly and became a leading voice for the monarchy. While the Prussian Assembly began its work, the Frankfurt congress hoped to create an all-German federal union, and events began with an aura of excitement and expectation. The National Assembly passed a series of bills and started drafting a constitution while Frankfurt witnessed one of the most prestigious gatherings of German intellectuals in history. All seemed to be going well until summer.[324]

Neither the members of the Prussian National Assembly nor the delegates of the German congress appeared to be able to make much pragmatic headway. The delegates were so intoxicated with democracy that they forgot to take practical actions. Unlike Great Britain or the United States, Germany had no long-term experience with democracy, and they were not quite sure how to implement the mundane functions of daily government. The National Assembly was quick to pass a series of democratic laws, and the poets of Frankfurt forged the intellectual basis for a free, united Germany. But, the diets could not restore order. The merchants of Berlin were much more concerned in forming a stable business climate than they were with fine tuning the constitution or German unification. When mobs appeared in the streets of Berlin in the summer of 1848, business people started worrying about profits. They wanted some form of rational order.[325]

The king observed the events and slowly assembled the forces he would need to regain power. A secret group of military and political advisors gathered around the king, gaining Frederick William's absolute trust. Informally,

---

[324]For a general summary see William Carr, *A History of Germany: 1815–1945* (2nd ed.) (New York: St. Martin's Press, 1979), 34–63.

[325]For analysis see P. H. Noyes, *Organization and Revolution: Working-Class Associations in the German Revolutions of 1848–1849* (Princeton, N.J.: Princeton University Press, 1966), 163 ff.

called the "Camarilla," the group began laying the ground work for the restoration of royal power. The summer of 1848 provided material for the Camarilla's script. Pointing to the inability of the National Assembly to restore order, they told Frederick William that he was obligated to act in order to protect citizens, and the deliberations in Frankfurt gave the Camarilla further ammunition. By late autumn the Frankfurt delegation had degenerated from political to academic debate while the Prussian National Assembly still failed to police Berlin. Frederick William's informal advisors urged him to move, and the king agreed with their advice. In November Frederick William took the offensive. He ordered the army back to Berlin.

The majority of officers were delighted as General Friedrich von Wrangel led the army through the city. Liberals in Berlin established a resistance front—even kidnapping Wrangel's wife and threatening to hang her—but they melted when Wrangel's hussars rode through the Brandenburg Gate. After Wrangel informed the king that his capitol was secure, Frederick William took control. He dissolved the National Assembly with Wrangel's troops at his side and began a counterrevolution. Not only did the army manage to restore royal power, but Frederick William sent troops throughout Germany. In May 1859, Prussian soldiers went to Dresden to secure the throne for the King of Saxony, and later two army corps under Prince William quashed a popular rebellion in Baden. Many of Germany's finest minds lost their lives in the struggle against William's army. The Frankfurt parliament disbanded with Frederick William's blessing, and, to top things off, the king delivered the coup de grace. Under the guidance of the Camarilla, he wrote his own constitution.[326]

Frederick William's constitution became law in 1850. Most of the document dealt with the structure of government, and the Camarilla assured that few of the king's powers were diminished. The National Assembly could be called for the purpose of raising revenue, but the king remained free to answer to no outside authority. Cabinet rulings had the power of law, and, while the Camarilla was essentially an illegal body, it continued to meet in spite of the constitution. The document also specifically dealt with the question of the military. Article 108 said that the army was the exclusive

---

[326]John R. Gillis, *The Prussian Bureaucracy in Crisis, 1840–1860: Origins of an Administrative Ethos* (Stanford University Press, 1971), 159–84. See summary of issues in Berdahl, 311–26.

servant of the king, not the state. Only the king could commission officers, and the officer corps was immune from the provisions of the constitution. Legally, the army was a police force, the prop behind the throne. Undoubtedly, the soldiers would have remained in this role had not foreign affairs called them to other duties.

Chapter 15

Unification

By the mid-nineteenth century the Prussian army had evolved into a domestic police force. High-ranking military officers, along with conservative members of the newly created Chamber of Deputies, viewed the army as a tool for maintaining internal political order. The officers even cautioned against foreign ventures when they felt such actions threatened their domestic role. When Austria, for example, challenged Prussian leadership in the German Diet in 1850, the officer corps backed away from any confrontation, fearing that foreign involvement might weaken the army's hold on internal political affairs. This happened again during the Crimean War when general officers urged the king to leave his troops at home. In 1859 Austria and France went to war in Italy, and the high command again opted for noninvolvement. The army seemed quite content to serve as a symbol of political stability. Undoubtedly, the army would have remained content in its domestic role for many years had not the foreign policies of Otto von Bismarck called Prussian forces to the field.

Bismarck and William I

Frederick William IV was never the same after the 1848 uprising. He remained bellicose, flamboyant, and quick with proud words, but the revolution left him shaken. The constitution he wrote satisfied neither the liberals nor the reactionaries, and he increasingly turned to the informal, unconstitutional Camarilla for political advice. At the same time, the lower house in the new Chamber of Deputies sought to expand the power of the middle class. Each issue brought confrontations between the conservative forces surrounding the king and the economic power of the middle class. All of this was too much for Frederick William, and the king slowly lost his ability to deal with the details. Sometime after 1855 the king's associates began to suspect that he was ill, and by 1858 it was clear that the king had lost his grasp of reality. The cabinet offered the throne to Frederick William's brother, and William accepted it as Prince Regent. Three years later Frederick William passed away and the Prince Regent became William I, King of Prussia at the age of 64.

The liberals, who knew of William's record in Baden during the uprising, placed few hopes in the new king. He was a tough, seasoned soldier, schooled in the profession of arms, and when he ascended to the throne, he had a military agenda. According to William, the most pressing political problem in Prussia dealt with the term of enlistments for Prussian soldiers. With the exception of the one- year student soldiers (*Einjahringe*), Prussian males were conscripted for a period of two years. After the initial enlistment period, discharged soldiers were to serve for a number of years in a reserve unit. This followed Stein's idea of the citizen soldier, but it was not in keeping with the ideology of the new king. William wanted to separate his soldiers from Prussian society, and two years did not give the army time to do it. After short-term training, William believed, recruits were nothing more than civilians in uniforms—precisely the idea that the reformers had in mind! These short-term soldiers were politically unreliable, and the *Landwehr* was of little comfort. William proposed a solution.[327]

The best way to increase the political loyalty of the army was to lengthen the term of enlistment and to further reduce the *Landwehr*. The stumbling block for William, quite ironically, was the new Hohenzollern-approved Prussian constitution. By its terms, the king could neither alter military structure nor pay for the army without the approval of the Chamber of Deputies. The upper house presented no problem because it was controlled by Junkers who supported the king, but the lower house was another matter. Composed of workers and middle-class business people, the lower house viewed the king's proposed changes with skepticism. They believed, quite correctly, that a three-year conscript gave conservatives a better tool for repressing the people. Since there seemed to be no immediate foreign threat, the lower house of the Chamber of Deputies wanted nothing to do

---

[327]Herbert Rosinski, *The German Army* (Washington, D.C.: The Infantry Journal, 1944), 55–56.

with William's reform package. They turned a cold shoulder to the king when the enlistment bill was introduced.[328]

The most influential soldier in William's army was General Albrecht von Roon, the Minister of War. As the legislative battle unfolded, Roon found himself in a most difficult position. He agreed with the king's desire to increase the time of active duty, but he also understood the constitutional power of the Chamber of Deputies. If the king pushed his idea too far, the deputies could close the purse. Spurred on by William, Roon introduced the entire military reform package in 1864, only to see it resoundingly defeated in the lower house. The king was infuriated and threatened to extend the term of enlistment without legislative approval, but Roon warned his sovereign that the deputies might respond by cutting off finances. The king would not be swayed, and he extended the enlistments. The Chamber of Deputies responded, just as Roon had predicted, leaving Prussia without an income. A dejected William decided to abdicate in favor of his son, a man immensely more popular with the liberal delegates. Anticipating the crisis, Roon turned to his arsenal for another political weapon. With no intention of failing either his king or his army, he sent a prearranged telegram to the Prussian ambassador in Paris.[329]

Otto von Bismarck-Schönhausen, not surprised by the political confrontation in Berlin, was anxiously awaiting word from Roon as he represented Prussia in Paris. He first came to the attention of the crown in 1849 when he championed the royal cause as a conservative delegate in the Prussian National Assembly. Polishing his skills and diplomatic views, Bismarck served in the German Congress from 1851 to 1859. He formulated a political philosophy in the Frankfurt-based diet that extended far beyond the simple reactionary policies of William and the army. Although opposed to middle-class liberalism, Bismarck realized that Germany was ideologically on the verge of unification. He began to think of the possibilities, not of a united liberal democracy, but of a Germany joined together under Prussian control. Germany would come together, Bismarck believed, if there were a foreign threat, and he left Frankfurt to find one. In 1859 he served as the Prussian

---

[328]Martin Kitchen, *A Military History of Germany: From the Eighteenth Century to the Present Day* (Bloomington, Ind.: Indiana University Press, 1975), 98 ff.

[329]Gordon A. Craig, *The Politics of the Prussian Army: 1640–1945* (London: Oxford University Press, 1968), 140–45.

ambassador to Russia, and by 1862 he received the most prestigious post outside of the king's cabinet, an appointment as ambassador to France. The selection had Roon's blessing, and the war minister kept in close contact with Bismarck, promising to call the ambassador to Berlin at the first sign of crisis.[330]

Roon's telegram, a promised signal telling Bismarck to return to Berlin, arrived in Paris on September 18, 1862. When Bismarck returned, Roon arranged a meeting with the king, and William was immediately impressed with the pragmatic, royalist loyalty exhibited by his ambassador to France. Bismarck seemed to be just what the government needed. He told the king he would find a solution to both the economic crisis and the military enlistment problem, and in exchange Bismarck expected to receive the reins of power. The king agreed. Withdrawing his offer of abdication, William appointed Bismarck as the Minister of Foreign Affairs and to the office of Prime Minister. With internal and foreign affairs under his domain, Bismarck had unprecedented power, and he immediately set about doing a rather impossible task, running a country without an income.[331]

Bismarck made no pretense of humility. In a speech before the Chamber of Deputies, he informed the representatives that the king no longer needed their authority to raise finances, and the shocked deputies were stunned to hear Bismarck rail at their petty democratic tomfoolery. Germany would be united, the new minister said, not by democracy, but by

---

[330]There is a plethora of English-language work on Bismarck. In terms of military affairs Ritter, vol. 1, compares Bismarck and Moltke as does Craig. A more recent essay, critical of past interpretations, is Dennis E. Showalter, "The Political Soldiers of Bismarck's Germany: Myths and Realities," *German Studies Review* (1) February 1994: 59–77. For standard biographies see Erich Eyck, *Bismarck and the German Empire* (New York: W. W. Norton and Co., Inc., 1968) (entertaining and easy to read); Lothar Gall, *Bismarck: The White Revolutionary* (2 volumes, trans. by J. A. Underwood) (London: Allen and Unwin, 1986); and Otto Pflanze, *Bismarck and the Development of Germany* (3 volumes) (Princeton, N.J.: Princeton University Press, 1990). Examples of earlier works are Moritz Busch, *Bismarck: Some Secret Pages of His History* (2 volumes) (New York: AMS Press, 1970; orig. 1898) (memoirs from a secretary); Emil Ludwig, *Bismarck: The Story of a Fighter* (trans. by Eden and Cedar Paul) (Boston: Little, Brown, and Co., 1927), ; and A. J. P. Taylor, *Bismarck: The Man and the Statesman* (London: Hamish Hamilton, 1955).

[331]Eyck, 57 ff.

blood and iron. William, infuriated because Bismarck implied that force would be used against other Germans, summoned his minister to demand an explanation, and the surprised Bismarck soothed his king's feelings. William, who wanted to discipline Bismarck, had no choice save to accept his minister's apology and ask him to continue in office. Prussia had a new leader, and the king could not afford to lose him. For the next eighteen years Bismarck guided the ship of state, and he became the most powerful minister to serve a Hohenzollern.

Bismarck ran over the Chamber of Deputies, challenging them at every step. Although his actions were controversial and often illegal, he raised revenues and ended the financial crisis. Whenever a liberal bureaucrat pointed to the illegal nature of his taxation policy, Bismarck fired the official, replacing him with a worker more sympathetic to the king. He invoked the king's military laws without waiting for legislative approval, and when the press criticized his actions, he imposed news censorship. Frederick William, the crown prince, openly challenged Bismarck's attempts to silence the press, writing an editorial in a local newspaper. Undaunted, Bismarck convinced the king to reprimand his own son. The new Prussia was to be Bismarck's Prussia.[332]

The unification of Germany under Prussia was the driving factor behind Bismarck's domestic and foreign policies. Without a unified country, Prussia would never be able to compete with the established world powers. Yet, there was one major factor inhibiting the unification of Germany. As long as Austria maintained a dominant role in the German diet, Vienna, not Berlin, would be the seat of power. Austria had divided loyalties. With one face turned toward the Balkans and the other looking north, the Hapsburgs ruled both Germans and a multitude of other ethnic groups in their conglomerated empire. If Austria were included in a unified state, Slavs would be incorporated in the new Germany, and, depending on how the borders were drawn, they might even outnumber the Germans. Bismarck knew that Prussia could only dominate when the Austrians and the non-German states of the Hapsburg empire were excluded from German affairs. This could mean only one thing; Austrian power had to be eliminated.[333]

---

[332]Pflanze, vol. I, 205–17. See also Craig, *Politics*, 161–66.
[333]Craig, *Politics*, 167 ff.

The 1864 War with Denmark

Bismarck's plans for Austria forced the army to meander away from its role of domestic police officer, but before he could move toward Vienna events to the north captured the nation's attention. The 1848 uprising had also swept across Denmark, and, unlike Prussia, the Danes developed their own democratic constitution. In the period following the revolution, Prussian soldiers attempted to restore the Danish government, but Danish nationalists not only blocked their efforts, they forced Prussia to recognize Danish influence in two southern provinces, Schleswig and Holstein. Schleswig, Germany's northernmost state, was linked to Denmark through a sizable Danish minority population, but the majority of its citizens were Germans. Holstein, on the other hand, was thoroughly German and a member of the German trading confederation. Although both provinces wanted to join Germany for economic reasons, the Danes maintained partial control. This testy situation lasted for over a decade until a change in the Danish government in 1864 brought the issue to a head. Danish liberals wanted to annex the two provinces, giving Bismarck an opportunity to focus Prussian attention on a foreign threat. When Denmark announced its plans for annexation, Bismarck responded with the threat of war and an appeal for military assistance throughout Germany.[334]

Despite Bismarck's bellicose actions, the Prussians were not quiet ready for war, especially a war that would involve their participation in a large coalition of German forces. Roon was beset with the type of bureaucratic problems that inundate coalition armies, and Bismarck, who had little patience for military discipline and protocol, was drawn into the military arena.[335] The most pressing problem dealt with command. Bismarck and Roon felt that a Prussian general should lead the all-German contingent, but the Austrians balked. Bismarck pointed out that Prussia was supplying most of the troops and that Schleswig and Holstein bordered Prussian territory. The Austrians finally agreed to serve under Prussian command, if the Prussians would select a seasoned leader. Roon selected Friedrich von Wrangel, the general who restored the Hohenzollerns in the 1848 revolution. Although Wrangel was acceptable to the Austrians, Bismarck had his doubts,

---

[334]Kitchen, 111.
[335]Gall, vol. I, 248.

believing the events of the world had bypassed the aged Wrangel. Bismarck was right, and the old general was soon at odds with the General Staff, Roon, the king, and his own deputy commander. The question of Bismarck's role further complicated the matter. Bismarck was the prime minister, but, as his friend Roon pointed out, he was *only* a major in the *Landwehr*. A major could not issue orders to a general, no matter what his civilian rank, and Roon asked that all of Bismarck's orders come through the king, rather than through his ministerial offices. William ordered his irritated prime minister to accept the arrangement.

To be sure, tensions between Bismarck and the military hierarchy increased throughout the build up. Bismarck focused on politics while the army centered on military affairs. Bismarck wanted to use the Schleswig-Holstein incident for a variety of political and military purposes aimed at bringing Germany under Prussian domination, but the generals thought only in military terms. Wrangel argued that his needs, not Bismarck's, should dominate Prussian decisions, and Wrangel's chief of staff, General Vogel von Falkenstein, was particularly obstinate. To his credit, William intervened. The king, experienced enough in military matters to know that such problems permeate war, made military operations subservient to political necessities. If Bismarck's political maneuvers allowed the army to achieve victory, the king said, the army had to be satisfied with its mission. Wrangel and Falkenstein felt betrayed, but the two friends, Bismarck and Roon, quickly achieved an understanding. In January 1864 Bismarck issued an ultimatum to Denmark with Roon's full approval. If the Danes refused to evacuate Schleswig and Holstein, the German states would go to war under Prussian leadership. The Danes refused.[336]

Plans for the war were drawn by the General Staff in Berlin, and since Prussia had no fleet, there were two options available for attacking Denmark. The first plan centered on the possibility of encircling the Danish army along the border. The General Staff referred to this as the *Kesselschlacht*, the battle of double encirclement, and it became increasingly popular among the staff officers. Yet, the *Kesselschlacht* would work only if the Danes tried to make a stand. At the first sign of Prussian forces, a prudent Danish general would abandon the border positions and retreat to strong fortresses inside Denmark. This scenario, the course of action considered

---

[336]Eyck, 87–89, and Craig, *Politics*, 181–86.

most likely by the General Staff, would force the army either to wait at the border or to pursue into Denmark. If Wrangel decided to pursue, the staff recommended bypassing the fortresses and invading Jutland, the main Danish province to the north of Schleswig. Wrangel favored the idea of invading Jutland.[337]

Despite the General Staff's optimism, Bismarck worried about the invasion plan. Austria, he believed, would fight to free German provinces, but it would not invade Denmark. Roon, believing that the army would be forced to move beyond the border, argued that all political meaning would be lost if German forces merely chased the Danes out of Schleswig. The Danes would run when the Prussians came, Roon said, but they would reoccupy the territory when the Prussians left. Bismarck appealed to the king, intimating that he would resign if the king endorsed the invasion of Jutland. Before the king decided, Roon sought a compromise. With two war plans in his hand, he ordered Wrangel to encircle the Danish army while it was in Schleswig. Officially, this canceled any anticipation of a Danish invasion. If Wrangel failed to surround the army, the question of Jutland would be revisited, but as military preparations entered their final phase, both Bismarck and the Austrians were satisfied with Roon's order. Wrangel prepared to attack.[338]

War began on February 1, 1864, but it did not follow plans. Wrangel did not, and perhaps could not, understand the concept of double envelopment. As soon as his forces crossed the border, the Danes scampered to the safety of secure positions, and by mid-February most of the Danish army was safely ensconced in the fortress of Düppel inside Denmark. A frustrated Wrangel ordered the invasion of Jutland on February 17, openly defying Berlin, and the Austrians, just as Bismarck had predicted, were shocked. They felt that the objective of the war had been reached when the Danes abandoned Schleswig, and Hapsburg generals stated that they would not invade Denmark, openly speaking of leaving the German alliance. Bismarck acted quickly, persuading the king to send General Edwin von Manteuffel, a trusted advisor of the king and Chief of the Military Cabinet, to Vienna. Manteuffel was well regarded in the Hapsburg court, and he soon calmed

---

[337]Curt Jany, *Die Königlich Preussischen Armee und das Deutsche Reichsheer 1807 bis 1914* (Berlin: Verlag von Karl Siegismund, 1933), 233–35.

[338]Craig, *Politics*, 187.

the Austrians by explaining that the Prussians intended only to isolate Düppel, not to launch a full-scale invasion of the Danish peninsula. At the same time, Bismarck asked the king to order Wrangel to cease operations in Jutland, save those around Düppel. William, angered by Wrangel's insubordination, personally reprimanded his general and ordered him to hold his position. A month later, when Wrangel still refused to obey Berlin, the king relieved him, giving temporary command of the army to Vogel von Falkenstein. Roon searched for a permanent replacement.[339]

Wrangel had been selected under Austrian pressure, and Roon wanted to replace him with a candidate able to work with the General Staff. William's nephew, Prince Frederick Charles, fit the bill. Competent, though a bit over-cautious, Frederick Charles was extremely popular with the common soldiers and a good leader. Roon also thought it best to replace the obstinate Falkenstein. Frederick Charles needed a more competent chief of staff, a deputy who would work as closely with the General Staff as his commander. Manteuffel and Roon searched the rolls of general officers when Manteuffel proposed a unique solution. The talents of the Chief of the General Staff, Manteuffel postulated, were being wasted in Berlin. If Roon wanted to improve communications with Berlin, who would be better to send than the chief? Roon liked the idea, and the king approved the selection. On May 11, General Helmuth von Moltke, the Chief of the Prussian General Staff, made arrangements to leave Berlin and join the army in Schleswig.

Helmuth von Moltke

Helmuth Karl Bernhard von Moltke was not a Prussian. Born in Denmark in 1800, he was the son of a Danish general, and he began his career in the Danish army, hoping to follow in his father's footsteps. In 1821 he received an offer from Berlin. Sensing greater opportunity, he left his homeland and joined the Prussian army where his intellect drew the attention of Gneisenau and Clausewitz. By 1823 he received an assignment to the *Kriegschule*. A royalist who never embraced the political ideology of the reformers, Moltke had no political problems in the post-Boyen era, and,

---

[339]See the cartoon cited by Pflanze, vol. I, 256. (Bismarck is trying to guide a train in tandem with the Austrian prime minister, but Wrangel is the erratic steam pipe.)

even though he disagreed with the politics of his mentors, he embraced their approach to organization. He saw the staff structure as the central key to victory in any future war. His teachers were impressed and believed that he would come to epitomize the staff officer.[340]

In 1834 Moltke journeyed through the Balkans, eventually moving on to the Middle East. His journal became popular reading in Prussia, but the prolific writer produced more than mere travelogue. While in Syria, he accidentally found himself in the middle of a fight between the Ottoman Turks and their insurgent Egyptian subjects. When the Ottoman sultan heard that a Prussian officer was in the area, he immediately offered Moltke a command. Although the war turned sour for the Turks, Moltke experienced combat firsthand, commanding the Turkish artillery for a number of months. He returned to Prussia to complete several staff assignments, and he was promoted to Chief of the General Staff in 1857. Reporting directly to Roon and Manteuffel, Moltke had the reputation of possessing one the keenest intellects in the army.[341]

Moltke made an impression on Manteuffel, Roon, and even Bismarck. His commanders took note of prolific writings in which Moltke explained the nature of war, extensive theories on railroads, industrial production, strategy, and tactics. Moltke theorized that railroad transportation could improve the speed of deployment, and, if properly employed, railroads gave armies a first-strike capability. He realized that machines had revolutionized the tactical aspects of combat, and that rapid firing breech-loading rifles gave the defense a tremendous advantage. In response, he devised a different style of offensive—something Basil H. Liddell Hart would later call "the indirect approach." In Moltke's new offensive, the purpose of an attack was to turn an enemy flank and surround an opposing army. This forced an enemy to attack the superior forces of the original attackers, while the new defensive positions gave the surrounding army a tactical advantage. Citing the ancient tactics of Hannibal and Scipio Africanus, Moltke referred to this style of fighting as the *Kesselschlacht*, the battle of double envelopment.

In May 1864 Moltke was still an untried intellectual in the king's eyes when he left Berlin to join Prince Frederick Charles at the siege of Düppel.

---

[340]See Walter Goerlitz, *The German General Staff: 1657–1945* (trans. by Brian Battershaw) (New York: Frederick A. Praeger, Publishers, 1962), 67–73.

[341]See Rosinski, 74 ff.

Great Britain was calling for a negotiated end to the conflict, and Bismarck wanted to bring some type of victory to the bargaining table. Since invasion of Jutland was out of the question, an assault on Düppel was the only alternative. Frederick Charles thought an assault on the fortress would be unwise, and Moltke, hating the idea of a frontal assault, agreed. William intervened. Knowing that the great powers would impose their will on Prussia without a victory, he ordered his nephew to give Bismarck a bargaining chip: Düppel was to be taken. Always able to understand an order, Moltke asked only for time to prepare.[342]

Moltke's preparation was brilliant, and his actions in Denmark allowed Bismarck to focus on the real objective of his belligerent policy, Austria. On April 18 the Prussian army took Düppel with minimal costs. Although the Prussians were extremely lucky, Moltke's reputation soared in Berlin, and Bismarck went to the peace table in a position of power. Bismarck wanted to keep the Danes off balance, while demonstrating to the German states that Prussia was capable of acting on its own. When the Danes balked at his initial peace offer, Bismarck sent Frederick Charles and Moltke on the offensive, and they quickly pushed the Danes back to the channel. Moltke planned a highly efficient assault on another Danish fortress at Alsen, demonstrating that the victory at Düppel was no fluke. Prussia could move where it wanted, with or without Austria. Denmark surrendered on July 12, and Moltke came back to Berlin not as an intellectual but as a bloodied commander with William's full confidence. All the while, Bismarck made overtures against Austria before the ink dried on the Danish treaty.

The Seven Weeks' War

Moltke would later write that war between Austria and Prussia was inevitable, but in 1865 he was not fully committed to the idea of armed conflict. Bismarck, however, always believed that Austria was Prussia's enemy. When the short Danish war concluded, the prime minister announced plans for the annexation of Schleswig and Holstein, knowing full well the impact such a policy would have in Vienna. Austria and the whole German Con-

---

[342]See Kitchen, 112; Gall, vol. I, 249; and Eyck, 118 ff. See also Heinrich Friedjung, *The Struggle for Supremacy in Germany: 1859–1866* (trans. by A. J. P. Taylor and W. L. McElwee) (New York: Russell and Russell, 1966; orig. 1897), 46–84.

federation recoiled. They were perfectly willing to fight for the freedom of the two provinces, but they were not willing to tolerate the aggrandizement of Prussia. Bismarck, feigning surprise, claimed that annexation was Prussia's right. Within months most of the German states rallied behind Austria, and Bismarck had his war—not only against Austria but against a whole host of German states. In February 1866, Moltke set about developing plans to destroy the Austrian army and its German allies.[343]

Two elements dominated Moltke's strategic thinking. First, most of the German states, including the major dominions of Baden, Hesse, Bavaria, Hanover, and Saxony, signed an alliance to support Austria. The Saxons fled to the south, but the other states assembled an army of 150,000 men in western Germany. If these forces united with Austria, Prussia would be greatly outnumbered. Moltke wanted to neutralize the west German forces before they could play a role, but he wanted to force them to capitulate after accepting minimal casualties. Killing fellow Germans could only lead to long-term hatred and animosity. Moltke wanted to defeat the German armies in such a way as to leave the door open for unification, and the *Kesselschlacht* seemed to be the obvious answer. When the German states assembled their army, Moltke hoped to surround them and ask for their surrender. Once victory was assured in the west, Moltke hoped to unleash the whole of the Prussian army against the Austrians somewhere in Bohemia, planning to eradicate the Austrian army in order to eliminate Hapsburg influence in German affairs. The *Kesselschlacht* was once again to be Moltke's tool, but in Bohemia his purpose would be destruction, not surrender.[344]

Railroad mobility played the key role in Moltke's thinking. While most of the world wagered on a Austrian victory, Moltke thought the railroads of northern Germany gave him an upper hand. Railroads allowed him to move faster than his enemies, and if the German armies could be quickly defeated in the west, Moltke gambled that all Prussian forces could be quickly transferred to the Austrian theater. He developed a daring strategy, all depending on the speed of rail transportation. Knowing that Austria would move into either Silesia or Bohemia at the beginning of hostilities, Moltke planned to

---

[343]Kitchen, 116.

[344]Larry H. Addington, *The Blitzkrieg Era and the German General Staff, 1865–1941* (New Brunswick, N.J.: Rutgers University Press, 1971), 5–7.

divide his forces into three armies to increase maneuverability. The Elbe Army was to be stationed on the far western flank. If the western German forces were defeated, the Elbe Army would move south into central Bohemia. Another force, the First Army under Frederick Charles, was positioned directly to the north of the Bohemian border. It was to march south and eventually link with the Elbe Army. The final force, the Second Army under Crown Prince Frederick William, was to move into Silesia to contain the Austrians in Bohemia. When the First Army moved, the Second Army was to cross the mountain passes and fight its way to the First Army's position. Moltke planned to separate the armies for movement but to have them assembled on the day of battle. Everything depended on the ability to move and supply the troops, and Moltke believed that the Prussian rail network could support such a feat.[345]

Moltke seized the initiative on June 16, 1866. Using the railroads for mobilization, he had gathered all his forces in an incredible twenty-one day period. He sent Vogel von Falkenstein, Wrangel's chief of staff in the Danish war, toward Hanover with 50,000 troops, while assembling three armies on the Bohemian border. Outnumbered, Falkenstein's troops advanced in a wide arc, and the Hanoverian commander, General Alexander von Arentschildt, mistakenly took this as a sign of weakness. Gathering as many of the German allies as possible, he struck Falkenstein at Lagensalza on June 27, directly in the center of the Prussian arc. Arentschildt's numerical superiority overwhelmed the lead elements of the Prussian forces in Lagensalza. Assuming that he had sent the Prussians reeling, he threw all available forces into the fracas. Even as the German armies were driving the Prussians from the city, however, Arentschildt became one of the first people outside of the Prussian army to learn the definition of *Kesselschlacht*. As the west German army moved forward, Falkenstein's flanking divisions closed in on Arentschildt's flanks and rear. Arentschildt turned from his attack to meet this new threat, only to realize that the Prussians were digging in. The Prussians held superior defensive positions, and Arentschildt immediately ordered a retreat. When his corps commanders informed him that Falkenstein was blocking his rear, Arentschildt realized that his entire army had been surrounded. Without the heart to fight other Germans to the finish, the

---

[345]Friedjung, 197–200.

west Germans surrendered on June 29, leaving Moltke free to concentrate on Bohemia.

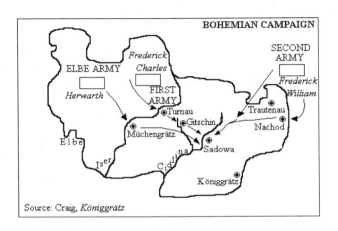

The main Prussian effort had begun on June 22, 1866. Frederick Charles and the First Prussian Army crossed the Bohemian border from the north, while the Elbe Army under General Karl Herwarth von Bittenfeld, reinforced from elements of Falkenstein's command a few days later, moved along the First Army's western flank.[346] Moltke planned to unite both armies in the vicinity of the Iser River, a stream running from north to south about midway through Bohemia, but Herwarth did not move as fast as Moltke desired. In addition, Frederick Charles stopped to rest on the second day of the campaign. Neither man understood Moltke's plan and the need for speed. On the Austrian side, Field Marshal Ludwig von Benedek also failed to grasp the situation. Commanding the Hapsburg forces as well as those Germans who had managed to flee south into Bohemia, Benedek could not believe that the Prussians had assembled so quickly. He decided to assemble his forces in southern Bohemia while establishing a thin defensive line under Saxon Crown Prince Albert along the Iser River. Not believ-

---

[346]Summary based on Gordon A. Craig, *The Battle of Köninggrätz* (London: Weidenfeld and Nicolson, 1965), 57–175. (Craig's work is the definitive English-language version of the battle.)

ing that the Prussians could be advancing in strength, Benedek gave Albert two corps, telling him to maintain his position on the Iser until the Austrian army was fully assembled.

Frederick Charles had one thing in common with his enemies. He believed that Moltke's main effort would be a straight shot at the Austrians with the First Army, and he had no idea why the Chief of the General Staff thought that speed was so essential. He wanted the army rested and fit for the upcoming battle with Benedek. After receiving frantic orders to move forward, Frederick Charles crossed the Iser on June 26, confused by Moltke's sense of urgency. Establishing a bridgehead in the village of Turnau, Frederick Charles set up observation points and started bringing the remaining elements of the First Army across the river. Albert's Saxons—just south of the First Army's position—held no love for Prussia, and they were spoiling for a fight. When Crown Prince Albert informed Benedek of Frederick Charles's position, he did so while launching an attack against the Prussians on the evening of the twenty-sixth.

Albert's attack was to be indicative of the campaign. As Albert's forces moved forward, local Prussian commanders seized the high ground to the north of Turnau, and lying on the ground the men of the First Army fired their breach-loading weapons into the advancing Austrians. The advantage was clear. The concealed Prussians fired from the prone position, but the Austrians—armed with muzzle loading rifles—had to stand while reloading. The Austrians were mowed down by an unseen enemy. As darkness enveloped the area, Albert was forced to retreat, and the road to Benedek was completely open. Frederick Charles, however, refused to take advantage of the situation. Moltke no doubt would send orders to move, but the First Army had just been tested in battle. Frederick Charles decided to rest his troops at Turnau.

Things were very different in the east. Prussian Crown Prince Frederick William started moving his Second Army through the Silesian mountain passes as soon as he received orders from Moltke. He knew that if he failed to join Frederick Charles, the Second Army would either be held in check in the mountain passes or isolated and destroyed in Bohemia. Everything depended on clearing the key passes and moving to southern Bohemia as swiftly as possible. The Prussian crown prince moved quickly, dividing his army into three corps to speed his movement through the mountain passes. Although his actions met with Moltke's approval, Frederick William did not know that a large Austrian army sat directly in the center of his divided forces. Moltke always allowed forces to separate for ease of movement while marching but never allowed them to disperse on the eve of battle. Unfortunately for the Prussians, the Second Army had quite inadvertently violated Moltke's dictum. When they found the Austrians, two of the Second Army's corps were able to join near the northern invasion route near Trautenau, but a third was completely cut off in the south at Nachod. The Second Army had been split.

The Second Army's first encounter with the Austrians came on June 27 when General Karl von Steinmetz, commanding the corps to the south of the crown prince's main position, ran into a strong Austrian force in the village of Nachod. If Steinmetz was surprised to find the Austrians in eastern Bohemia, the Austrians were equally mystified by the advance of a strong Prussian force from Silesia. Although he was isolated, Steinmetz took advantage of the situation. Bringing every available unit to the front, he ordered a full-scale assault on Nachod. The Austrians were forced to retreat, and Benedek sent no reinforcements to the area. He believed that the main attack was coming from Frederick Charles and had no intention of wasting resources on a diversionary assault from Silesia.

Things did not go as well to the north. The next day, June 28, the other segment of Frederick William's army was driven back to its base in Silesia, leaving Steinmetz on his own. With the majority of the Prussians either stuck in the mountains or regrouping in Silesia, a grim Frederick William ordered his troops to the defensive. He decided that he would hold the mountain passes, hoping that Steinmetz could somehow escape. Realizing the gravity of the situation, Frederick William's generals ordered troops to dig in and await the Austrian onslaught. It never came. Still believing the Silesian effort to be a feint, Benedek thought that he had merely driven a diversionary force away. He ordered local commanders to cease all attacks on the twenty-eighth, telling them to march west toward Prince Albert on the twenty-ninth. The relieved—and astounded—Prussians watched the Austrians melt away without a fight. Frederick William followed the retreating Austrians, moving to the west of Nachod. By evening Frederick William was able to assemble his entire army, and Steinmetz was no longer on his own.

Helmuth von Moltke had a more strategic view of the situation. Commanding the entire operation miles away by telegraph, he could see the potential trap springing. After Herwarth's Elbe Army linked with Frederick Charles, Moltke gave Frederick Charles command of both armies and sent him a strongly worded telegram telling the general that the king would like to see the First Army moving with all possible haste. Frederick Charles took the hint. Moving like a man possessed (after the king's name had been invoked), Frederick Charles drove Prince Albert from the Iser at Müchengrätz. Without waiting for his forces to assemble, he moved forward, meeting the Austrians again at Gitschin on June 28. At Gitschin, however, Saxon Crown Prince Albert was somewhat more successful than in previous days. Stopping the Prussians dead in their tracks, he sent word to Benedek, asking permission to press the attack. Benedek would have agreed a few days earlier, but now he had reports of thousands upon thousands of Prussian troops moving from Silesia. Realizing that the Second Army was no diversionary force, Benedek denied Albert's request, telling him that all forces should be assembled as quickly as possible. He ordered Albert to retreat to the main Austrian position near Königgrätz.

In the waning days of June, Austrian Field Marshal Benedek thought the fight for Bohemia was lost. With Frederick Charles advancing to his front and Frederick William coming on the flank, Benedek's corps commanders wanted Emperor Franz Joseph's permission to move the army to Austria. Franz Joseph would hear nothing of such defeatism, and he ordered Benedek to make a stand. The order initially depressed the imperial field marshal. Believing the emperor could not see the hopelessness of his situation, Benedek surveyed the ground north of Königgrätz, and much to his surprise he found the ground to be defendable. When Albert arrived on July 1, Benedek concentrated all his forces along the Bistritz River, and Albert approved. Becoming cautiously optimistic, Benedek replaced his corps commanders, save Prince Albert, with men who spoke of victory. Frederick Charles was ready to attack, but the Austrians had relatively equal strength and held the high ground. If he could defeat Frederick Charles in a defensive struggle, Benedek thought, he could turn and strike Frederick William. He explained the plan to the delight of his corps commanders, and the Austrians enthusiastically began fortifying their positions with earth works. As the Austrians were digging in, however, a train carrying King William and Bismarck arrived at Frederick Charles's headquarters. Moltke was with them.

The Battle of Königgrätz

As Moltke explained the Prussian attack to William, Bismarck, and the commanding generals, Frederick Charles did not completely understand the Chief of the General Staff.[347] To Frederick Charles, *Kesselschlacht* meant that the First Army would surround Benedek and force the Austrians to submit. Moltke brushed this idea aside, telling Frederick Charles that First Army would "absorb" the Austrian army in a frontal assault and that the Elbe Army would be detached to move along the Austrian left. Herwarth would turn the flank and move to the Austrian rear, while Frederick William—who was miles away—would converge on the other flank. Frederick Charles was skeptical. He asked Moltke about his own flank and why the frontal assault was necessary when Moltke's very doctrine denounced such tactics. Moltke patiently explained that Frederick Charles was simply to tie the Austrian army to a single position. The flank was no problem, Moltke said, because the Second Army would attack the Austrian right and move to the rear to link with the Elbe Army. Frederick Charles did not understand this at all. Second Army was more than twenty miles away, and it would not arrive in time to fight. He hoped the Chief of the General Staff knew what he was doing.

It is quite possible, indeed, quite probable, that the three Prussian army commanders did not understand what Moltke had in mind, and, for that matter, neither the king nor Bismarck had a clear picture. The battle of double encirclement was clear, but gathering three separated armies during the course of a battle on a single day seemed an impossible task. Frederick Charles and Herwarth grumbled as they left Moltke's briefing, looking at the Austrian positions in the waning hours of July 2. The Austrian front was guarded by the Bistritz River. The river was fortified by a series of villages running from Nechanitz on the southwest, through Sadowa, and into the village of Benatek on the northeastern flank. Each village was fortified so that Austrian riflemen could fire from excellent cover. Sadowa was flanked by two small forests, Holawald on the south and Swiepwald to the north. The bulk of the Austrian army waited on the high ground behind the fortified river bank. According to Moltke's plan, Frederick Charles was to cross the river near Sadowa, take the fortified positions, and eventually take the

---

[347]Ibid., 99 ff.

high ground. Frederick Charles did not believe this to be the wisest of all
plans, and it certainly did not seem like a battle of encirclement.

General Herwarth liked the plan even less. His Elbe Army was assigned
the task of moving through Nechanitz and sweeping to the Austrian rear. As
he looked toward the fortified positions that fell in Frederick Charles's sector, Herwarth believed that the First Army would be massacred on the banks
of the Bistritz. And, if the Elbe Army actually moved through Nechanitz, as
Moltke had ordered, his troops would be surrounded and destroyed after
Benedek eliminated Frederick Charles. Herwarth publicly promised to obey
Moltke but privately planned to leave most of his Elbe Army at Nechanitz
so that he could retreat if Frederick Charles was destroyed. Moltke was a
paper soldier, Herwarth thought, and did not understand that this was a suicidal attack. In addition, the Second Army was too far away. It would be no
help on the day of battle. Despite the talk of encirclement, Herwarth believed that Moltke had accidentally planned a head-on attack into a heavily
fortified position.

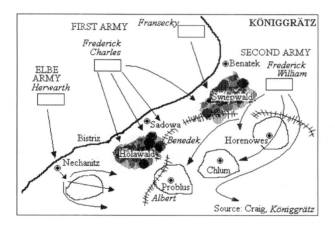

On the other side of the hill, Benedek shared the same impression. The
Austrian staff assured him that Frederick William's Second Army was too
far away to march into battle, and the Prussians had given every intention
that they planned to attack frontally the next day. The Austrian position was
sound. Most of the front was covered by natural defensive barriers, and the

open positions were reinforced with trenches. The villages that dotted the Bistritz had loopholes chiseled out of the walls and houses. The line was backed by strong infantry forces on the hills behind the Bistritz, and each avenue of approach was covered by rifled Austrian artillery. The cannoneers had already sited their weapons and determined the azimuths of fire. They needed only to point in the prearranged direction and discharge for deadly accuracy. Should the Prussians attempt to cover the left flank, they would encounter infantry. If they tried to sweep from the right, they would be met by a cavalry reserve. The position seemed virtually impregnable, unless the Second Army arrived. That, Benedek's staff promised, could not happen.

Frederick Charles may not have liked Moltke's plan, but he was an obedient soldier. At 7:15 A.M. on July 3 he moved the First Army to the Bistritz and began crossing in the vicinity of Sadowa, the very heart of Benedek's strength. As he expected, things did not go well. His troops made some headway, moving into the small forest of Holawald. The experience, however, soon turned into an absolute nightmare. Austrian gunners, having craftily hidden from Frederick Charles's view, had a direct line of fire toward the Prussian troops. The 4th Infantry Division moved into the Holawald only to be met by the direct, deadly cannon fire. With no artillery of their own, the Prussians could not return fire, and the Prussians attempted to advance, only to be gunned down. Survivors crawled behind trees, their comrade's dead bodies, and anything that seemed to promise some form of cover while the artillery shells burst among them. By mid-morning Frederick Charles's attack was going nowhere, and he begged Herwarth to begin his own attack to relieve some of the pressure.

General Herwarth was no doubt moved by Frederick Charles's appeal, but he had no intention of placing the Elbe Army on what he believed to be a sacrificial altar. Prince Albert's Saxons held the ground around Nechanitz in front of the Elbe Army, and they were giving a good account of themselves. Herwarth engaged them cautiously. He managed to dislodge the first rank of Saxon defenses, and he sent some troops forward. Nevertheless, he kept the majority of his forces in position to retreat. When Frederick Charles's plea came, Herwarth sympathized, but he maintained his position. Frederick Charles, seeing that the Elbe Army was holding its position on the Bistritz while his own soldiers were being pounded by Austrian artillery, decided to take action. In desperation, he appealed directly to the king, and William, who came forward to have a look at the Bistritz, was moved by his nephew's cry for help. Sending personal word to Herwarth, King William urged the general forward. Herwarth would not move for Moltke, but, like most Prussian officers, he was willing to fall on his sword for his sovereign. Shortly before noon the Elbe Army launched a devastating assault on Nechanitz.

The only brief bright spot for the First Army appeared around 10:30 A.M. Lieutenant General Eduard von Fransecky's 7th Division managed to take the village of Benatek on the Austrian right. The Austrians were swept back into the trees of Swiepwald, and Fransecky continued the attack, moving beyond the village. The forest of Swiepwald was thicker than Holawald, and Fransecky's men were able to find cover. Unable to dislodge Fransecky with artillery, the Austrians launched a series of infantry assaults through the woods. Although the 7th Division held, each attack drove it back toward the village of Benatek. By early afternoon, Fransecky managed to hold a thin strip on the edge of the woods, but a strong attack could dislodge the 7th Division. The First Army was in a deadlock all along the front, just as Frederick Charles had feared.

Only Moltke seemed unworried. Shortly after noon, the Prussian headquarters was flooded with reports of terrible casualties, and every unit in the First Army was pinned down. The 4th Division was in pitiable disarray, and Fransecky's 7th Division was holding on for dear life. Herwarth was finally making a little headway, but all attention was focused on the First Army. Frederick Charles seemed to be getting nowhere, and hundreds of Prussians were being massacred in the process. Bismarck turned to Moltke, asking if the situation was acceptable. Moltke brushed him off. Later, when the king asked the same question, Moltke gave a more military response. As bad as the slaughter was, Moltke explained, Frederick Charles had completely "absorbed" the Austrian front. While Frederick Charles could not advance, the Austrians could not disengage. The crown prince would arrive shortly with the Second Army, Moltke told his sovereign, and it would move to Benedek's rear. This would doom the Austrian army. All eyes turned in the direction of the crown prince's suspected route of march. Everything depended on his arrival.

The fight of Fransecky's 7th Division seemed to underscore the situation. The king, unlike his generals, believed Moltke's plan to be viable. If his Chief of Staff recommended holding the positions along the Bistritz without reinforcements, the king would follow the recommendation, but it was not an easy decision. Around noon, an urgent message came from Fransecky. The Austrians were preparing yet another counterattack, and Fransecky begged the king for more troops so that he could hold Swiepwald. The king looked at Moltke. The general told William that he knew Fransecky personally, and that Fransecky would not fail. He will hold the Swiepwald, Moltke calmly stated. William raised his eyebrows in thought (maybe prayer), then he denied the request for reinforcements. William was willing to gamble everything on Moltke and the arrival of his son's Second Army.

The king's wager was well placed. The Second Army had been on the move since early morning, and Crown Prince Frederick William knew that Moltke needed his forces. He also knew that if he moved fast enough, he could attack by early afternoon. Quite by chance, Frederick William had placed the majority of his artillery, under the command of Prince Kraft zu Hohenlohe-Ingelfen, near the front of his columns. As the army marched at a furious pace, Hohenlohe's artillery wagons led the way. When the Second Army neared the battle, Hohenlohe pointed his scouts toward a big clump of trees. He ordered the artillery to move in that position, and the infantry followed suit. This action fortuitously concentrated the artillery and the infantry in one spot one the Austrian flank—the position Frederick William was trying to strike.

Helmuth von Moltke, looking through field glasses, saw the Second Army on the Austrian flank in the early afternoon. The Chief of the General Staff broke into a rare smile as he turned to his king, assuring William that all would go well. In the Austrian headquarters, other generals scanned the right flank with field glasses, but there was little they could do save to watch in horror. The Austrian line was fully engaged, and the Elbe Army was enveloping its left flank. Having moved most of his troops forward to counterattack, Benedek had nothing left to meet the Prussian threat to his right. His modest flank was designed to hold Frederick Charles, not an entire army. Benedek's woes were compounded at 3 o'clock when Hohenlohe's artillery moved to the hills around Horenowes, opening a hellish barrage that reached the very heart of the Austrian positions. Unable to disengage, Benedek feared that his entire army would be lost.

To their credit, the Austrians did not panic. Benedek realized that the hill at Chlum had become the key to escape. One of the highest points in the Austrian rear, the hill offered a chance for the Austrians to rally and reestablish their line. If he could hold Chlum, or even delay the Prussians there, Benedek might be able to extricate the army and save it from complete destruction. If he lost it, however, his army was doomed. Frederick William's infantry took the hill after a brief fight, only to be met by a fierce Austrian counterattack. Despite being gunned down by Prussian breach-loading rifles and Hohenlohe's accurate artillery fire, the Austrians charged the hill repeatedly. By 4:30 their tenacity paid off. Although they could not move into the village, Hohenlohe's ammunition was depleted and he was forced to pull the artillery back. It was the only break the Austrians needed. Benedek struck with his cavalry reserve and pulled all the available infantry out of the area. About the same time, Frederick Charles received a new order; the First Army was ordered to advance. The Austrians had abandoned the defensive positions along the Bistritz.

The final chapter of the battle ended with a confused cavalry action between Chlum and Königgrätz. As the Austrians retreated, Benedek sent his cavalry forward as a rear guard. For the next few hours until darkness, Prussian and Austrian cavalry rode at one another on a scale that rivaled Liebertwolkwitz. The Austrian horse bested their Prussian rivals, but to no avail. Each small victory was quickly equalized by the deadly infantry fire from the Prussian needle nose rifles. The Austrian cavalry allowed Benedek's army to escape, and that was all it could do. Benedek knew that the army was defeated. His actions at Chlum saved the lives of many men, but he could not make another stand. On July 5 he received permission to negotiate an armistice.

Most contemporary military observers believed that the Austro-Prussian War would last quite some time. Emperor Napoleon III of France, for example, thought Prussia's invasion of Bohemia signaled the start of a second Seven Years' War. Instead, the war lasted seven weeks, and on August 23 Bismarck accepted a treaty from Vienna. There were small territorial concessions as Schleswig, Holstein, and Hanover fell to Prussia, and Austria was forced to pay a small indemnity. But Bismarck's greater concern came with Austria's role in German politics. The treaty formally excluded Austria from German affairs. Prussia was now the dominant member of the German Confederation and in control of most of northern Germany. The Catholic south was not excited about this prospect, but they could no longer turn to Vienna for help. Bismarck had only one question remaining: Would France allow Germany to unite under Prussian control? He would answer the question with blood and iron.

Chapter 16

Empire

In 1871 Prussia entered its final war as a self-contained, autonomous state. It was the end of an amazing transformation. The Great Elector and his corps of military commanders could have never envisioned the powers and abilities of the new Prussian army. With the capacity for mass mobilization, the strength of industrial power, and the speed of rail transportation, Moltke led what could arguably be called the best army in Europe. Only one nation was in a position to challenge that claim: France. Basking in the glory of its second Napoleonic empire, France was *the* continental power of the era, and few serious military analysts believed that Prussia could ever successfully challenge its position. Bismarck believed, however, that France would never accept a unified Germany and that Prussia would be forced to challenge its old nemesis, if Germany were to free itself from the fear of France. Moltke felt the same way, and he was confident in the army's ability to defeat France. Data compiled by the Prussian General Staff suggested that the French army was a paper tiger, and if war came, many members of the General Staff believed, the Prussian army would emerge as the victor.

The Diplomatic Preparations for War

During the night of July 3, 1866, a sullen Bismarck rode through the muddy roads around Königgrätz searching for a place to sleep. He was shaken by what he saw. The fields were covered with pain—wounded men writhing in agony among the lifeless forms of other men. Most of the wounds were ghastly, but Bismarck, who had never seen such destruction, would be soothed by the long-term outcome. Within weeks Prussia was clearly in control of the German Confederation, and the German states were solidifying under the Hohenzollern throne. Yet, the job was not finished. Many individual German states simply refused to see themselves as vassals to a larger country, and nowhere was this feeling more evident than in the

317

south. Southern Germany, an area that differed from the north in culture and dialect, had traditionally been a hotbed for Hapsburg loyalty, and it had always remained in the Roman Catholic fold. To many southerners, unification in the German Confederation was a northern affair. Southern Germans, although supporting the new confederation, were not quite ready to accept Prussia leadership completely.[348]

Bismarck was concerned by the lack of cohesion in the south, but the situation appealed to Charles Louis Napoleon III, the emperor of France. Questions about southern loyalty kept the Prussians from becoming too strong, and Napoleon III was content to have the German states play their power against one another. Napoleon's European military plans no longer included a desire to annex the Rhineland, and, so long as Germany remained politically fragmented, Prussia offered no resistance to France's interests overseas. Nevertheless, a powerful Germany would change the equation. Should German unification pose a threat to France, Napoleon might be forced to act. Indeed, most Germans believed that Napoleon would strike as soon as Germany galvanized beyond the loose bonds of confederacy. Accepting this state of affairs as an outgrowth of hundreds of years of animosity, Bismarck believed in the French threat, but he also felt France's position presented an opportunity for Prussian diplomacy. If France raised its sword against the German Confederation, it could well become the catalyst to unite all the German states. Officially, however, France and Prussia were united by treaty. If Bismarck wanted to cast Napoleon in an adversarial role, he needed to kindle a fire that would reverse the friendly relations between the two countries.[349]

The spark came in July 1870 when the last of the Spanish Hapsburgs died leaving no heir to the throne. As Spain searched the royal houses of Europe for a new leader, Bismarck hit upon an idea that would inflame the French nationalists in Napoleon's government. He proposed a Hohenzollern candidate, Prince Leopold zu Sigmaringen-Hohenzollern, as the successor to the Spanish throne. The Spanish parliament embraced the idea, and Prince Leopold was delighted. William also liked the thought of expanding

[348]William Carr, *A History of Germany: 1815–1945* (2nd ed.) (New York: St. Martin's Press, 1979), 110–17.

[349]Erich Eyck, *Bismarck and the German Empire* (New York: W. W. Norton and Co., Inc., 1968), 163–68.

Hohenzollern influence to the Iberian peninsula, but he was careful to keep a low profile lest he rouse the ire of his fellow monarchs. Bismarck expected these reactions, but his real target was Paris. He knew full well that Napoleon would never allow France to be flanked by two Hohenzollern thrones. Napoleon III had gained the French throne in 1848 during the street revolution. By 1852 he had garnered enough political support to proclaim himself emperor, and he maintained power by appeals to fervent nationalism. His supporters would never accept a Hohenzollern on the Spanish throne. When Leopold's candidacy was announced in the summer of 1870, there was no joy in Paris. French nationalists spoke of war.[350]

The French parliamentary reaction to Bismarck's proposal was swift and bellicose. Without waiting for word from Berlin, they ordered a partial mobilization of the French army in early July. The French press took up the cry, demanding a march on Berlin. King William was absolutely stunned. Unaware of Bismarck's plans, the king wanted nothing to do with war. His wife spoke to Leopold, urging him to withdraw from contention for the Spanish throne, and William breathed a sigh of relief when Leopold quietly removed his name from the list of possible successors. Assuming that the issue was settled, the king left for a holiday to the hot baths of Bad Ems.

Despite William's perception, the issue was far from settled. Paris was aflame with rhetoric, and demands for belligerency multiplied. The French press and right-wing nationalists wanted something more significant than Leopold's withdrawal. They wanted an apology from William and an assurance that Prussia would never again seek the Spanish throne. France's honor had been tinged, they said. In response, Napoleon sent the French ambassador to see William at Bad Ems, but the old Prussian king only wanted to escape the pressures of office. He was in no mood to see the ambassador. When the ambassador demanded to see the king, William refused an audience, telling the ambassador to talk to his ministers. After the French envoy left, William thought it would be wise to inform Bismarck of the incident, and he sent a lengthy telegram to Bismarck telling him of the visitation. The king did not know that Bismarck would edit the document.[351]

---

[350]Martin Kitchen, *A Military History of Germany: From the Eighteenth Century to the Present Day* (Bloomington, Ind.: University of Indiana Press, 1975), 125.

[351]For good narratives see Carr, 119 ff., and Eyck, 169 ff.

Bismarck was dining with Moltke and Roon when the king's telegram arrived. As he read William's narration, Bismarck realized that he could exclude a few key phrases, giving the Bad Ems telegram an entirely different flavor. By carefully editing sentences, Bismarck made it look as though the king had done something more than tell the French ambassador not to interrupt his bath. Bismarck made it appear as though King William had severed diplomatic relations with France. Although nothing of the sort had happened, Bismarck released the modified telegram to the press. It hit like a bombshell. Paper after paper printed inflammatory excerpts from the message, spinning sensationalized headlines first across Germany, then to France, and then throughout the world. The story took on a life of its own. When outraged French parliamentary representatives assembled in Paris, they were caught in the heated frenzy. In no mood for rationalization, they demanded military action. On July 15, 1870, they voted to fund a full military mobilization, and on July 19 they declared war, never suspecting that they had been duped.

Moltke's Operational Plan

Moltke, Roon, and Bismarck convinced the king to mobilize German forces in late July. The French had a head start, but Moltke knew that it would be quite some time before Napoleon could gather his scattered forces. His studies of the French rail system told him something more. The Prussian General Staff was aware that modern armies were moved and supplied by rail. In eastern France there were only two facilities capable of sustaining the French army, Metz and Strasbourg, and no other railheads could accommodate large numbers of troops or serve to supply the French army. Although the French simply believed they could march on Berlin, Moltke planned to assembled three large German armies along the west German railroad network and move on the French railheads. With his reputation firmly in hand after the Austro-Prussian War, Moltke had become William's formal military advisor. When he unveiled the plans to strike Metz and Strasbourg, the king endorsed the concept.[352]

---

[352]Larry Addington, *The Blitzkrieg Era and the German General Staff, 1865–1941* (New Brunswick, N.J.: Rutgers University Press, 1971), 7–11.

Source: Howard, *Franco-Prussian War*

Organizing three German armies was a sensitive task. Rival German forces had been slaughtering one another a mere four years earlier, but the General Staff counted on the fear of France to force unity. Moltke took the matter a step further. Like Napoleon III, the Chief of the General Staff was not completely trusting of the southern Germans, and he suspected that many hearts were not in the cause. Rather than letting each state fight as an individual entity, Moltke allowed individual units to keep their identification and command structures, but he completely integrated them within the Prussian army. For all intents and purposes, the German army became subservient to the Prussian army. The situation was not without its ironies. Crown Prince Albert of Saxony, the man who offered such strong resistance to Frederick Charles in the Seven Weeks' War, now commanded a corps for Frederick William. Although hundreds of Prussians had died at his hands, Albert marched with the Prussians a scant forty-eight months after König-grätz. The efficiency of the mobilization was another issue. By the end of July Moltke had 1,800,000 men under arms, with 460,000 of them in the Rhineland. The ability to assemble such large numbers was a tribute to the General Staff. If the army would fight as a German force, Moltke had the upper hand.[353]

---

[353]See Curt Jany, *Die Königlich Preussische Armee und das Deutsche Reichsheer*

Moltke's battle plan was based on the experience in Austria. He hoped to move quickly, keeping the units dispersed until the day of battle. Knowing that the French would be concentrated in Metz and Strasbourg, Moltke planned to surround the majority of French forces close to the border and destroy them. This would open the road to Paris, and by quickly taking the city Moltke hoped to put the king in a position where he could dictate peace terms. To accomplish these objectives, Moltke divided the troops in the Rhineland into three armies. Crown Prince Frederick William—popular in the south—received command of the Third Army because it had a large concentration of southern Germans, including Prince Albert's 12th Saxon Corps. Frederick Charles assumed command of the Second Army, and the smaller First Army went to Karl von Steinmetz, the seventy-year-old hero of Nachod. Both the Crown Prince and Frederick Charles proved to be a bit too cautious, and Frederick Charles planned to seek the glory he felt denied him at Königgrätz. Steinmetz, unconvinced that Moltke had any role in field operations, soon became a thorn in the Chief of Staff's side. Moltke could not afford such attitudes. His plan was only as strong as the willingness of the three army commanders to cooperate.[354]

Both the Germans and the French had organizational problems. Despite the well-ordered Prussian rail system, supplies often failed to reach German units in the field. Material moved to the rail yards, but the army lacked the vehicles necessary to move stocks to individual units. While Moltke was frustrated by logistical failures, his problems were infinitesimal when compared to those of the French. Napoleon III had no plan other than that of his War Minister Edmond Leboeuf, who claimed that the French army would march to Berlin. French regiments moved at a snail's pace as railroads jammed and supplies moved to the wrong destinations. Battalions marched away with improper uniforms and, on at least one occasion, the wrong shoes. French staff work was absolutely atrocious. The weary troops marched and countermarched until they were exhausted, and Leboeuf insisted on fortifying every sector of the 200-mile Franco-German border. Not

---

1807 bis 1914 (Berlin: Verlag von Karl Siegismund, 1933), 262–67.

[354]Michael Howard, *The Franco-Prussian War: The German Invasion of France, 1870–1871* (London: Routledge, 1988), 57–62. (The summary of the campaign is based on Howard's narrative. Like the books by Duffy and Craig, this is a "standard-setting work." It is the definitive English-language text on the Franco-Prussian War.)

only was this physically impossible, but the troops were perilously stretched in a thin line. Supplies were almost non-existent, and the common soldiers began to lose faith in their officers. Some of the unsupplied troops became openly defiant while others rebelled. To make matters worse, many of the troops began to forage from the countryside, stripping their own land bare. Moltke had only to contend with standard operational problems. As the French army gathered in late July, it was obvious that they were completely unprepared for war.[355]

The Guns of August

The chain of command and hierarchical structure of the French army was horrid. Officially, Napoleon and his War Minister Leboeuf ran all operations, but local commanders were forced to assume their own authority, while units along the border became openly insubordinate. No single political philosophy dominated France's approach to the war, and as various factions publicly quarreled, events quickly dissolved into a hodgepodge of individual actions. On August 2, General Charles Frossard, Napoleon's military tutor, let the frustrating situation get the best of him. Without waiting for either orders or support from other units, he defiantly marched across the Rhine to seek the Germans. He was surprised at what he found. After taking the town of Saarbrücken without too much resistance, he found himself in the midst of a massive German build up. While the French press hailed Saarbrücken as the first step toward victory, Frossard sent urgent word to Leboeuf asking for help. His pleas for help fell on deaf ears. Napoleon and Leboeuf were more concerned with their own organizational problems than with Frossard's dilemma in the Rhineland.[356]

At first, Frossard's sudden move threw Moltke into a panic. Assessing the situation from Berlin, the Chief of the General Staff believed that he had badly underestimated the strength of the French army. Hastily assembling plans for defensive operations, Moltke grew calmer when the French showed no signs of supporting Frossard's initial attack. When Frossard's forces began to dig in, Moltke ordered Frederick William's Third Army to

---

[355]For assessment of railroads see Arden Bucholz, *Moltke, Schlieffen, and Prussian War Planning* (New York: Berg Publishers, Inc., 1991), 41 ff.

[356]Howard, 82–89.

attack them on August 4. The crown prince complied, slipping to the south of Frossard's position in an effort to encircle the French troops and running headlong into a French force commanded by General Patrice MacMahon. Realizing that Frederick William was attacking at full strength, MacMahon retreated to a strong position around the area of Fröschwiller, placing his troops on high ground above a river valley. Frederick William assaulted the position on August 6 in the first major action of the war. Casualties were heavy on both sides, but the French could not stop the Third Army. By evening MacMahon withdrew, giving Prussia its first victory. [357]

Moltke wanted something more than a local victory. As he examined the situation, he saw that Frederick William's Third Army had moved through the French southern flank. While MacMahon retreated from Fröschwiller, Frossard was entrenched in front of Steinmetz's First Army, and the majority of the French army sat dormant to Frossard's north. If Frederick William could move behind the French forces, Moltke thought it would be possible to trap the entire French army between Frederick William and the First and Second Prussian Armies. Unfortunately, Moltke's army commanders did not grasp the point. Frederick William decided that it would be much wiser to pursue MacMahon than to move behind him, while Steinmetz defiantly resisted any attempt to place First Army in a blocking position. Instead of serving as the hinge on Moltke's trap, the insolent Steinmetz ordered all his units to attack Frossard without waiting for orders.

Frossard was completely unaware of MacMahon's retreat to his south. Knowing that he had encountered the bulk of the Prussian build up, he moved his forces to a series of hills around the village of Sprichen. Steinmetz saw this as the opportunity to attack. Ignoring Moltke's orders to hold his position, Steinmetz abandoned his sector and crossed in front Frederick Charles's Second Army. The Second Army commanders were infuriated to see Prussian troops moving across their fields of fire, but there was nothing they could do. Steinmetz attacked Sprichen on August 6 without support and soon learned that the French were intent on holding their positions.[358]

Although Steinmetz attacked with every available unit, Frossard would not be moved. Never one to admit defeat, Steinmetz sent a message to Second Army demanding reinforcements. Frederick Charles was beside him-

---

[357]Ibid., 99–107.
[358]Ibid., 91–95.

self. Steinmetz's movement had completely blocked his front, effectively removing him from contact with the enemy, and now his only option was to support Steinmetz by a direct attack on fortified positions. He reluctantly committed Second Army, forcing Frossard slowly to give ground. By evening the sector was in Prussian hands, and Steinmetz sent Moltke a wire telling him of the unwanted victory. Exasperated, Moltke watched the French retreat from the trap that he had been trying to spring. He asked King William formally to reprimand Steinmetz.

Despite Moltke's misgivings, the French were badly shaken, and August 6 had been a dark day for the French army. Frederick William had achieved a complete victory at Fröschwiller, while Steinmetz and Frederick Charles had dislodged the troops at Sprichen. Napoleon III and his generals began to realize the poor quality of both the French army and French military technology. French generals had hoped that their new machine gun, the mitrailleuse, would stop attacking Prussian infantry in their tracks, but the highly acclaimed weapon had been of little use to local commanders. The cannon were worse. The bronze muzzle-loading guns of the French were no match for the rifled steel Krupp breech-loaders of the Prussians. The only bright spot was the French rifle, the chassepot. Its range was twice that of the Prussian needle gun, but Napoleon could see no bright spot on August 6. His major problem was not technology. Napoleon's enigma was that the French were not organized to fight a modern war.

To his credit Napoleon attempted to take decisive steps. He reduced Leboeuf to a corps commander, a level that better suited the former war minister, and he divided the army into two commands. Patrice MacMahon, the general bloodied by Frederick William at Sprichen, organized all the southern forces in the newly christened Army of Alsace. The northern forces, the Army of Lorraine, went to Marshal François Achille Bazaine, a combat commander with an excellent reputation earned in Mexico. Napoleon ordered both armies to retreat, hoping to gain time to reorganize the entire front, but reorganization accomplished little. Both Bazaine and MacMahon were expected to maintain their original jobs as corps commanders while commanding several corps in their new armies. In addition, neither general was given a staff to assist with duties. Napoleon also made no provision for coordination between the two new armies, so both men acted independently. Within a week, Napoleon realized that his organizational solution was not sufficient. In frustration, he relinquished command to Bazaine on August 12.[359]

While Bazaine's position grew increasingly hopeless, Moltke's frustrations blossomed. To begin, Steinmetz went into a pouting fit. Stung by the king's reprimand, the general stopped sending information to either his fellow army commanders or to Moltke. This caused Frederick Charles to lose all patience with Steinmetz, and he refused to support any actions from the First Army. To the south, Frederick William's Third Army was completely in the dark. Satiated with his victory at Sprichen, the crown prince saw no reason to move with vigor, and when he did advance, he did not turn to surround the French forces. To make matters worse, it started raining. The roads turned to mud, and troops were constantly soaked to the bone. Some of the Prussian commanders believed they were losing the war.

As usual, Moltke developed a plan. Since his commanders had destroyed the original hopes for double encirclement, the Chief of Staff ordered his generals to do something they would understand. All three Prussian armies were to launch a general frontal advance. Steinmetz, skeptical of Moltke, slowed the First Army in defiance, and Frederick Charles moved the Second Army forward to get away from Steinmetz. Frederick William took a different approach. Believing that his purpose should be to lead an advance into France, he moved forward with such speed that he soon outstripped the other two armies. The crown prince wanted to find MacMahon and destroy him before Frederick Charles and Steinmetz could take away his thunder.

The results would have been almost comical had Moltke not realized the situation developing before him. Steinmetz, holding back in the north, anchored First Army in front of the French position. Frederick Charles, marching quickly to get away from Steinmetz, moved into the center, forcing the French to turn their flank. Frederick William's Third Army left everybody behind and advanced far to the rear of the French positions. As a result, MacMahon and Bazaine were completely separated, and Frederick William ended up in the rear of Bazaine's position. Moltke had not planned such a deployment, but he was certainly willing to take advantage of it.

Moltke's opportunity for encirclement started on August 15 when Steinmetz ran into Bazaine at the fortress of Metz. Even though Steinmetz had

---

[359]Ibid., 124–25.

a chance to trap Bazaine's Army of Lorraine in the city, he decided not to attack. As a result, Bazaine's troops were able to retreat, but when they did so, they encountered Prussian troops the next day in the vicinity of Mars-la-Tour. Bazaine's men had come to grips with General Constantin von Alvensleben, commander of the III Corps in the Second Army, by far the most aggressive general in Frederick Charles's charge. Alvensleben, unaware that his isolated corps had found the entire Army of Lorraine, ordered all his units into action. The first attacks were successful, forcing the French from the road they were using to retreat, yet Alvensleben detected a problem. Each time he assaulted, he seemed to encounter a fresh supply of French reinforcements. By midmorning it was obvious that III Corps had bumped into a major French army. Alvensleben told his officers that their only chance for survival was to convince the French that III Corps was the entire Second Army, so each unit launched a full blown attack. In the meantime Alvensleben sent a frantic appeal to Frederick Charles for help.[360]

An aggressive French commander would have attacked. Surprised by the battle at Mars-la-Tour and Alvensleben's tenacity, Bazaine had no intention of attacking; he ran. As Frederick Charles frantically established a thin line to support Alvensleben, Bazaine retreated to the east toward Metz. Since Steinmetz had not entered the city, Bazaine knew that the fortress was still intact, and he thought it would be prudent to establish a defensive line to the west of Metz, using the fortress as a base for defensive operations. Accordingly, on August 17, Bazaine moved his troops away from Frederick Charles to a defensive position a few miles away. That same evening Moltke arrived from Berlin to take personal command of the war.

The Battle of Gravelotte-St. Privat

The Army of Lorraine comprised five corps, and its defense-minded commander arrayed the troops along an eight-mile front to the west of Metz.[361] The southern flank was anchored by the Moselle River where rolling hills protected the entrenched infantry. Charles Frossard commanded a

---

[360]See also J. F. C. Fuller, "Gravelotte-Saint Privat," in Cyril Falls (ed.) *Great Military Battles* (London: Spring Books, 1969), 192–203. (This is a coffee table book, well-known for its art work and illustrations.)

[361]Based on Howard, 167–82.

corps in this sector. Just to his north, in another good topographical defensive position, sat the corps of Edmond Leboeuf, the former war minister. The terrain changed on Leboeuf's right north of the Leipzig farm, shifting from rolling hills to a gentle slope. The ground was ideal for the long-range fire of the mitrailleuse and the chassepot. Two corps completed the line in this area, the IV Corps of General de Ladmirault in the south and Marshal Canrobert's VI Corps stretching north to St. Privat. Canrobert's flank was exposed, but a seasoned African veteran, General Bourbaki, commanded a reserve corps behind Leboeuf. If the Prussians tried to sweep across the exposed flank, Bourbaki could seal it. Given his mental state, Bazaine had placed his troops in a strong position.

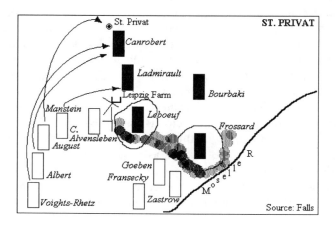

Moltke did not have a complete picture of Bazaine's strong line, but he understood the overall situation. Steinmetz, albeit unwillingly, blocked the French route of escape. If Second Army were to swing north, Moltke could spring the trap. Elements from Steinmetz's First Army were in contact with the southern French flank along the Moselle, so Moltke believed that Second Army should move north, a move around the French left to surround Bazaine. The problem was that Moltke did not know where the French line came to an end. Therefore, he ordered Frederick Charles to move to the north until he could locate the northern French flank. Steinmetz was to assist Second Army with the holding action by detaching a corps to occupy the French in the hills along the Moselle. With the front thus engaged, Moltke hoped that the northern portion of Frederick Charles's army would go around the flank to surround Bazaine.

The battle began in the early morning hours of August 18. Frederick Charles moved his forces northward in a wide swinging arc, and Steinmetz, ignoring Moltke's orders, tried to break through along the Moselle. Instead of tying down the French forces, Steinmetz sent the First Army's VII Corps, two divisions under General Zastrow, straight toward Frossard's trenches. Zastrow received a nasty surprise. Most of Frossard's men were hidden in the wooded hills, and Zastrow's corps was subjected to merciless chassepot fire. Usually, rifled Prussian artillery sent French infantrymen running, but trenches and trees negated the effect of the Prussian bombardment. Frossard's men held firm. By midmorning, Zastrow was stuck, his troops exchanging fire with unseen enemies. Steinmetz decided to launch General Goeben's VIII Corps in support, and he removed them from Second Army without authority. Goeben, who had officially been transferred to Second Army, knew better than to cross Steinmetz. He attacked, only to become bogged down in a situation similar to that of Zastrow. The French refused to be moved.

With Goeben insubordinately removed, Frederick Charles had five corps at his disposal. They included: Constantin von Alvensleben's III Corps, Constantin von Voights-Rhetz's X Corps, Manstein's IX Corps, Prince August's Prussian Guards, and Saxon Crown Prince Albert's XII Saxon Corps. As Steinmetz slammed into the southern flank, Frederick Charles marched north with Manstein leading the way and Prince August and Prince Albert covering his rear. Alvensleben maintained light contact with the French front to hold the troops in place while Voights-Rhetz remained in reserve. At 10 A.M. Manstein reported that he had reached the northern flank. Frederick Charles ordered Manstein forward, sending Albert and August further north with orders to turn and encircle the French. The maneuver would have served to surround Bazaine's army had Manstein not made a mistake. Instead of hitting the northern flank, Manstein marched directly into the center of the French line.

As Manstein went forward into murderous chassepot fire, he soon realized that he was nowhere near the French flank. He hurriedly sent word to Frederick Charles, but it was too late. Within minutes, Manstein was fully engaged, and Alvensleben fell in on Manstein's right flank to support his efforts. The open ground in the French center provided an excellent advantage for the French riflemen. While their bronze cannon could not compete with the accurate fire from the Prussian artillery, their chassepots held the Prussians at bay. Manstein had no choice, save to call his artillery to counter French rifle fire. August and Albert were able to avoid the fray, but by early afternoon, the majority of Frederick Charles's corps were engaged along the enemy line. The French were holding their own.

By mid-afternoon August's scouts brought word that the French flank actually ended in the area of St. Privat. After notifying Frederick Charles, August sent his artillery forward while Canrobert's Frenchmen sought cover behind a stone fence surrounding the village. Prince August launched a series of assaults, but the French fought for every inch of ground. The situation was growing critical for Moltke. With August held in check in the north and the center degenerating into an artillery duel, Frederick Charles was not moving. The southern flank, on the other hand, turned into a complete disaster. Steinmetz's troops were pinned down, unable to move since morning. The insubordinate general was demanding reinforcements, fearing that his command might be destroyed. Only Voights-Rhetz remained free to move, but Frederick Charles needed to keep his corps in reserve. The situation was ripe for a French counterattack.

Fortunately for Moltke, the thought of taking the offensive was as remote to Bazaine as interplanetary travel. Frederick Charles, cut from a different cloth than Bazaine, decided to place all his efforts on the northern flank. If the flank could be turned, the French position would weaken. He urged Albert to move forward with all possible haste, and he ordered August to turn all available fire on St. Privat. At the same time he ordered an all-out barrage on French gun emplacements all along the front. The tactic worked. By 5 P.M. St. Privat was weakening and the French artillery had been knocked out of the battle. Moltke was relieved. Although he had no more units to reinforce Frederick Charles, the time for a French counterstrike had passed.

There was still time for tragedy, however. The Prussian Guards were in position just below St. Privat, and at 6 P.M., for some reason, Prince August decided to launch them toward the town. General von Pape, the Guards' commanding officer, was disgruntled. The French controlled the gentle slope in front of the village, and their position was not sufficiently weakened, he told Prince August. The Prince responded by ordering Pape to weaken it. Accordingly, the fathers, sons, and brothers of Prussia's aristocracy lined up in parade-ground fashion and began marching forward. On the slope outside St. Privat French Marshal Canrobert's riflemen lay prone with their chassepots pointed toward their enemies. Their rifles had an effective range of 600 yards, and, since they were breech-loaders, the French riflemen needed not to stand to fire. They could remain hidden while firing at fast as they could load. As the Prussian Guard moved forward, the French opened sustained accurate fire.

It took twenty minutes. At first, the Guard infantrymen held their ranks while the officers on horseback were methodically sent to the dust. The French riflemen then turned their attention to the ranks. After a few hundred yards, the Guards wavered and spread into a line formation. Somehow, they kept coming, but in a few minutes even the line disappeared. No Prussian soldier made it closer than 600 yards to the wall of St. Privat. The survivors crawled or ran back to the Prussian lines. In the course of only twenty minutes the Guards suffered over 8,000 casualties. Many veterans began weeping, and within days tears ran all the way back to the homes of Prussia.

The Crown Prince of Saxony was more fortunate. At 7 P.M. Prince Albert's XII Corps completely dislodged all the troops on the French northern flank. At 7:30 Frederick Charles committed Voights-Rhetz and the survivors of the Guard to a second assault on St. Privat. Unlike the previous assault, this time St. Privat had been blasted by Prussian artillery. The French flank turned, and Canrobert reluctantly ordered his men to retreat. As the French units moved back, they became confused and mingled with the reserves who had come to help. Within minutes the retreat turned into a confused mob; then panic set in. The French were defeated.

Back at his command post, Moltke was only aware of the human meat grinder along the Moselle. With no word from Frederick Charles, Moltke only hoped that Steinmetz could hold the southern flank, but there was little reason for optimism. The French were in control, and Steinmetz had been unable to move for almost seven hours. At 5 P.M. Eduard von Fransecky arrived with his II Corps. Steinmetz immediately asked for them, and, despite Moltke's objection, the king approved. Fransecky went forward to be met not by the French but by Zastrow. The French fire had been devastating all day, and, as Fransecky approached, Zastrow's men found they could take no more. They broke and ran. Fransecky's men composed themselves and advanced. Preparing to retake the vacated area, his troops sent several hundred rounds of accurate fire into Zastrow's old positions and prepared to assault them. It was a mistake. Zastrow had gone, but Goeben's men were still there. Fransecky had fired on fellow Prussians. The remainder of Goeben's corps panicked, and, throwing their rifles aside, they ran for the rear. Some even ran past the king and Moltke. As Fransecky dug in for an expected French attack, Moltke feared that the day was lost.

The expected counterattack never came. At 9:30 a despondent William reluctantly gave Moltke permission to attack the next day, but there was no elation in Prussian headquarters. The king wanted to retreat, and, too tired to argue with Moltke, he simply gave in and approved another attack. The cease fire sounded, and the Prussians and the other German soldiers retired to lick their wounds. The casualty reports—growing to 20,000—seemed unending, and news of the Guards devastated the king. The mood lifted a little two hours later. At midnight, an excited messenger found a tired Moltke and gave him a report from Frederick Charles. Moltke read the message, hardly able to grasp its contents. Elements of the Second Army were behind the original French line. Frederick Charles reported no resistance; the French were running in a disorganized mob toward Metz. Victory, the message told Moltke, belonged to Prussia.

The Battle of Sedan

Marshal Patrice MacMahon realized both the political and military dangers facing the French government after the defeat of Gravelotte-St. Privat. As Bazaine withdrew to Metz, Parisians demanded action, and political resistance to Napoleon's military government increased. There was talk of a popular uprising among leftists, and, unless the army achieved some victory on the battlefield, there was a good chance that the French government would topple. Since victory could not come from Bazaine, MacMahon represented all of Napoleon's hopes. Moving to Châlons, Mac-Mahon reorganized his forces and increased the ranks. Assembling 120,000 men in four corps, MacMahon and an ailing Napoleon marched toward the Meuse on August 21 in an effort to relieve Bazaine. The French press, anxious to report a victory, published reports of MacMahon's strength and probable routes of march. Moltke soon had as much information about the new Army of Châlons as the French General Staff.

Moltke was quick to react to the situation. Now assured that his pan-Teutonic army would fight as a unified force, he reorganized his army. Leaving Frederick Charles to besiege Metz, he detached elements of the Second Army to create a new force, the Meuse Army. Assigning the courageous but cantankerous Steinmetz to Frederick Charles's new command, Moltke awarded leadership of the new Meuse Army to his former adversary, Albert the Crown Prince of Saxony. He ordered Albert to join Frederick William's Third Army and continue the advance into France. Their object was to find MacMahon's army and destroy it.

Frederick William was the first to find MacMahon. Although he still did not completely understand the *Kesselschlacht,* he did know how to find, surround, and destroy an enemy on a localized front. As the Army of Châlons marched in the direction of Metz, Frederick William delivered a severe blow to it at Nouart on August 29. Stunned, MacMahon altered his route, turning north toward the Belgian border. Prince Albert, who understood Moltke's view of grand encirclement, took the Meuse Army in MacMahon's direction, envisioning another trap. On August 30 Albert surprised the French at Beaumont, forcing MacMahon to move further north. Moltke ordered another strike the next day, realizing that he might be able to position MacMahon between Frederick William and Albert. The plan worked, and MacMahon soon found himself in Sedan, a small town near the Belgian border, eighty miles away from Metz. Frederick William sat on his southern positions, while Albert covered his eastern flank. Moltke moved his headquarters to the German positions outside Sedan on August 31 to assess the situation.[362]

The French and the Germans realized the potential outcome of the situation at the same time. Moltke saw that if Frederick William could cover the west, the French would be trapped against the Belgian border. MacMahon also saw the problem, and he promised his staff that the French army would move out of Sedan within a day. On the evening of August 31, one of the French staff officers objected. If they waited to move, the officer told MacMahon, it would be too late. Indeed, the French officer was prophetic. Moltke, taking personal command of the situation, moved Frederick William and Albert toward the flanks of the French position. Sitting in a valley beneath a semicircle of hills on the Meuse River, Sedan was tailor-made for

---

[362]Ibid., 184–203.

Moltke's plan. If Moltke took the high ground, the French would be forced to attack in order to break out. By the evening of the thirty-first, Albert's Meuse Army held the hills on the eastern flank, while Frederick William was a few hours' march from the hills to the west.[363]

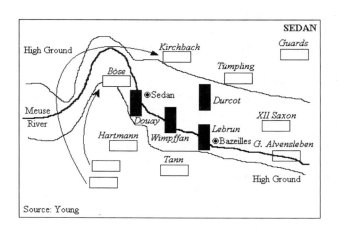

Source: Young

Moltke had several corps to close the position. Frederick William's army was positioned with two corps—the I and II Bavarians—forward, and two more units—the XI and V Corps—ready to sweep the left flank to take the hills. The II Bavarian Corps under General von Hartmann held the south, while General von der Tann's I Bavarians were positioned to Hartmann's right. The XI Corps (General von Böse) and the V Corps (General von Kirchbach) were stationed behind the II Bavarians, ready to swing around the flank. Prince Albert's Meuse Army was also in an excellent position. He placed his own XII Saxon Corps on the heights overlooking Sedan from the east, flanking them with General von Tüpling's VI Saxons and General Gustav von Alvensleben's IV Prussians. Holding the Prussian Guards in reserve, Moltke referred to the position as a mousetrap. A French officer trapped in Sedan was more graphic, saying that the French army was

---

[363]Peter Young, "Von Moltke at Sedan," in Young (ed.), *Great Generals and their Battles* (New York: The Military Press, 1984), 86–103. (This is another coffee table book with excellent illustrations and good commentary.)

in a chamber pot and that the German army was about to defecate. His crude assessment was correct.[364]

MacMahon never intended to fight at Sedan.[365] After the debacle at Beaumont, he had hoped only to rest and refit. His troops were arranged in a triangular defensive position in the Meuse River valley, but the post was designed to be only temporary. Up before dawn on September 1, MacMahon issued marching orders, but before his units could assemble, the first booming echoes of the German cannon sounded through the valley. The French position burst open in black smoke and pain as messengers brought news that the eastern flank was under attack. MacMahon hastily called his commanders for a conference.

At 4 A.M. Generals Böse and Kirchbach, commanding the XI and V corps in Frederick William's Third Army, slipped from behind the two forward Bavarian corps and crossed the Meuse to move to the western flank of the French position. Before they reached their objectives, however, Tann's I Bavarians opened fire from the eastern flank. Tann's bombardment set the forward French positions in Bazailles ablaze, while the surprised Böse and Kirchbach wondered what to do. Kirchbach's men reached their objective first, but Böse was out of position. Breaking ranks to hasten their pace, Böse's XI Corps began running to the hills—the location Frederick William had assigned. On the French side of the line, Böse's broken ranks drew the attention of French cavalrymen in Sedan. As Böse's men ran toward their objective, the French launched a cavalry charge.

In past times, the French onslaught would have devastated broken infantry, but rifle fire had changed the nature of battle. The scattered Prussian units had no need to rely on a protective bayonet wall, and when Kirchbach's V Corps moved into to cover Böse's left, Prussian riflemen mowed down the French cavalry. Astounded by their inability to break their enemy, the French continued to reassemble and charge, only to be sent back by accurately aimed bullets from Dryse rifles. Watching from the hills, King William—perhaps recalling the Prussian Guard at St. Privat—commented sadly on the courage of the French cavaliers. Regardless, cavalry charges could not stem the well aimed shots of Prussian soldiers. Thousands of

---

[364]The officer was General Drucot who said, *"Nous sommes dans un pot de chambre et nous y serons emmerdés."*

[365]Howard, 203–23.

French soldiers were killed, including the French cavalry commander who had his jaw shot away.

In the meantime, French artillery responded to the Bavarian artillery fire, but Tann's rifled guns had the advantage. When a puff of smoke betrayed a French artillery piece's position, rifled Krupp guns knocked the cannon out. As MacMahon watched his artillery being systematically consumed by Bavarian fire, shell fragments suddenly sent the French commander to the ground with splinters in his leg. Unable to continue, he passed command to General Auguste Ducrot. As if the situation were not bad enough, this sparked a controversy. General Emmanuel de Wimpffen, recently arrived from Paris, produced papers giving him charge of the French army. After a brief debate, Ducrot agreed to let Wimpffen command. As Ducrot prepared to return to his corps, Wimpffen condescendingly informed him that any form of retreat was out of the question. Ducrot did not let the remark pass. You will be lucky, he told MacMahon's arrogant replacement, if you have an opportunity to retreat.

Moltke did everything possible to assure that the French would not escape. With the Meuse Army on the east, Böse and Kirchbach held the west. Tann's artillery fire increased, and the Meuse valley turned into a smoking hell. French cavalry attacks increased, not for the purpose of breaking the infantry but to find away out of the valley. By noon Wimpffen was enclosed in an iron ring.

With his troops in position, Moltke did not want to waste German blood by moving forward too quickly. He methodically inched the infantry forward under the protective bombardment of artillery batteries, hoping the French would wear themselves out by launching counterattacks. The French did not disappoint him. Deciding that his only option was to strike south, Wimpffen asked Napoleon III to lead the breakout. The emperor declined. Undaunted, Wimpffen moved all available forces to the south about 1 P.M. and smashed headlong into Hartmann's II Bavarians. Furious fighting raged for two hours, and Hartmann sent for reinforcements, believing that his line was about to break. But he overestimated his enemy. The French were completely spent. Around 5 o'clock Wimpffen gathered about 1,000 men and tried to break out through Tann's I Bavarian Corps. The assault surprised Tann, but he repelled it. The French refer to the attack as the battle of the last cartridges, and, indeed, it was. Wimpffen ran out of ammunition. As evening settled on the field, Napoleon realized the cause had ended. He personally surrendered to William.

Bombardment and Guerrilla War

The remaining portion of the Franco-Prussian War was not indicative of Moltke's *Kesselschlacht*, and the German-speaking forces could no longer properly be called "Prussian." For the first time, it was proper to speak of a German war and a German army. Born from a military revolution in the seventeenth century, the Prussian army faded from history, and Sedan became a German victory, not simply a battle belonging to Prussia. As the spoils were gathered from Sedan, the German army moved south to Paris, arriving on September 19, 1870, to find a new French army—a popular army created by the people of Paris. Untrained, ill-equipped, and hardly professional, the people's army offered to defend Paris with their lives. Moltke understood the ferocity of such defense, and rather than send his troops into a costly street-by-street battle, he surrounded the city.[366]

Moltke's original goal was to impose a peace treaty after defeating the French national forces, but continued fighting confused German foreign policy. Metz surrendered on October 27, freeing all the German army for the siege of Paris, but the French fought on. They raised guerrilla forces, striking Moltke's supply lines, and they recruited new popular armies to the south of Paris and in Lorraine. Frustrated German troops resorted to harsh measures, not knowing how to counter guerrillas.[367] In the end, however, conventional tactics won. The French came in substantial numbers, but they could not organize under the sights of German guns. By December the people of Paris were starving, and Moltke believed that a bombardment was in order. Although Bismarck disagreed, the German guns began firing into Paris on January 5, 1871.

---

[366]See Gordon A. Craig, *The Politics of the Prussian Army: 1640–1945* (London: Oxford University Press, 1968), 206–15.

[367]German soldiers either murdered and executed French civilians as a matter of course in retaliation for attacks on German soldiers. American General Philip Sheridan, dining with Bismarck, encouraged the Germans to take such harsh actions to make the French civilians "feel" the war. Otto Pflanze, *Bismarck and the Development of Germany, Volume I, The Period of Unification, 1815–1871* (Princeton, N.J.: Princeton University Press, 1990), 483.

Starved into submission, Paris surrendered on January 28, after all relief attempts failed. Several fortresses along the Rhine continued to resist, including the garrison at Belfort. After a 105-day siege, Belfort had become a symbol to the French people, similar to Gneisenau's stubborn resistance at Kolberg in 1807. Moltke did not want a costly assault, and Bismarck suggested letting the garrison surrender with full military honors. It was a sound compromise. Although the French had a moral victory, resistance at Belfort came to an end on February 15. The savage guerrilla war wound down in the spring, and the new country of Germany signed a peace treaty with France at Frankfurt on May 10, 1871. The Franco-Prussian War was over, but the bitterness between Germany and France was not.

Militarism was to be part of the new German order. One could argue that the Great Elector never intended this, and it is certainly demonstrable that the reformers did not want a military state. Yet, the irony of history is that the pan-Germanic unification movement was closely linked to the conservative monarchialism of the Prussian army. Nationalism and military unification went hand in hand. The Prussian army—the triumphant conservative army of 1819 and 1848—defined the new state. The newly created Great German General Staff was politically popular, giving the army much to say about the future, and the situation did not change until March 1945.

Nothing symbolized this relationship more than the Treaty of Frankfurt, the settlement ending the Franco-Prussian War. Bismarck wanted France to keep its dignity and possibly to seek partnership with the newly established Germany. The General Staff could not envision friendship with France. While Bismarck focused on political pragmatism, Moltke and his generals asked the king to both punish France and annex portions of French territory. In the end, Bismarck lost. The French Rhenish provinces of Alsace and Lorraine became part of Germany, not because Germany had a political need to control the area, but because the General Staff wanted the road system for future wars. Bismarck objected, arguing that such an action would only deepen French animosity and increase the possibilities of future confrontations. His insight went unheeded, but the issue was raised again in August 1914 and September 1939.

The End of the Prussian Army

For all practical purposes, the history and traditions of the Prussian army continued in the German army. Prussian military methods governed the new army as well as its General Staff. Officially, the Prussian army came to an end on January 18, 1871, at Versailles. Bismarck achieved his dream as all the German states joined together in a confederation, the Second Empire, or *Reich*. William was selected as the king among kings and given the ancient Roman—and Hapsburg—title of "emperor" or "*Kaiser*." There is a marvelous engraving by Anton von Werner that shows the newly crowned emperor being saluted by his ministers and generals. This book is designed for American college students, and it is fitting to end with a classroom comment on the picture.

As American students look at copies of the engraving in history books or encyclopedias, some of them recognize Kaiser William. Many more recognize Otto von Bismarck, who stands close to the newly crowned emperor wearing a white coat. A German king stands next to William as a whole host of generals and princes raise their swords or helmets in salute of the new government. Standing beneath William, raising his helmet in salute, is a general most of my students do not recognize. He appears to be a stereotypical German militarist—minus the monocle. He was the outcome of a tradition and the prop behind the imperial throne. More importantly, his army allowed William to stand and receive the crown. The man, of course, is Helmuth von Moltke, Chief of the Prussian General Staff.

It is a glorious picture, a military celebration, but students would be wise to consider some of the unseen events in the painting. There is no rendition of starving, lost children inside a city under siege. The celebration does not portray a Scharnhorst dying from an infected wound or a Seydlitz expiring from syphilis. The men in the picture are not hungry, thirsty, or cold. They are dry. The backdrop is a palace, not a field covered with men and human limbs. There is no smell of death. The lives forever shattered from news of death and destruction are nowhere to be seen. There are no tears. The picture does not and cannot portray the words of dying soldiers, words that always remain the same regardless of nationality. As a German officer once said, neither uniform nor age matter to the dying. Just before his last breath, the officer said, a soldier always cries for his mother. Such is the nature of military glory.

Works Cited

Most of the works cited here can be found in many American college libraries. While the list is by no means definitive, it is designed to give English-speaking readers an introduction to Prussian military history.

For more in-depth reading, the following books are recommended:

Ferdinand Schevill, *The Great Elector.*
Robert Ergang, *The Potsdam Führer.*
Christopher Duffy, *Frederick the Great.*
Robert Asprey, *Frederick the Great.*
F. N. Maude, *The Jena Campaign* (may be difficult to find).
Peter Paret, *Clausewitz and the State.*
Dennis Showalter, *Railroads and Rifles.*
Gordon Craig, *The Battle of Königgrätz.*
Michael Howard, *The Franco-Prussian War.*

Two other works round out this recommended list. Michael Howard's *War in European History* examines the social meaning of military conflict. One of the most important books in military history is John Keegan's *The Face of Battle.* Although Keegan's purpose is not to discuss the Prussian army, he captures the meaning of battle for individual soldiers.

Works cited in the text

Addington, Larry H. *The Blitzkrieg Era and the German General Staff, 1865–1941.* New Brunswick, N.J.: Rutgers University Press, 1971.
Ashley, Maurice. *A History of Europe: 1648–1815.* Englewood Cliffs, N.J.: Prentice-Hall, Inc., 1973.
Anderson, M. S. *Eighteenth-Century Europe: 1713–1789.* London: Oxford University Press, 1968.

Aron, Raymond. *Clausewitz: Philosopher of War.* Translated by Christine Booker and Norman Stone. Englewood Cliffs, N.J.: Prentice-Hall, Inc., 1985.

Asprey, Robert B. *Frederick the Great: The Magnificent Enigma.* New York: Ticknor and Fields, 1986.

Atkinson, C. T. *A History of Germany: 1715–1815.* New York: Barnes and Noble, Inc., 1908.

Axtmann, Roland. " 'Police' and the Formation of the Modern State: Legal and Ideological Assumptions on State Capacity in the Austrian Lands of the Hapsburg Empire: 1500–1800," *German History* (1) February 1992: 39–61.

Beiser, Frederick C. *Enlightenment, Revolution, and Romanticism: The Genesis of Modern German Political Thought, 1790–1800.* Cambridge, Mass.: Harvard University Press, 1992.

Beloff, Max. *The Age of Absolutism: 1660–1815.* London: Hutchinson University Library, 1964.

Berce, Brian and Jonathan R. White, "Roadblocks to Democracy: The Prussian Army as a Police Force in the 1848 Revolution," *Academy of Criminal Justice Sciences Annual Conference*, Chicago, 1994.

Berdahl, Robert M. *The Politics of the Prussian Nobility: The Development of a Conservative Ideology, 1770–1848.* Princeton, N.J.: Princeton University Press, 1988.

Black, Jeremy. *The Rise of the European Powers: 1679–1793.* London: Edward Arnold, 1990.

Bucholz, Arden. *Moltke, Schlieffen, and Prussian War Planning.* New York: Berg Publishers, Inc., 1991.

Bunney, Richard. *The European Dynastic States: 1494–1660.* Oxford: Oxford University Press, 1991.

Busch, Moritz. *Bismarck: Some Secret Pages of His History, Volumes I-II.* New York: AMS Press, 1970; orig. 1898.

Carlyle, Thomas. *History of Friedrich II of Prussia Called the Great*, John Clive (ed.). (Abridged). Chicago: University of Chicago Press, 1969; orig. 1900.

Carr, William. *A History of Germany: 1815–1945.* 2nd ed. New York: St. Martin's Press, 1979.

Carsten, F. L. *The Origins of Prussia.* Oxford: Oxford University Press, 1954.

———. "The Rise of Brandenburg," in G. N. Clark et al. (eds.), *The New Cambridge Modern History, Volume IV, The Ascendancy of France: 1648–88.* London: Cambridge University Press, 1961.

Catt, Henri de. *Frederick the Great: The Memoirs of His Reader: 1758–1760, Volumes I-II.* Translated by F. S. Flint. London: Constable and Company, Ltd., 1916.

Chandler, David. *The Art of War in the Age of Marlborough.* New York: Hippocrene, 1976.

Chandler, David G. *The Campaigns of Napoleon.* New York: Macmillan, 1966.

Churchill, Winston S. *Marlborough: His Life and Times.* New York: Charles Scribner's Sons, 1968.

Clausewitz, Karl von. *On War.* Translated by Michael Howard and Peter Paret. Princeton, N.J.: Princeton University Press, 1984.

Craig, Gordon A. *The Politics of the Prussian Army: 1640–1945.* London: Oxford University Press, 1968.

———. *The Battle of Köninggrätz.* London: Weidenfeld and Nicolson, 1965.

Delbrück, Hans. *History of the Art of War within the Framework of Political History, Volumes I-IV.* Translated by Walter J. Renfree, Jr. Westport, Conn.: Greenwood Press, 1985; orig. 1920.

Demeter, Karl. *The German Officer-Corps in Society and State: 1650–1945.* Translated by Angus Malcolm. London: Weidenfeld and Nicolson, 1965.

Dorwart, Reinhold August. *The Prussian Welfare State: Before 1740.* Cambridge, Mass.: Harvard University Press.

———. *The Administrative Reforms of Frederick William I of Prussia.* Cambridge: Harvard University Press, 1953.

Dowd, David L. *The French Revolution.* New York: Harper and Row, 1965.

Doyle, William. *The Oxford History of the French Revolution.* New York: Oxford University Press, 1989.

Droz, Jacques. *Europe Between Revolutions: 1815–1848.* Translated by Robert Baldick. New York: Harper Torchbooks, 1967.

Duffy, Christopher. *Siege Warfare: The Fortress in the Early Modern World, 1494–1660.* London: Routledge and Kegan Paul, 1979.

———. *The Fortress in the Age of Vauban and Frederick the Great: 1660–1789.* London: Routledge and Kegan Paul, 1985.

————. *The Army of Frederick the Great.* New York: Hippocrene Books, Inc., 1974.

————. *Frederick the Great: A Military Life.* London: Routledge, 1988.

Dwyer, Philip G. "The Politics of Prussian Neutrality: 1795–1805," *German History* (12) October 1994: 351–73.

Easum, Chester V. *Prince Henry of Prussia: Brother of Frederick the Great.* Westport, Conn.: Greenwood Press, 1971; orig. 1942.

Elting, John R. *Swords Around a Throne: Napoleon's Grand Armee.* New York: The Free Press, 1988.

Ergang, Robert. *The Potsdam Führer: Frederick William I, Father of Prussian Militarism.* New York: Octagon Books, 1972; orig. 1941.

Eyck, Erich. *Bismarck and the German Empire.* New York: W. W. Norton and Co., Inc., 1968.

Falls, Cyril. *Great Military Battles.* London: Spring Books, 1969.

Faucitt, William. *Regulations for the Prussian Infantry.* New York: Greenwood Press, 1968; orig. 1759.

————. *Regulations for the Prussian Cavalry.* New York: Greenwood Press, 1968; orig., 1757.

Feuchtwagner, E. J. *Prussia: Myth and Reality: The Role of Prussia in German History.* Chicago: Henry Regnery Co., 1970.

Ford, Franklin L. *Europe: 1780–1830.* London: Longmans, Green, and Co. Ltd., 1970.

Ford, Guy Stanton. *Stein and the Era of Reform in Prussia: 1807–1815.* Gloucester, Mass.: Peter Smith, 1965.

Forest, Alan. *Soldiers of the French Revolution.* Durham, N.C.: Duke University Press, 1990.

Freytag, Gustav. "The German Catastrophe," (orig. 1882) in Theodore K. Rabb, (ed.) *The Thirty Years' War.* Lexington, Mass.: D. C. Heath and Co., 1972.

Friedjung, Heinrich. *The Struggle for Supremacy in Germany: 1859–1866.* Translated by A. J. P. Taylor and W. L. McElwee. New York: Russell and Russell, 1966; orig. 1897.

Frischauer, Paul. *Prince Eugene, 1663–1736: A Man and a Hundred Years of History.* New York: William Morrow and Company, 1934.

Fuller, J. F. C. *The Conduct of War, 1789–1961.* Minerva Press, 1968.

————. "Gravelotte-Saint Privat," in Cyril Falls (ed.) *Great Military Battles.* London: Spring Books, 1969.

Gall, Lothar. *Bismarck: The White Revolutionary, Volumes I-II.* Translated by J. A. Underwood. London: Allen and Unwin, 1986.

Gaxotte, Pierre. *Frederick the Great.* Translated by R. A. Bell. New Haven, Conn.: Yale University Press, 1942.

Gillis, John R. *The Prussian Bureaucracy in Crisis, 1840–1860: Origins of an Administrative Ethos.* Stanford University Press, 1971.

Goelitz, Walter. *The German General Staff: Its History and Structure: 1657–1945.* Translated by Brian Battershaw. New York: Frederick A. Praeger, Publishers, 1962.

Gooch, G. P. *Frederick the Great: The Ruler, the Writer, the Man.* Hamden, Conn.: Archon Books, 1947.

Hale, J. R. *War and Society in Renaissance Europe: 1450–1620.* New York: St. Martin's Press, 1985.

Hamilton-Williams, David. *Waterloo New Perspectives: The Great Battle Reappraised.* New York: John Wiley and Sons, Inc., 1993.

Harris, R. W. *Absolutism and Enlightenment: 1660–1789.* New York: Harper and Row, 1966.

Hassall, Arthur. *The Balance of Power: 1715–1789.* London, Rivingtons, 1925.

Haythornwaite, Philip and Brian Fosten. *Frederick the Great's Army 2 Infantry.* London: Osprey, 1991.

Henderson, Nicholas. *Prince Eugene of Savoy.* New York: Frederick A. Praeger, 1964.

Henderson, W. O. *Studies in the Economic Policy of Frederick the Great.* London: Frank Cass, 1963.

Hobsbaum, E. J. *The Age of Revolution: 1789–1848.* New York: Mentor, 1962.

Hofschoer, Peter. *Leipzig 1813.* London: Osprey, 1993.

Hogg, O. F. G. *Artillery: Its Origin, Heyday, and Decline.* Hamden, Conn.: Archon Books, 1970.

Holborn, Halo. *A History of Modern Germany.* New York: Alfred A. Knopf, 1964.

Howard, Michael. *War in European History.* New York: Oxford University Press, 1987.

———. *Clausewitz.* New York: Oxford University Press, 1988.

———. *The Franco-Prussian War: The German Invasion of France, 1870–1871.* London: Routledge, 1988.

Howarth, David. *Waterloo: Day of Battle.* New York: Antheneum, 1968.

Hubatsch, Walther. *Frederick the Great of Prussia: Absolutism and Administration.* London: Thames and Hudson Ltd., 1975.

Jany, Curt. *Geschichte der königlich Preussischen Armee bis zum Jahre 1807, Volumes I-III.* Berlin: Verlag von Karl Siegismund, 1928–1929.

————. *Die Königlich Preussische Armee und das Deutsche Reichsheer 1807 bis 1914.* Berlin: Verlag von Karl Siegismund, 1933.

Johnson, Hubert C. *Frederick the Great and His Officials.* New Haven, Conn.: Yale University Press, 1975.

Jones, Archer. *The Art of War in the Western World.* Urbana, Ill.: University of Illinois Press, 1987.

Kafker, Frank A. and James M. Laux (eds.), *The French Revolution: Conflicting Interpretations.* New York: Random House, 1968.

Kaiser, David. *Politics and War: European Conflict from Philip II to Hitler.* Cambridge, Mass.: Harvard University Press, 1990.

Keegan, John. *The Face of Battle: A Study of Agincourt, Waterloo, and the Somme.* New York: Penguin, 1978.

Kitchen, Martin. *A Military History of Germany: From the Eighteenth Century to the Present Day.* Bloomington, Ind.: Indiana University Press, 1975.

Kleinschmidt, Harald. "Studien zum Quellenwert der deutschsprachigen Exerzierreglements vornehmlich des 18. Jahrhunderts (III)," *Zeitschrift für Heereskunde* 51 (1987): 162–64.

Knopp, Werner. *In Remembrance of a King: Frederick II of Prussia, 1712–1786.* Bonn, Inter Nationes, 1986.

Koch, H. W. "Brandenburg-Prussia," in John Miller (ed.), *Abosultism in Seventeenth Century Europe.* New York: St. Martin's Press, 1990.

Kossman, E. H. "The Dutch Republic," in G. N. Clark et al. (eds.). *The Cambridge Modern History, Volume IV, The Ascendancy of France 1648–88.* Cambridge: Cambridge University Press, 1961.

Kranzberg, Melvin (ed.). *1848: A Turning Point.* Boston: D. C. Heath and Co., 1959.

Lachouque, Henry. *Napoleon's Battles: A History of His Campaigns.* Translated by Roy Monkcom. London: George Allen and Unwin, Ltd., 1966.

Lauisse, Ernest. *The Youth of Frederick the Great.* Translated by Mary Bushnell Colemen. New York: AMS, 1972; orig. 1892.

Lehmann, Max. *Scharnhorst, Volumes I-II.* Leipzig: Verlag von S. Hirzel, 1887.

Levinger, Matthew. "Hardenberg, Wittgenstein, and the Constitutional Question in Prussia: 1815–22," *German History* (3) October 1990: 257–77.

Liddell Hart, Basil H. *Strategy.* New York: Praeger, 1967.

Linnebach, Karl. *König Friedrich Wilhelm I und Fürst Leopold zu Anhalt-Dessau.* Berlin: B. Behr's Verlag, 1907.

Little, H. M. "Thomas Powell and Army Supply, 1761–1766," *Journal of the Society for Army Historical Research,* 1987 (26): 92–104.

Longman, F. W. *Frederick the Great and the Seven Years' War.* New York: Charles Scribner's Sons, 1926.

Ludtke, Alf. *Police and State in Prussia: 1815–1850.* New York: Cambridge University Press, 1989.

Ludwig, Emil. *Bismarck: The Story of a Fighter.* Translated by Eden and Cedar Paul. Boston: Little, Brown, and Co., 1927.

Luvaas, Jay. *Frederick the Great on the Art of War.* New York: Free Press, 1966.

Lynn, John A. "The Pattern of Army Growth, 1445–1994," in John A. Lynn (ed.), *Tools of War: Instruments, Ideas and Institutions, 1445–1871.* Urbana, Ill.: University of Illinois Press, 1990.

Manchester, William. *The Arms of Krupp.* Boston: Little, Brown, and Co., 1968.

Mann, Golo. *The History of Germany Since 1789.* Translated by Marian Jackson. New York: Frederick A. Praeger, Publishers, 1968.

Maude, F. N. *The Jena Campaign, 1806.* London: Swan Sonnenschein & Co., 1909.

Maurice, C. Edmund. *Life of Frederick William: The Great Elector of Brandenburg.* London: George Allen and Unwin, Ltd., 1926.

McKay, Derek. *Prince Eugene of Savoy.* London: Thames and Hudson Ltd., 1977.

Meinecke, Friedrich. *The Age of German Liberation: 1795–1815.* Translated by Peter Paret and Helmuth Fischer. Berkeley, Calif.: University of California Press, 1977.

Menne, Bernhard. *Blood and Steel: The Rise of the House of Krupp.* New York: Lee Furman, Inc., 1938.

Mitford, Nancy. *Frederick the Great.* New York: Harper and Row, Publishers, 1970.

Montross, Lynn. *War through the Ages.* New York: Harper and Brother Publishers, 1944.

Niderost, Eric. "Desperate Victory Overshadowed," *Military History* (4) August 1987: 19–25.

Noyes, P. H. *Organization and Revolution: Working-Class Associations in the German Revolutions of 1848–1849.* Princeton, N.J.: Princeton University Press, 1966.

Palmer, Alan. *Frederick the Great.* London: Weidenfeld and Nicolson, 1974.

Paret, Peter. *Clausewitz and the State.* London: Oxford University Press, 1976.

———. *Yorck and the Era of Prussian Reform.* Princeton, N.J.: Princeton University Press, 1966.

Parker, Geoffrey. *The Thirty Years' War.* London: Routledge and Kegan Paul, 1984.

———. *The Military Revolution: Military Innovation and the Rise of the West, 1500–1800.* Cambridge: Cambridge University Press, 1988.

Pengel, R. D. and G. R. Hurt, *Prussian Cavalry, Seven Years War.* Birmingham, England: 1981.

Pepper, Simon and Nicholas Adams, *Military Architecture and Siege Warfare in Sixteenth-Century Siena.* Chicago: University of Chicago Press, 1986.

Pflanze, Otto. *Bismarck and the Development of Germany, Volumes I-III.* Princeton, N.J.: Princeton University Press, 1990.

Polišenský, J. V. *The Thirty Years' War.* Translated by Robert Evans. Berkeley, Calif.: University of California Press, 1971.

Postgate, Raymond. *Story of a Year: 1848.* New York: Oxford University Press, 1956.

Ranke, Leopold von. *Memoirs of the House of Brandenburg during the Seventeenth and Eighteenth Centuries, Volume I.* New York: Greenwood Press, 1968; orig. 1849.

Reddaway, William F. *Frederick the Great and the Rise of Prussia.* New York: G. P. Putnam's Sons, 1911.

Reiners, Ludwig. *Frederick the Great: A Biography.* Translated by Lawrence P. R. Wilson. New York: G. P. Putnam's Sons, 1960.

Reinke, Herbert. "The State, the Military, and the Professionalization of the Prussian Police in Imperial Germany," in Clive Emsley and Barbara Weinberger (eds.), *Policing Western Europe: Politics, Professionalism, and Public Order, 1850–1940*. New York: Greenwood Press, 1991.

Riemann, Karl-Heinz. "Graf Yorck von Wartenburg—preussicher General zwischen feudalem Konservatismus und Patriotismus," *Militärgeschichte* (6) 1987: 585–91.

Ritter, Gerhard. *Frederick the Great: A Historical Profile*. Translated by Peter Paret. Berkeley, Calif.: University of California Press, 1970.

———. *The Sword and the Scepter: The Problem of Militarism in Germany, Volume I, The Prussian Tradition: 1740–1890*. Translated by Heinz Norden. Coral Gables, Fla.: University of Miami Press, 1969.

Robertson, Priscilla. *Revolutions of 1848: A Social History*. Princeton, N.J.: Princeton University Press, 1952.

Ropp, Theodore. *War in the Modern World*. Durham, N.C.: Duke University Press, 1959.

Rosinski, Hebert. *The German Army*. Washington, D.C.: The Infantry Journal, 1944.

Rothenburg, Gunther E. "Maurice of Nassau, Gustavus Adolphus, Raimondo Montecuccoli, and the 'Military Revolution' of the Seventeenth Century," in Peter Paret et al. (eds.). *Makers of Modern Strategy: From Machiavelli to the Nuclear Age*.

Savory, Reginald Arthur. *His Britannic Majesty's Army in Germany during the Seven Years' War*. Oxford: Clarion Press, 1966.

Schevill, Ferdinand. *The Great Elector*. Chicago: University of Chicago Press, 1947.

Schroeder, Paul W. *The Transformation of European Politics: 1763–1848*. New York: Oxford University Press, 1994.

Schweizer, K. W. "The Bedford Motion and the House of Lords Debate, 5 February 1762," *Parliamentary History*, 1986 (5): 107–23.

Seeley, J. R. *Life and Times of Stein, or Germany and Prussia in the Napoleonic Age, Volume I: 1757–1807*. London: Cambridge University Press, 1878.

Shanahan, William O. *Prussian Military Reforms: 1786–1813*. New York: AMS Press, 1966; orig. 1945.

Sheehan, James. *German History: 1770–1866*. New York: Oxford University Press, 1989.

Sheehan, James J. *German Liberalism in the Nineteenth Century.* Chicago: University of Chicago Press, 1978.

Showalter, Dennis E. "Hubertusberg to Auerstadt: The Prussian Army in Decline?" *German History* (3) October 1994: 308–33.

————. *Railroads and Rifles: Soldiers, Technology, and the Unification of Germany.* Hamden, Conn.: Archon, 1976.

————. "The Political Soldiers of Bismarck's Germany: Myths and Realities," *German Studies Review* (1) February 1994: 59–77.

Snell John L. *The Democratic Movement in Germany, 1789–1914.* Chapel Hill, N.C.: University of North Carolina Press, 1976.

Sonnino, Paul. *Louis XIV and the Origins of the Dutch War.* Cambridge: Cambridge University Press, 1988.

Steward, John Hall. *A Documentary Survey of the French Revolution.* New York: MacMillan, 1951.

Stoye, John. *Europe Unfolding: 1648–1688.* London: William Collins Sons, 1954.

Strachan, Hew. *European Armies and the Conduct of War.* London: George Allen and Unwin, 1983.

Sutherland, D. M. G. *France 1789–1815: Revolution and Counter Revolution.* New York: Oxford University Press, 1985.

Sweet, Paul R. *Humbolt: A Biography.* Columbus, Ohio: Ohio State University Press, 1978.

Symcox, Geoffrey. (ed.) *War, Diplomacy, and Imperialism: 1618–1763.* New York: Walker and Company, 1974.

Taylor, A. J. P. *The Course of German History.* New York: Capricorn Books, 1962; orig. 1946.

————. *From Napoleon to the Second International: Essays on Nineteenth-Century Europe.* New York: Penguin, 1994.

Treitschke, Heinrich von. *History of Germany in the Nineteenth Century, Volumes I-VII.* Translated by Eden and Cedar Paul. New York: McBride, Nast, and Co., 1915–1919; orig. 1884.

Tuttle, Herbert. *History of Prussia, Volumes I-IV.* New York: AMS Press, 1971; orig. 1884.

Van Creveld, Martin. *Command in War.* Cambridge, Mass.: Harvard University Press, 1986.

————. *Technology and War: From 2000 B.C. to the Present.* New York: The Free Press, 1989.

Verney, Peter. *The Battle of Blenheim.* London: B. T. Batsford Ltd., 1976.

Weigley, Russell F. *The Age of Battles: The Quest for Decisive Warfare from Breitenfeld to Waterloo.* Bloomington, Ind.: Indiana University Press, 1991.

Weller, Jac. *Weapons and Tactics: Hastings to Berlin.* New York: St. Martin's Press, 1966.

White, R. J. *Europe in the Eighteenth Century.* New York: St. Martin's Press, 1965.

White, Charles Edward. *The Enlightened Soldier: Scharnhorst and the Militärische Gesellschaft in Berlin, 1801–1805.* New York: Praeger, 1989.

Wolf, John B. *Toward a European Balance of Power: 1620–1715.* Chicago: Rand McNally and Co., 1970.

Wood, Anthony. *Europe: 1815–1945.* New York: David McKay Co., Inc., 1964.

Young, Peter. "Von Moltke at Sedan," in Young (ed.), *Great Generals and their Battles.* New York: The Military Press, 1984.

Zeller, G. "French Diplomacy and Foreign Policy in the European Setting," in G. N. Clark et al. (eds.). *The Cambridge Modern History, Volume IV, The Ascendancy of France: 1648–88.* Cambridge: Cambridge University Press, 1961.

Index